T0134839

Studies in Systems, Decision and Control

Volume 187

Series editor

Janusz Kacprzyk, Polish Academy of Sciences, Warsaw, Poland
e-mail: kacprzyk@ibspan.waw.pl

The series "Studies in Systems, Decision and Control" (SSDC) covers both new developments and advances, as well as the state of the art, in the various areas of broadly perceived systems, decision making and control–quickly, up to date and with a high quality. The intent is to cover the theory, applications, and perspectives on the state of the art and future developments relevant to systems, decision making, control, complex processes and related areas, as embedded in the fields of engineering, computer science, physics, economics, social and life sciences, as well as the paradigms and methodologies behind them. The series contains monographs, textbooks, lecture notes and edited volumes in systems, decision making and control spanning the areas of Cyber-Physical Systems, Autonomous Systems, Sensor Networks, Control Systems, Energy Systems, Automotive Systems, Biological Systems, Vehicular Networking and Connected Vehicles, Aerospace Systems, Automation, Manufacturing, Smart Grids, Nonlinear Systems, Power Systems, Robotics, Social Systems, Economic Systems and other. Of particular value to both the contributors and the readership are the short publication timeframe and the world-wide distribution and exposure which enable both a wide and rapid dissemination of research output.

More information about this series at http://www.springer.com/series/13304

Damian Piotr Muniak

Regulation Fixtures in Hydronic Heating Installations

Types, Structures, Characteristics and Applications

Damian Piotr Muniak
Department of Thermal Power Engineering
Cracow University of Technology
Kraków, Poland

ISSN 2198-4182 ISSN 2198-4190 (electronic)
Studies in Systems, Decision and Control
ISBN 978-3-030-13216-3 ISBN 978-3-030-03128-2 (eBook)
https://doi.org/10.1007/978-3-030-03128-2

Softcover re-print of the Hardcover 1st edition 2019

This Springer imprint is published by the registered company Springer Nature Switzerland AG
The registered company address is: Gewerbestrasse 11, 6330 Cham, Switzerland

Contents

Symbols

A	Surface area of the inner cross section of a pipe or valve carrying the medium, (mm^2, m^2)
A_i	Surface area of the inner cross section of a pipe carrying the medium in the *i*-th section of the circuit, (m^2)
A_M	Active surface area of the valve membrane, (mm^2, m^2)
A_x	Surface area of the medium flow cross section at the position of the valve closing/throttling element corresponding to partial opening, (mm^2, m^2)
A_{100}	Surface area of the medium flow cross section at the position of the valve closing/throttling element corresponding to full opening, (mm^2, m^2)
a	Valve regulated section authority, (–), or the slope factor of the pump throttling characteristic, (–)
a'	Valve superior role, (–)
a_c	Valve total authority, (–)
$a_{c,100}$	Total authority of the valve closing element for full opening of both regulation stages, (throttling and closing sections), (–)
$a_{c,x}$	Total authority of the valve closing element for a given pre-setting, (–)
$a_{dł}$	Valve throttling criterion, (–)
a_m	Share of local hydraulic resistances in total hydraulic resistance of a section of the pipework, (–)
a_w	Valve inner authority, (–)
$a_{w,100}$	Inner authority of the valve closing element at full opening, (–)
$a_{w,min}$	Minimum value of the valve closing element inner authority, (–)
$a_{w,I,min}$	Minimum value of the throttling element authority (full opening) of a valve with two adjustable sections of the medium flow, (–)

$a_{w,max}$	Maximum value of the valve closing element inner authority, (–)
$a_{w,x}$	Inner authority of the valve closing element for a given pre-setting, (–)
a_z	Valve outer authority, (–)
$a_{z,100}$	Valve outer authority at full opening, (–)
$a_{z,min}$	Minimum value of the valve outer authority, (–)
$a_{z,x}$	Valve outer authority at partial, *x*-th-degree opening for a given pre-setting, (–)
$a(n_i)_{c,min}$	Function describing the total authority variability depending on the pre-setting of a valve with one adjustable section of the medium flow, for no constraints on the range of the closing/throttling element travel, (–)
$a(n_i)_{w,min}$	Function describing the inner authority variability depending on the pre-setting of a valve with one adjustable section of the medium flow, for no constraints on the range of the closing/throttling element travel, (–)
$[a(n_i)_{w,I,min}]_{Xp}$	Function describing variability in the throttling element inner authority depending on the thermostatic valve pre-setting, for the closing element position corresponding to a given proportional range X_p, for no constraints on the range of the throttling element travel, (–)
$[a_{w,I,min}]_{Xp}$	Minimum value of the thermostatic valve throttling element authority (no constraints on the range of travel) for the closing plug lift corresponding to a given proportional range X_p, (–)
$[a_{z,min}]_{Xp}$	Thermostatic valve outer authority for the closing element position corresponding to a given proportional range *Xp*, for no constraints on the range of the throttling element travel, (–)
c	Equal-percentage factor of the valve primary closing characteristic, (–)
c_w	Specific heat of water, (J/(kg deg))
C_1, C_2, C_3	Coefficients for the *Colebrook-White* formula, (–)
d	Inner diameter of the pipe or element carrying the medium, (m, mm)
d_i	Inner diameter of a pipe in the *i*-th section of the pipework, (m, mm)
dh_x	Elementary increment in the valve opening degree, (m, mm)
$d\dot{V}_X$	Elementary increment in the medium volume flow, (m^3/h, m^3/s), (–)
e	Relative roughness of the pipe inner surface, (–)
e_{gr}	Boundary value of the pipe inner surface relative roughness, (–)
F_F	Spring tension force, (N)
F_M	Force on the valve membrane and plug, (N)
g	Gravitational field acceleration, (m/s^2)

h_{100}	Position of the valve closing/throttling element corresponding to full relative opening, (mm, m)
h_{max}	Position of the valve closing/throttling element corresponding to full absolute opening, (mm, m)
h_N	Closing plug lift over the valve seat corresponding to the nominal proportional range, e.g. for $X_p = 2K$, (mm, m)
h_x	Position of the valve closing/throttling element corresponding to partial opening, (mm, m)
h	Position of the thermostatic head follower, (mm, m)
h_h	Height of the liquid column, (m)
H	Heat loss coefficient, (W/K)
k_z	Valve amplification factor, (–)
k_m	Thermostatic head amplification factor, (mm/K, m/K)
$k_{m,\%}$	Thermostatic head percentage amplification factor, (%)
k_s	Room amplification factor, (K/W, °C/W)
$k_{s,\%}$	Room percentage amplification factor, (%)
k	Mean absolute roughness of the pipe inner surface, (m, mm)
k_v	Flow factor, $m^3/(h\ bar^{0.5})$, m^3/h (according to the simplification of units adopted in practice)
$k_{v,z}$	Valve flow factor, ($m^3/(h\ bar^{0.5})$, m^3/h)
$k_{v,rz}$	Flow factor real value obtained by means of experimental measurements, (m^3/h)
$k_{v,ob}$	Circuit flow factor, ($m^3/(h\ bar^{0.5})$, m^3/h)
$k_v(n_i)$	Function describing the flow factor variability depending on the pre-setting of a valve with one adjustable section of the medium flow (no possibility of a current increase in the flow above the value resulting from the pre-setting), ($m^3/(h\ bar^{0.5})$, m^3/h)
$k_{v,k}$	Valve body flow factor, ($m^3/(h\ bar^{0.5})$, m^3/h)
$k_{v,k+I,i}$	Resultant flow factor of the valve body and throttling element for the *i*-th pre-setting of the valve, ($m^3/(h\ bar^{0.5})$, m^3/h)
$k_{v,reg}$	Flow factor of the valve current-regulation element, ($m^3/(h\ bar^{0.5})$, m^3/h)
$k_{v,reg,100}$	Flow factor of the current-regulation element of a fully open valve, ($m^3/(h\ bar^{0.5})$, m^3/h)
$k_{v,reg,x}$	Flow factor of the valve current-regulation element for partial opening corresponding to the pre-setting, ($m^3/(h\ bar^{0.5})$, m^3/h)
$k_{v,x}$	Valve flow factor for partial opening corresponding to the pre-setting or to an intermediate position of the closing element, ($m^3/(h\ bar^{0.5})$, m^3/h)
k_{v0}	Valve flow factor for full (mechanical) closing, ($m^3/(h\ bar^{0.5})$, m^3/h)
$k_{v,100}$	Flow factor of a fully open valve, ($m^3/(h\ bar^{0.5})$, m^3/h)

k_{vN} Flow factor of a thermostatic valve for the closing element opening corresponding to proportional range $X_p = 2K$, ($m^3/(h\ bar^{0.5})$, m^3/h)

k_{vs} Mean flow factor of the tested series of fully open valves, ($m^3/(h\ bar^{0.5})$, m^3/h)

$[k_v(n_i)]_{Xp}$ Function describing the flow factor variability depending on the pre-setting of the thermostatic valve throttling element, for the closing element position corresponding to a given proportional range X_p, (m^3/h, $m^3/(h\ bar^{0.5})$)

$[k_{v,x}]_{Xp}$ Thermostatic valve flow factor for a given pre-setting, at the closing plug lift corresponding to a given proportional range X_p, (m^3/h, $m^3/(h\ bar^{0.5})$)

$[k_{v100}]_{n_i}$ Flow factor of a thermostatic valve, or a double-regulation manual valve, for the throttling element pre-setting, at the closing plug maximum lift, (m^3/h, $m^3/(h\ bar^{0.5})$)

$[k_{vs}]_{Xp}$ Mean flow factor of the thermostatic valve series for the maximum pre-setting, at the closing plug lift corresponding to a given proportional range X_p, (m^3/h, $m^3/(h\ bar^{0.5})$)

l_i Length of a straight pipe section in the i-th section of the pipework, (m)

$l_{z,i}$ Equivalent length of the i-th local obstacle, (m)

m Correction exponent of a straight pipe hydraulic characteristic, (–)

$\dot{m}$ Working medium mass flow, (kg/s, kg/h)

$\dot{m}_i$ Working medium mass flow through the i-th element of the pipework, (kg/h, kg/s)

$\dot{m}_X$ Working medium mass flow for the valve partial opening, (kg/s, kg/h)

n Correction exponent of the pipework characteristic, (–)

n_i i-th value of the valve pre-setting, (–)

n_{max} Maximum value of the valve pre-setting, (–)

p_A Pressure on the secondary side of the valve membrane, (bar, Pa)

p_{atm} Atmospheric pressure, (bar, Pa)

p_E Pressure on the primary side of the valve membrane, (bar, Pa)

p_h Hydrostatic pressure, (bar, Pa)

P_{el} Electric power needed to drive the pump, (W)

$\dot{Q}_i$ Thermal power in the room, (W)

$\dot{Q}_N$ Nominal thermal power in the room, (W)

$\dot{Q}_X$ Thermal power corresponding to the valve partial opening, (W)

$\dot{Q}_{ob}$ Thermal power supplied in the circuit, (W)

q_{mmax}	Maximum mass flow—the highest possible mass flow of water (fully open valve) at differential pressure of 10kPa, (kg/h)
$\dot{m}_{\mathrm{N}}$, q_{mN}	Nominal mass flow—characteristic mass flow for the intermediate position of the thermostatic head selector (20–24 °C). In the case of a valve with the possibility of pre-setting, this is the mass flow through a valve with no pre-setting, (kg/h, kg/s)
q_{ms}	Characteristic flow established for $X_p = 2\mathrm{K}$ and differential pressure of 10kPa at any value of the thermostatic head setting, (kg/h)
R_{i}	Unit linear pressure losses in the *i*-th straight section, (Pa/m)
Re	Reynolds number, (–)
Re_{gr}	Reynolds number boundary value, (–)
$r_{\mathrm{c,i}} = r_{\mathrm{l,i}} + r_{\mathrm{m,i}}$	Total resultant hydraulic resistance of the pipework *i*-th section, ((h^2 bar)/m^6, (Pa s^2)/m^6)
$r_{\mathrm{c,100}}$	Total hydraulic resistance of a given section of the pipework and of a fully open valve, ((h^2 bar)/m^6, (Pa s^2)/m^6)
$r_{\mathrm{c,x}}$	Total hydraulic resistance of a given section of the pipework and of a valve for partial opening, ((h^2 bar)/m^6, (Pa s^2)/m^6)
r_{I}	Hydraulic resistance of the valve throttling element, ((h^2 bar)/m^6, (Pa s^2)/m^6)
$r_{\mathrm{I,0}}$	Hydraulic resistance of the valve throttling element at the valve minimum opening, ((h^2 bar)/m^6, (Pa s^2)/m^6)
$r_{\mathrm{I,100}}$	Hydraulic resistance of the valve throttling element at the valve full opening, ((h^2 bar)/m^6, (Pa s^2)/m^6)
$r_{\mathrm{I,i}}$	Hydraulic resistance of the valve throttling element at the opening position determined by the *i*-th opening degree corresponding to a given pre-setting, ((h^2 bar)/m^6, (Pa s^2)/m^6)
r_{II}	Hydraulic resistance of the valve closing element, ((h^2 bar)/m^6, (Pa s^2)/m^6)
$r_{\mathrm{II,100}}$	Hydraulic resistance of the valve closing element for the full available range of travel, ((h^2 bar)/m^6, (Pa s^2)/m^6)
r_{k}	Hydraulic resistance of the valve body, ((h^2 bar)/m^6, (Pa s^2)/m^6)
$r_{\mathrm{k+I,100}}$	Total hydraulic resistance of the valve body and the valve throttling element at the valve full opening, ((h^2 bar)/m^6, (Pa s^2)/m^6)
$r_{\mathrm{k+I,i}}$	Total hydraulic resistance of the valve body and the valve throttling element at a given opening degree corresponding to the pre-setting, ((h^2 bar)/m^6, (Pa s^2)/m^6)
$r_{\mathrm{l,i}}$	Hydraulic resistance of straight sections of pipes in a given part of the pipework related to the volume flow of the medium, ((h^2 bar)/m^6, (Pa s^2)/m^6)

$r_{m,i}$ Hydraulic resistance of the *i*-th local obstacle, ((h^2 bar)/m^6, (Pa s^2)/m^6)
r_i Hydraulic resistance of the *i*-th element of the pipework related to the mass or the volume flow, (–)
r_{ob} Hydraulic resistance of the circuit, ((h^2 bar)/m^6, (Pa s^2)/m^6, kPa/(dm^3/s)^2)
r_{reg} Hydraulic resistance of the valve current-regulation section of the medium flow, ((h^2 bar)/m^6, (Pa s^2)/m^6)
$r_{reg,100}$ Hydraulic resistance of the valve current-regulation section of the medium flow for full opening, ((h^2 bar)/m^6, (Pa s^2)/m^6)
$r_{reg,x}$ Hydraulic resistance of the valve current-regulation section of the medium flow for a given opening degree, ((h^2 bar)/m^6, (Pa s^2)/m^6)
r_{str} Hydraulic resistance of all elements of a given part of the pipework excluding the valve, ((h^2 bar)/m^6, (Pa s^2)/m^6)
r_z Hydraulic resistance of the valve, ((h^2 bar)/m^6, (Pa s^2)/m^6)
$r_{z,100}$ Hydraulic resistance of the valve at full opening of both elements of regulation, ((h^2 bar)/m^6, (Pa s^2)/m^6)
$r_{z,x}$ Hydraulic resistance of the valve for a given degree of opening of both elements of regulation, ((h^2 bar)/m^6, (Pa s^2)/m^6)
$[r_{z,100}]_{Xp}$ Hydraulic resistance of the valve at full opening of the thermostatic valve throttling element for the closing plug lift corresponding to a given proportional range X_p, ((h^2 bar)/m^6, (Pa s^2)/m^6)
t_i Conventional air temperature in the room, (°C)
t_e Conventional outdoor temperature, (°C)
$t_{i,k}$ Conventional final air temperature in the room, (°C)
t_p Working medium temperature at the radiator outlet, return temperature, (°C)
t_z Working medium temperature at the radiator inlet, supply temperature, (°C)
t_{zad} Temperature set at the temperature regulator selector, (°C)
t_c Temperature measured by the temperature sensor, (°C)
$\dot{V}$ Working medium volume flow, (m^3/h, m^3/s)
$\dot{V}_{100}$ Working medium volume flow through the valve at the valve full (100%) opening, (m^3/h, m^3/s)
$\dot{V}_k$ Working medium volume flow through the valve body, (m^3/h, m^3/s)
$\dot{V}_z$ Working medium volume flow through the valve, (m^3/h, m^3/s)
$\dot{V}_i$ Working medium volume flow through the *i*-th section of the circuit, (m^3/h, m^3/s)

$\dot{V}_{\mathrm{x}}$ — Working medium volume flow through the valve at the valve partial opening, ($\mathrm{m^3/h}$, $\mathrm{m^3/s}$), or volume flow for a given intermediate working point on the pump throttling characteristic, ($\mathrm{m^3/h}$, $\mathrm{m^3/s}$, $\mathrm{dm^3/s}$)

$\dot{V}_{\mathrm{max}}$ — Working medium maximum volume flow through the pump, ($\mathrm{m^3/h}$, $\mathrm{m^3/s}$, $\mathrm{dm^3/s}$)

$\dot{V}_{\mathrm{i}}$ — Working medium volume flow through the i-th element of the pipework, ($\mathrm{m^3/h}$, $\mathrm{m^3/s}$, $\mathrm{dm^3/s}$)

$\dot{V}_{\mathrm{N}}$ — Working medium volume flow through the valve at the closing element opening corresponding to proportional range $X_p = 2\mathrm{K}$, ($\mathrm{m^3/h}$, $\mathrm{m^3/s}$)

$\dot{V}_{\mathrm{k+I,i}}$ — Working medium resultant volume flow through the valve body and the throttling element for the i-th pre-setting, ($\mathrm{m^3/h}$, $\mathrm{m^3/s}$)

$[\dot{V}_{100}]_{Xp=2}$ — Working medium volume flow at the thermostatic valve throttling element full opening and proportional range $X_p = 2\mathrm{K}$, ($\mathrm{m^3/h}$, $\mathrm{m^3/s}$)

$\dot{V}_{\mathrm{ob}}$ — Working medium volume flow in the circuit, ($\mathrm{m^3/h}$, $\mathrm{m^3/s}$)

w — Working medium mean velocity in the cross section of a pipe or an element carrying the medium, (m/s)

w_{i} — Working medium mean velocity in the cross section of a pipe in the pipework i-th section, (m/s)

$\dot{V}_1, \dot{V}_2, \dot{V}_3, \dot{V}_3, \dot{V}_4$ — Working medium volume flow, for example, working points, ($\mathrm{m^3/h}$, $\mathrm{m^3/s}$)

X_{p} — Proportional range of the thermoregulator, (K, °C), or of the automatic balancing valve, (kPa, bar)

$X_{\mathrm{p,max}}$ — Maximum proportional range of the thermoregulator, (K, °C)

$X_{\mathrm{p,s}}$ — Proportional range of the room (room thermal characteristic), (W)

Z_{i} — Pressure losses on local obstacles in the i-th section of the circuit, (bar, Pa)

Greek Symbols

α — Valve discharge coefficient, (–)

$\Delta k_{\mathrm{v,max}}$ — Maximum error in the flow factor determination, ($\mathrm{m^3/h}$)

Δn — Change in the number of the pump impeller rotations, (rpm)

Δp_1 — Pressure value from the thermoregulator measurement performed according to relevant standards, $\Delta p_1 = 10\mathrm{kPa}$, or pressure for an example working point, (bar, Pa)

Δp_2	Pressure loss for the nominal flow or for the flow characteristic of valves with a possibility of pre-setting with no pressure losses in the regulated section, read from the $\Delta p = f(q_m)$ characteristic of a fully open valve and corresponding to the maximum flow, (kPa) or pressure for an example working point, (bar, Pa)
Δp_3, Δp_3, Δp_4	Pressure, for example, working points, (bar, Pa)
Δp_0	Working medium pressure loss on the valve, according to the flow factor definition, $\Delta p_0 = 1\text{bar} = 10^5\text{Pa}$
$\Delta p_{c,i}$	Working medium total pressure loss in the pipework i-th section, (bar, Pa)
Δp_{I}	Working medium pressure loss on the valve throttling element, (bar, Pa)
$\Delta p_{I,i}$	Working medium pressure loss on the valve throttling element at the opening position determined by the i-th opening degree corresponding to a given pre-setting, (bar, Pa)
$\Delta p_{I,0}$	Working medium pressure loss on the valve throttling element at its minimum opening, (bar, Pa)
$\Delta p_{I,100}$	Working medium pressure loss on the valve throttling element at its full opening, (bar, Pa)
Δp_{II}	Working medium pressure loss on the valve closing element, (bar, Pa)
$\Delta p_{II,100}$	Working medium pressure loss on the valve closing element for its full available range of travel, (bar, Pa)
Δp_{k}	Working medium pressure loss on the valve body, (bar, Pa)
Δp_{i}	Working medium pressure loss caused by the pipework i-th element, (bar, Pa)
$\Delta p_{k+I,i}$	Working medium total pressure loss on the valve body and the valve throttling element at a set opening degree corresponding to the i-th pre-setting, (bar, Pa)
$\Delta p_{l,i}$	Working medium pressure loss on a straight section in the pipework i-th section, (bar, Pa)
$\Delta p_{m,i}$	Working medium pressure loss on local obstacles in the pipework i-th section, (bar, Pa)
Δp_{ob}	Pressure loss in the circuit, (bar, Pa)
Δp_{cz}	Active (differential) pressure in the circuit, (bar, Pa)
Δp_{reg}	Working medium pressure loss on the valve current-regulation section of the medium flow, (bar, Pa)
$\Delta p_{reg,100}$	Working medium pressure loss on the valve current-regulation section of the medium flow for full opening, (bar, Pa)
$\Delta p_{reg,x}$	Working medium pressure loss on the valve current-regulation section of the medium flow for a given opening degree, (bar, Pa)
Δp_{str}	Working medium pressure loss on all elements of a given part of the pipework excluding the valve, (bar, Pa)

$\Delta p_{str,100}$ Working medium pressure loss on all elements of the circuit (pipework) excluding the valve for the valve full opening, (bar, Pa)
$\Delta p_{str,x}$ Working medium pressure loss on all elements of the circuit (pipework) excluding the valve for a given degree of opening of both regulation elements of the valve, (bar, Pa)
Δp_{z} Working medium pressure loss on the valve, (bar, Pa)
$\Delta p_{z,0}$ Working medium pressure loss on the valve for full closing of both regulation stages (throttling and closing section), (bar, Pa)
$\Delta p_{z,100}$ Working medium pressure loss on the regulation valve for full opening, (bar, Pa)
$\Delta p_{z,x}$ Working medium pressure loss on the regulation valve for a given degree of opening of both regulation stages (throttling and closing section), (bar, Pa)
$\Delta p_{z,max}$ Maximum error in the measurement of the pressure drop on the valve, (bar)
$\Delta\Delta p$ Change in the working medium pressure, (bar, Pa)
$[\Delta p_{z,100}]_{Xp}$ Working medium pressure loss on the thermostatic valve at full opening of the throttling element for the closing plug lift corresponding to a given proportional range X_p, (bar, Pa)
Δp_{x} Pressure for a given intermediate working point on the pump throttling characteristic, (bar, Pa)
Δp_{max} Maximum pressure produced by the pump, (bar, Pa)
Δt_{i} Change in conventional temperature of air in the room, (°C)
Δt_{w} Water cooling in the radiator, (°C)
$\Delta \dot{V}_{max}$ Maximum error in the working medium volume flow measurement, (m^3/h)
$\Delta \dot{Q}_{i}$ Change in the room thermal power, (W)
$\Delta \dot{Q}_{i,\%}$ Percentage change in the room thermal power, (W)
ΔP_{el} Change in the electric power needed to drive the pump, (W)
$\Delta \dot{V}$ Change in the working medium volume flow, (m^3/h, m^3/s)
λ_{i} Coefficient of linear pressure losses in the pipework *i*-th section, (–)
λ Coefficient of linear pressure losses, (–)
μ Dynamic viscosity of the medium, (Pa s)
v Kinematic viscosity of the medium, (m^2/s)
$\eta_{c,i}$ Total efficiency of the pump, (–)
ρ_{i} Working medium density at set values of temperature and pressure in the pipework *i*-th section, (kg/m^3)
ρ_{o} Working medium density according to the flow factor definition, (kg/m^3)
ρ Working medium density, (kg/m^3)

δ	Percentage error in the circuit balance, (%)
τ	Time, (s)
ζ	Coefficient of local pressure losses, (–)
ζ_i	Coefficient of local pressure losses in the pipework *i*-th section, (–)

Introduction

The book *Regulation Fixtures in Hydronic Heating Installations: Types, Structures, Characteristics and Applications* is devoted to issues of the selection, the principle of operation and the types and structures of regulation valves in hydronic heating systems. It also deals with hydraulic problems of the valves operation.

The publication presents a broad discussion of the types and kinds of regulation fixtures used in heating installations at present and describes practical aspects of their selection. Reasons are indicated for the application of specific types of valves in certain installation points, and their impact on the other elements of the system is described. The analysis is supplemented with connection diagrams, figures and photographs of cross sections of real valves. Such an approach facilitates the understanding of the principle of operation of individual elements.

A considerable part of the book is devoted to the problem of the working medium pressure losses in the installation pipe system, the hydraulic resistance and the valve co-operation with the pipework and the regulation elements, which is the fundamental aspect of the valve operation. The most common hydraulic characteristics encountered in the theory of regulation are presented together with an extensive mathematical basis. This provides a solid foundation for the presentation of a mathematical model that enables an analysis of the effects of the installation control by means of regulation valves and of their impact on the operating parameters of the other elements of the system. A lot of attention is given to the notion of the valve *authority*, as one of the main parameters determining the process of regulation by means of a valve. An extensive theoretical basis is presented together with a detailed mathematical analysis. The algorithms are compared to those used in the engineering theory and practice so far, indicating the differences, the reasons behind them and their consequences for the heating installation regulation process. Moreover, the book offers a novel and original analytical method of the regulation valve sizing. It also presents a novel and original technical solution in the form of a double-regulation valve free from the limitations having a negative impact on the quality of the hydraulic circuit regulation process, which are typical of the known and popular structures currently in use.

The book is closed with a chapter presenting computational examples. The examples are specially constructed to offer great theoretical value but also to relate to practical problems of the regulation valve operation in a heating installation and of the valve co-operation with the regulated objects, i.e. the room radiators.

The publication is intended for teaching specialists, designers, heating installation makers and operators, as well as scientists and authors of computer programs used for the heating system thermal and hydraulic balancing. It will also prove useful for students of specialities such as environmental engineering, power engineering.

Chapter 1
Aim of the Process of a Hydronic Heating Installation Control and Regulation

Control is the process of using input signals to affect the controlled object so that the output signals or the object status should reach the desired value. Input signals are referred to as the input function, whereas output signals—as the response. The water mass flow rate and the radiator power can serve as examples of the input function and the response, respectively.

The control action is realized through a control system/controller. The control process can be realized in an open-loop or a closed-loop control system. In the former, the input parameter (or a group thereof) is set so that, at the controlled object known response characteristic (static or dynamic regulation characteristic), the desired value or status can be obtained at the outlet, taking no account of the variable impact of disturbing signals produced for example by the environment in which the object operates. In other words, the input signal is independent of the output signal. Closed-loop control differs from the open-loop system in that the controller receives additional information about the object status or about the process output values. The information is used to correct the input value systematically to bring the output value to the desired level. Such actions are referred to as feedback. In practice, if the control system has a feedback loop, i.e. if the control system is a closed-loop one, it is referred to as a regulation system. The controlled object is then called the regulated object and the controller is referred to as the regulator.

No-feedback systems are scarce in practice because they do not provide a mechanism for correcting the object status if the object or the controller operation is disturbed.

Feedback can be positive or negative. Negative feedback is the most common self-regulation mechanism both in nature and technology. It occurs if the disturbance causing deviation of the process output values or of the object status from the set (desired) values generates actions leading to a change in the output values or in the object status in the opposite direction (hence "negative"), i.e. the actions aim to eliminate (compensate for) the consequences of the deviation. The effect of negative feedback is that the parameter oscillates around the set value. In the case of positive feedback, if disturbance arises in the system, the system attempts to change the output

D. P. Muniak, *Regulation Fixtures in Hydronic Heating Installations*, Studies in Systems, Decision and Control 187, https://doi.org/10.1007/978-3-030-03128-2_1

values or the object status according to the disturbance direction and sense (hence "positive"). The effect of positive feedback is that deviation increases.

Open-loop (no feedback) and closed-loop (with feedback) control systems can be illustrated by the example of a heating installation with the radiator manual or automatic valves. If the room and the radiator static thermal characteristics are known [1, 2], it is possible to determine the supply water parameters, i.e. the supply water mass flow rate and temperature at which the radiator power guarantees the desired level of the room temperature. In this case, the radiator manual valve is enough to set the water mass flow at the water given temperature. A structure like that is therefore an open-loop system, and it is able to ensure the set value of the room temperature taking account of specific and known external conditions (e.g. the temperature of the environment). However, if disturbance occurs, e.g. a window is opened, the room temperature drops, i.e. an unwanted change occurs in the output parameters. An automatic regulation valve attempts to maintain the initial set temperature by opening itself accordingly to increase the water mass flow through the radiator and, in this way, raise the radiator heat output, i.e.—raise the temperature in the room. This is due to the information the valve receives as feedback from its sensor that the temperature level has dropped below the set value. The information is a part of the feedback loop.

In the case of the radiator manual valve, the control process can also have the nature of automatic regulation because the temperature sensor (and the regulator) can be the person in the room, who systematically and continuously corrects the valve setting to maintain the desired temperature level. Therefore, the distinction between manual and automatic regulation is mainly arbitrary because in both cases the process has the same character and "manual control" simply means that the regulator is a human.

The aim of the heating installation control and regulation process is to guarantee and keep at a certain level or prevent an excessive rise or drop in the following most common parameters:

- the room temperature. This is the basic parameter subject to regulation. The parameter value is stabilized and must be maintained at a level not lower than the defined one. By definition, a heating installation is intended to ensure a specific value of the heated room temperature, which is one of the basic parameters of thermal comfort, as stipulated by relevant legal requirements [1–3].
- temperature of the heating medium return to the boiler or another heat source. This parameter is usually not subject to stabilization. The only restriction is that it must not fall below a certain value because in some situations, especially in the case of boilers fired with solid fuels, a drop of the medium return temperature below a certain level is harmful (the issue is discussed in Sect. 2.3.3).
- temperature of the outside of the surface radiator, e.g. the underfloor radiator. This parameter is usually not subject to stabilization. The only restriction is that it must not rise above a certain value, which is related to permissible temperatures of surfaces that a human foot can come into contact with [1, 2].

It should be added that control and regulation systems most often do not have a direct impact on the quantities mentioned above. They affect them indirectly—through other parameters. In the case of the room temperature regulation and stabilization, these parameters are: the medium mass/volume flow rate and the medium temperature. The medium mass/volume flow rate is regulated by means of local throttling of pressure, e.g. using regulation valves, or through a global change in it, e.g. using pumps with the impeller adjustable rotational speed. The medium temperature is regulated for example by changing the amount of fuel fed into the boiler or by changing the amount of combustion air. For this reason, the control system has to include elements which are able to convert the value of one parameter, e.g. the measured-target parameter, to another, which is used to regulate the target parameter. The target object of regulation is de facto the room with all the elements of equipment and not the radiator because the regulated parameter is the room temperature.

References

1. Muniak, D.: Grzejniki w wodnych instalacjach grzewczych. Dobór, konstrukcja i charakterystyki cieplne Radiator in hydronic heating installations. Structure, selection and thermal characteristics), WNT/PWN, Warszawa (2015)
2. Muniak, D.: Radiators in hydronic heating installations. Structure, selection and thermal characteristics). Springer, Berlin (2017)
3. Rozporządzenie Ministra Infrastruktury z dnia 12 kwietnia 2002 r. w sprawie warunków technicznych, jakim powinny odpowiadać budynki i ich usytuowanie (Dziennik Ustaw Nr 75, Poz. 690), z późniejszymi zmianami (Regulation of the Minister of Infrastructure of 12 April 2002 on the technical conditions to be met by buildings and their location, Dz.U. (Journal of Laws) 02.75.690 as amended)

Chapter 2
Role, Types and Structure of the Heating Installation Regulation Valves

This chapter presents selected issues concerning regulation valves used in heating installations. It offers a description of the valves function, structure and principles of operation.

Regulation valves, or the heating installation regulation fixtures in general, constitute a group of basic devices intended for hydraulic, as well as (indirectly) thermal, balancing of heating circuits. Hydraulic balancing consists in such selection of pipe diameters and settings of the regulation fixtures that equilibrium is achieved between the medium pressure losses and available pressure in each circuit for a given value of the medium flow rate. Thermal balancing is related to such selection of heat receivers (radiators) that their heat output equals the design demand for thermal power in individual rooms.

Hydraulic and thermal balancing can occur for both static and dynamic operating conditions. The distinction should be understood as one between steady states, where the quantities and parameters describing the installation operation are constant in time (e.g. the medium flow rates in circuits), and unsteady states, where the quantities vary over time. To make matters simple, it may be stated that the distinction concerns systems with a constant and changeable flow rate of the medium. Nevertheless, design conditions that determine selection of individual devices concern the installation steady-state operation. The task of automatic regulation fixtures is continuous monitoring and current regulation of set operating parameters of the installation, circuit or part thereof. Manual regulation devices aim to balance heating circuits hydraulically in steady-state conditions (no continuous regulation). Both regulation types—manual and automatic—can be used for the installation any part or element. Due to some legal regulations (e.g. in Poland it is [42]) and reasons related to use and practice, the devices can generally be divided into regulation valves at heat receivers and balancing valves installed in other sections of the installation.

Regulation valves can also be divided with regard to their structure and type of the regulating element. The most common heating installation valves are plug valves, where regulation is the effect of the plug reciprocating motion. The regulating element

D. P. Muniak, *Regulation Fixtures in Hydronic Heating Installations*, Studies in Systems, Decision and Control 187, https://doi.org/10.1007/978-3-030-03128-2_2

is called a plug due to its characteristic shape. Ball regulation valves are also used, but these are less frequent. In this case, regulation is the effect of rotation of a ball with cut out orifices that the medium flows through in the valve body.

2.1 Regulation Valves at the Heating Installation Receivers

Regulation valves at the heating installation receivers, e.g. radiators, fan coils, etc., belong to the most essential elements of the heating installation fixtures. They feature a wide variety of design solutions but serve very similar functions, as specified below:

- balancing pressure losses with active pressure in the installation controlled circuits for a given value of the medium mass/volume flow,
- ensuring the design value of the medium mass/volume flow, according to the results of calculations of the process of the installation hydraulic and thermal balancing,
- ensuring the possibility of automatic current regulation, i.e. of the room temperature regulation by changing the radiator heat output in certain situations required by law (e.g. in Poland it is document [42]).

Before thermostatic heads and valves entered the market, manual radiator regulation valves had been in common use as the basic elements of the heating installation regulation fixtures. Depending on the country, there were different technical requirements regulating their proper use, as stipulated in relevant documents. In Poland for example, before European standards were implemented, they had been regulated by Polish standards. One of them was standard PN-M-75009:1991 [36]. The standard set out required hydraulic characteristics, parameters, as well as terminology and methodology of measurements. According to it, the following are distinguished for one-way valves:

- first-stage regulation, also known as initial, assembly-related hydraulic regulation of the central heating installation, consisting in steady balancing of pressure losses of all flows through individual branches and radiators with active pressure at design (computational) mass/volume flows of the heating medium, which is carried out by means of throttling elements in double-regulation valves. Related thereto is the throttling characteristic, i.e. characteristic of the regulation first stage, which is the hydraulic characteristic of a double-regulation valve defining the flow factor k_v dependence on the throttling element position at full opening of the regulating or closing element. First-stage regulation is related to determining the value of what is referred to as pre-setting.
- second-stage regulation, i.e. current hydraulic regulation of the central heating installation, consisting in introducing periodical changes in the heating medium flow rate in flows through individual heat receivers by means of regulating elements included in radiator valves or through individual risers or branches by means of closing elements of straight-through valves. Related thereto is the opening (closing or flow) characteristic, i.e. characteristic of the regulation second stage. Formally,

this is a relation between the relative surface area A_x/A_{100} of the medium flow cross-section and the valve plug relative lift h_x/h_{100}. Moreover, this is a commonly accepted method of determining the flow factor k_v dependence on the regulating/closing element position at a given position of the throttling element because the characteristics have the same shape.

Depending on the regulation valve structure and function, different regulation characteristics can be distinguished corresponding to the valve specific regulating element or the function thereof. It should also be remembered that in some cases the same terms are not identical, e.g. if a manual regulation valve is compared with a radiator thermoregulator (a thermostatic head connected to a regulation valve).

2.1.1 The Radiator Manual Regulation Valves

The stipulations included in the standards mentioned above were the effect of research necessitated by ever increasing requirements posed at the time for heating installations being energy-consuming objects that decide about thermal comfort in utilized rooms. The importance and use of radiator valves evolved significantly over the years. As early as at the end of the 19th century, valve structures characterized by a proportional (linear) regulation characteristic were known. The first comprehensive research in this field was carried out by *Ambrosius* in 1919 [3]. At the turn of the 20th century, an idea was developed in Europe of so-called double regulation, which enables both the balance of hydraulic resistances and current regulation of flow rates. This type of valves, as devices with two adjustable sections of the medium flow, were also made in Poland before the 1940s. Nevertheless, in the first half of the 20th century the main task of radiator valves was to make it possible only to cut off the water flow completely. The flow rate regulation or ensuring an appropriate regulation characteristic were things that were hardly ever taken into consideration as the purpose of the valve application. This was mainly due to the fact that generally the only installations operating at the time were natural (gravitational) circulation systems, where low hydraulic resistances are required to ensure correct operation. The regulation capacity of valves used at the time was rather limited in such conditions. Apart from that, no reliable mathematical models describing the regulation processes taking place in installations were known. Another essential problem was the difficulty in precise making of the flow-throttling elements with adequate repeatability. The situation started to change after the second world war. Pumps designed especially to satisfy the heating installation needs became available. The theory of regulation methods appeared and more attention was paid to the throttling technique and regulation fixtures [19, 23]. In Poland, the changes occurred in the 1960s and 1970s, when the serious drawbacks of gravitational installations were increasingly perceived, and the old systems started to be replaced with pump installations. It turned out that the fixtures manufactured at the time failed to meet any requirements concerning the regulation quality and costs related to the pump installation operation [12, 19]. The effect of that was common application of orifices (small round plates with a central hole with an appropriate diameter) mounted in the radiator connectors as assembly elements balancing the circuit flow resistances. From the operating point of view, this solution had one major downside—any correction to throttling in the system involved the orifice replacement, which meant a partial shutdown of the installation and, quite frequently, emptying a part thereof. More importantly, it often turned out that due to the limited minimal diameters of the orifices and the considerable differences between subsequent values in the diameters series of types, it was impossible to balance hydraulic resistances in circuits and ensure design operating parameters. Moreover, due to the deformation of the closing characteristic, the solution had a very unfavourable impact on the regulating capacity of the valve itself. In fact, in this way it became a cutting off element only. For this reason, as the problems could not be eliminated anyway, attempts were made to analyse the possibility of replacing the old solutions with orifices integrated permanently with the radiator valve [44, 46]. This option, however, never reached the stage of mass production. Another solution, developed by *Kwapisz* and *Piłatowicz,* was also discussed in the 1960s. It concerned a valve with a replaceable orifice [16, 22], selected depending on the required value of pressure to be throttled. The two scientists put forward a solution in the form of setting a different geometry of the closing plug for a given valve model depending on the share of the selected orifice hydraulic resistance/pressure losses. Depending on the share of the resistance, the regulation characteristic of the closing element (the valve plug) displayed a different deformation degree. A correction to the plug geometry was to ensure that the characteristic shape should remain unchanged [16, 23]. Neither of the solutions, though, went into mass production or practice. The structural and operating usefulness of the radiator double-regulation valves available at the time: the *M-3172* (straight-run) valve and the *M-3174* (angle) valve was also rather poor [12, 19]. First-stage regulation was realized by a cylindrical shutter placed beyond the regulating element to create an additional regulated section of the medium flow. The closing element itself—the plug—had a flat shape corresponding to the original linear regulation characteristic. The combination of the two aspects resulted in very bad regulating features. In addition, the valves did not ensure proper tightness and their parameters were characterized by considerable spreads compared to their nominal values. They were also often made based on different drawings and different specifications of their principal elements [12]. The available single-regulation valves such as the *M-3173* (straight-run) valve and the *M-3175* (angle) valve, which were intended for single-pipe gravitational installations mainly, also displayed poor regulating qualities due to the unfavourable shape of the plug. The research carried out in Poland in the 1960s and 1970s aimed to modernize current solutions and develop the radiator valve appropriate structure. Another goal was to create an opportunity to achieve the main targets: adequate current regulation of the flow rate and balance of hydraulic resistances, i.e. to implement the principle of double regulation. In 1970, a prototype of a manual double-regulation radiator valve was developed and made in Krakowskie Zakłady Armatur, in co-operation with ZBiD/COBRTI "Instal". The valve symbol was *M-3176* and it was intended for pump installations. The valve angle figure was also developed with the symbol *M-3177*. The double-regulation

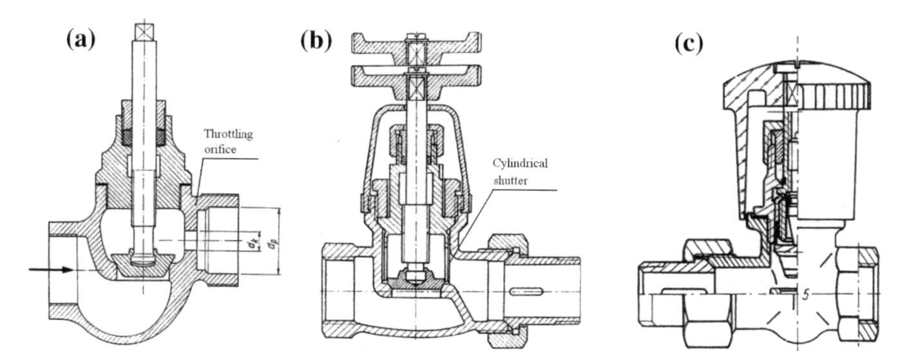

Fig. 2.1 Polish historical double-regulation radiator valves: **a** valve with an orifice integrated within the valve body; **b** *M-3172* valve with a cylindrical shutter; **c** *M-3176* valves with a reduction in the plug lift [16, 19]

concept was realized in the valves through a reduction in the lift of a single plug profiled for an equal-percentage/linear characteristic. The results of experimental testing and relevant parameters of the valve *M-3172* and *M-3176* models can be found in [16, 19, 23]. The structures of the *M-3176* and *M-3172* valves and of the valve with an integrated orifice are presented in Fig. 2.1.

Foreign radiator regulation valves could also be found on the market at the time but for systemic reasons their availability was rather limited. Heating installation fixtures using the principle of double regulation were offered by *Herz*, *Tour Agenturer*, *Dicon* and *Goeke* for example. The first two companies realized the principle by a reduction in the lift of a single plug, like in the case of valve *M-3176*. The *Herz* valve was additionally adapted to co-operate with either a manual or a thermostatic head. The *Goeke* company used the solution in the form of a cylindrical shutter, like in the valves already considered in Poland as imperfect. But it also drew on an interesting mechanism of a curvilinear shape of the single plug in the valve model named *Optimal*, whose characteristic feature was shaping the plug for an asymmetric profile. The plug rotation around the axis in relation to the liquid outflow channel caused a change in the free cross-section and, consequently, a change in hydraulic resistance and throttling, whereas current regulation was realized through a reciprocating motion. This is a favourable solution because it integrates the first and the second stage of regulation in one adjustable section of the medium flow, like in the variant discussed above. It has a positive impact on regulation properties because—as indicated in Sect. 4.1—it does not decrease the value of inner *authority*. Here, by contrast to the solution with a reduction in the plug lift, the closing element range of travel can be independent of the pre-setting. Owing to that, the structure is also favourable in terms of co-operation with thermostatic heads, as it ensures a constant value of what is referred to as proportional range X_p. It is also possible to shape the plug in such a manner that a favourable equal-percentage closing characteristic is ensured. Figure 2.2 presents cross-sections of the valves made by the companies mentioned above.

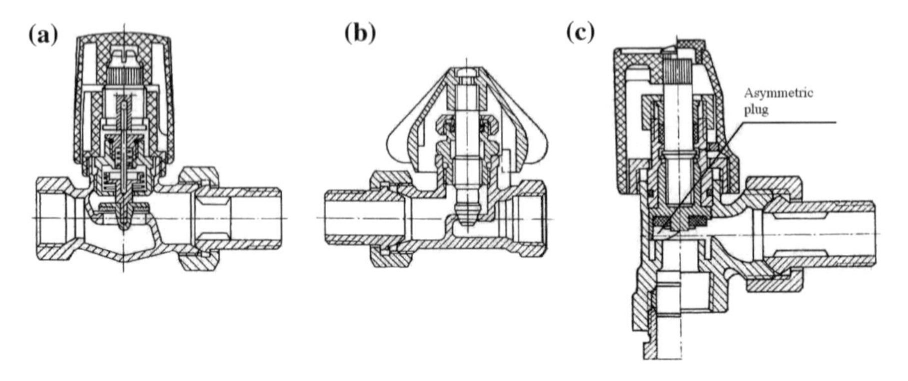

Fig. 2.2 Historical structures of radiator regulation valves: **a** *Herz* valve with a thermostatic head; **b** *RVO* valve made by *Tour Agenturer*; **c** *Optimal* valve made by *Goeke* [16]

2.1.2 The Radiator Automatic Thermoregulators

Most EU countries have regulations concerning thermal performance of residential buildings and the use of the radiator thermoregulators as devices that make it possible to rationalize the consumption of energy for heating purposes. According to the current legal state in Poland [42], in certain situations the radiator regulation valve (in buildings excluding e.g. single-family houses) must ensure automatic regulation of the room temperature. The valve satisfies this requirement in co-operation with the following elements:

- thermostatic heads,
- electronic and electric heads,
- heads with thermal heads/actuators.

Considering that the most popular and historically first solution in common use was the thermostatic head, the valves are now referred to as thermostatic valves. This term is not completely appropriate because a radiator valve is not an element of automatic regulation and itself it does not contain any thermostatic elements. As a result, it cannot change the medium flow rate automatically. However, the term is generally accepted and used despite the fact that according to standard EN 215:2004 [14] a *thermostatic valve* is defined as a combination of a regulation valve and a thermostatic head.

The idea of a radiator thermostatic thermoregulator has been known for dozens of years. In terms of automatics, it is a proportional, self-actuated temperature regulator operating continuously in an automatic system to keep the pre-set value at a constant level, with a negative feedback. Self-actuation means that the device does not need any extra energy apart from that taken from the regulated object. Proportional regulation (defined in automatics as *P regulation*) means that a change in the head (the valve follower, cf. Fig. 2.3a) response is proportional to the change in measured temperature. Radiator thermoregulators are used to compensate for momentary oscillations

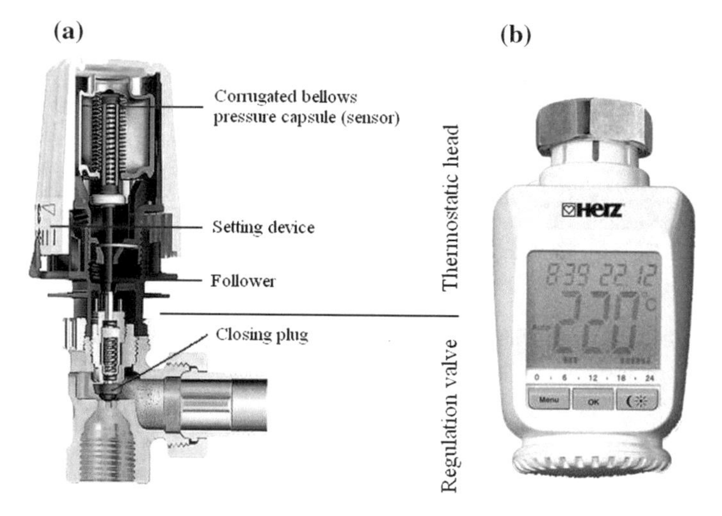

Fig. 2.3 Radiator thermostatic thermoregulator (**a**) and the *Herz ETKF* radiator electronic head (**b**) [6, 7]

in the demand for thermal power in a given room, arising e.g. due to unexpected heat gains from solar exposure, humans, electric appliances, etc., to ensure the room constant temperature. By doing so they also reduce the consumption of energy for heating purposes and favour its rational use. The savings generated by the occurrence of "free" heat gains and the reduction in the room overheating vary from case to case. They depend on the building thermal properties, the climatic zone, the installation operating parameters, etc. It is estimated that the savings may reach almost 20% [48, 49].

A typical thermoregulator is composed of two elements:

- the control unit (controller, regulator referred to as the head), which measures the temperature in the room as the regulated quantity and compares it to the pre-setting value, triggering the reciprocating motion of the follower controlling the valve,
- the executive unit in the form a of a single- or double-regulation radiator valve.

The two elements mentioned above—the regulation valve and the head, which together make up a *thermostatic unit* (also referred to as *thermoregulator*), are to ensure and maintain an appropriate temperature level in the heated area. It should be noted that the room adequate temperature is one of the main components of thermal comfort [33, 34]. Figure 2.3a presents a typical radiator thermoregulator as a combination of a regulation valve with an integrated thermostatic head and an electronic head mounted on the valve.

The principle of a thermostatic head operation is based on the phenomenon of thermal expansion, (vapour) pressure or the volumetric change caused by a change in the state of aggregation of the substance used to fill the temperature sensor, and the process conversion to a reciprocating motion of the follower controlling the valve plug. The head and the valve are selected for the design operating conditions, where

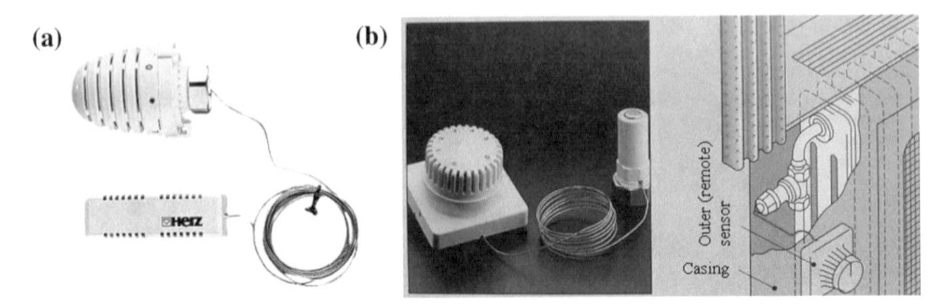

Fig. 2.4 Thermostatic head with an outer sensor (**a**); thermostatic head with an outer sensor and setting device (**b**) [1, 2, 7]

temperature t_{zad} set on the head setting knob is equal to temperature t_c measured by the sensor and corresponding to a given room temperature t_i. For such conditions, the valve plug position meets the design assumptions for which the required volume flow of the medium is supplied, ensuring the radiator required heat output. If temperature t_c measured by the head sensor changes, the volume or pressure of the substance filling the sensor will change too, changing the follower position and—thereby—altering the position of the valve plug. For example, if temperature t_c measured by the head sensor rises above set temperature t_{zad}, the capsule will expand, the head follower will come out and the valve plug will close. This will cause a decrease in the working medium mass/volume flow and a reduction in the radiator heat output. Consequently, the room temperature will be lowered. If the temperature measured by the head sensor drops, the process runs the other way round. It should be remembered that though the thermoregulator capability to reduce the medium flow rate is in fact unlimited (to zero), the possibility of raising it above the design value may be substantially restricted. This is the effect of two factors: the relative opening degree of the valve closing plug for a given proportional range X_p (for which the required position of the valve plug is specified and which informs how much the element may be opened above the value) and the value of what is referred to as the valve total *authority* discussed in detail in Sect. 4.1. For most radiator thermostatic heads, the temperature of 20 °C corresponds to number 3 on the setting knob, temperature 16 °C—to number 2, and temperature 24 °C—to number 4. In practice, there is always a difference between the room temperature t_i and the temperature measured by the head, as described further below. For this reason, for a long time the scale of values to be set on the head has not been a temperature one.

There are now several basic structural thermostatic head solutions:

- with a sensor and the setting device built into the head (cf. Fig. 2.3a),
- with a remote (outer) sensor connected to the head setting device using a capillary (cf. Fig. 2.4a),
- with a remote (outer) setting device connected to the sensor using a capillary,
- with a remote (outer) head, i.e. a remote sensor and setting device connected to the valve base containing the follower using a capillary (cf. Fig. 2.4b).

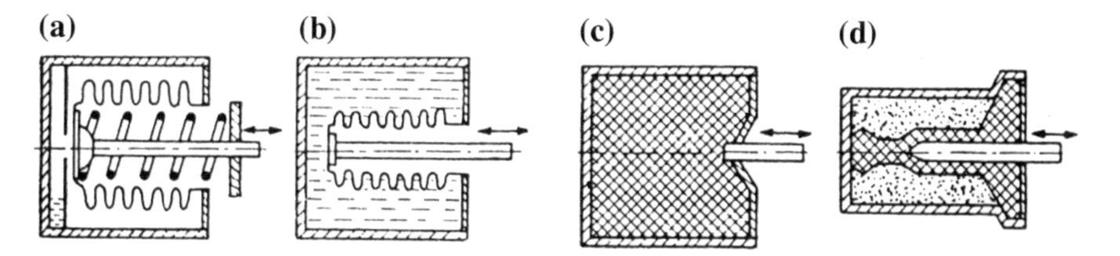

Fig. 2.5 Thermostatic head sensors: **a** vapour sensor; **b** liquid sensor; **c** sensor with a solid body; **d** sensor utilizing the change in the substance state of aggregation [18]

A thermostatic head with an outer sensor, as the second most popular solution, is shown in Fig. 2.4a. Figure 2.4b presents a remote head (with an outer sensor and setting device) installed in a typical way on the radiator.

Moreover, there are a few kinds of the thermostatic head sensors differing from each other by the type of the substance used in them and depending on the physical phenomenon their operation is based on. Using this criterion, the following kinds of thermostatic heads can be distinguished:

- heads based on thermal expansion of a liquid,
- heads based on thermal expansion of a solid body,
- heads based on changes in saturated vapour pressure over the surface of a liquid,
- heads based on changes in the substance volume occurring with a change in the state of aggregation (melting or solidification).

The design solutions of the sensors are presented in Fig. 2.5.

In the case of liquid or gaseous mediums, the sensor is built using a metal pressure capsule or an elastic membrane/bellows which get deformed due to changes in the medium pressure or volume and transform the process into reciprocating motion of the follower controlling the valve. The thermostatic liquid is usually a mixture of hydrocarbons, silicone oil, alcohols or other substances characterized by high thermal expansion. Mixtures of heavy hydrocarbons are frequently used in gaseous sensors. A rise in temperature involves evaporation of the medium liquid fraction and an increase in pressure until a new state of the mixture thermodynamic equilibrium is reached. Generally, there is no increase in the medium volume, but in a closed space a rise in pressure occurs. The sensors are equipped with springs balancing the arising force. They change their shape setting the follower position according to the set temperature value.

If a solid body is utilized, the sensor is made as a uniform block of silicone rubber or plastic which is covered with a moveable lid with the follower connected thereto.

If the sensor makes use of the phenomenon of the change in the substance state of aggregation, an elastic membrane or bellows is used like in the case of liquid sensors. For the thermoregulator to operate correctly, it is necessary to select a substance for which the change in the state of aggregation occurs smoothly in the required temperature range that coincides with the range of the head regulation and not abruptly for a single temperature value only. This is the feature of some mixtures of technical wax.

The sensor and the sensor design selection is dictated by a few basic parameters. The thermostatic regulator should be characterized by possibly high sensitivity to temperature changes to ensure considerable changes in the head follower position for a given change in the medium temperature, i.e. a high value of what is referred to as the head amplification factor k_{m}. At the same time, the sensor mass should be as small as possible to ensure a low value of thermal capacity. It is also favourable for the ratio between the sensor heat transfer surface area and mass to be possibly high. Combined with the sensor small mass, this favours reaching small values of thermal inertia and ensures a quick response to changes in ambient temperature. However, although a small mass of the sensor substance usually means low thermal inertia, this unfortunately also involves low sensitivity to temperature changes.

The most popular devices nowadays are liquid sensor thermostatic heads, which use the phenomenon of the working medium thermal expansion. The solution is fairly cheap and the most universal. The pressure difference arising on the valve plug always[1] creates a force on the head follower which has a sense opposite to the force produced by the sensor. Due to low compressibility of the liquid, liquid sensors are rather insensitive to changes in the force. Moreover, a liquid sensor head performs well in each of the thermostatic head design solutions mentioned above.

The option with a vapour sensor based on the vapour pressure phenomenon is noticeably more expensive, but it has some advantages compared to classical thermostatic heads using a liquid. The basic ones are the smaller hysteresis zone, being the effect of a considerable reduction in friction between moving elements, the small mass of the substance and the low thermal inertia, resulting in a low time constant and a short time of response, and the high value of the obtained amplification factor k_{m}, which is approximately twice higher compared to liquid sensors. However, gas sensors are not advantageous in the case of heads with an outer sensor. The hysteresis zone may be considerably larger then.

Sensors based on the principle of the solid body thermal expansion or those utilizing the change in the substance state of aggregation are cheaper solutions compared to the ones discussed earlier due to the simplicity of their design and making. For this reason, they were fairly popular in the 1970s. However, they are characterized by much larger hysteresis zones because of additional forces arising due to friction between the thermally expandable substance and the casing. Moreover, this phenomenon increases the sensors wear and failure rate.

Sensors making use of a solid body are quite easy to make, but they also have rather serious limitations. The solid body low thermal expansion makes it necessary to use a considerable volume of the substance to ensure an appropriate value of the amplification factor. This involves a rise in the thermoregulator thermal inertia and

[1] It is most often recommended that the medium should flow in under the plug so that the pressure difference threon should not generate a force closing the valve but one that opens it. In terms of usability, this is more favourable because if the medium flows in the system, the valve will not close by itself, and it will open if there is no setting element or if a failure occurs.

Fig. 2.6 Historical radiator thermoregulators, including the first to be made (far left) [5]

time constant. Moreover, the solution is rather sensitive to changes in pressure on the valve plug. The force generated by the pressure difference arising on the head follower, which makes the follower get into the sensor, shifts the static characteristic into the range of higher temperatures. Due to that a considerable difference is created between the temperature set on the head and the measured temperature value, and oscillations in this difference occur depending on the pressure difference on the valve plug and on active pressure in the installation given point. Moreover, the closing point is shifted into the range of higher temperatures.

The first thermostatic valve was constructed by *Mads Clausen*, the founder of the *Danfoss* company, in 1943 [22]. Within the structure, the valve was integrated with the sensor. In 1965, the *Danfoss* company was again the first to propose a solution with a separate thermostatic head mounted on the regulation valve. The devices, together with subsequent versions of the company thermostatic heads, are presented in Fig. 2.6.

In Poland, the first structure of the radiator thermoregulator came into being in 1970 [16, 20, 21]. It was a single-regulation valve co-operating with a head equipped with an integrated vapour pressure sensor. It was the *Jerzy Kwapisz OBD-SPEC* design drawing on foreign solutions. Apart from that, in the early 1970s in the *Mera* Industrial Research Institute for Automation and Measurements and in the *Predom-Dezamet* Research & Development Centre the *RCO 701* thermoregulator was developed with a head utilizing a wax sensor. The structure was quite untypical, with a remote sensor (unlike the *OBD-SPEC* design) and with a gear lever acting as an amplifier and connecting the head follower to the single-regulation valve [16]. The device is presented in Fig. 2.7.

The *RCO 701* model saw the day of serial production, but later on practice revealed its many defects, which resulted in the device withdrawal and development of the *RCO 708* prototype. The new structure was typical. It had a head with an integrated wax sensor and its target application was to co-operate with a double-regulation valve with a double plug with a lift of 2.5–3 mm. The 708 valve was developed in the late 1970s in co-operation between *Instal Warszawa* and *Zakłady Metalowe im. T. Dębala* in Nowy Dąb. This double-regulation valve was intended to co-operate

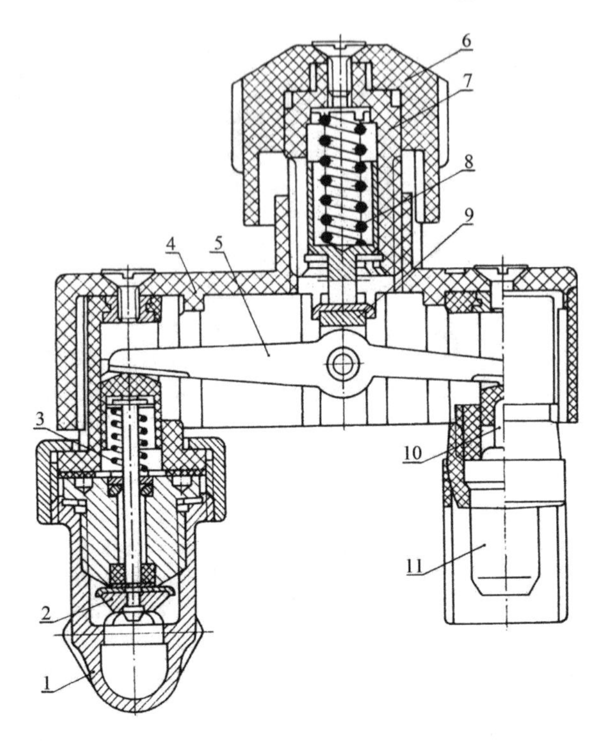

Fig. 2.7 View of the *RCO 701* radiator thermoregulator [16]

with either a manual or a thermostatic head. The double-regulation idea was realized in it through an outer closing plug (regulation second stage) integrated with an inner throttling element (regulation first stage). However, the thermoregulator did not go into serial production. The structures are illustrated in Fig. 2.8.

The standardization process carried out for radiator thermostatic regulators resulted in, then obligatory, Polish standard PN-M-75010:1991 [37] being a translation of the German standard DIN 215 [15]. The standard current version is EN 215:2004 [14]. The document concerns the radiator thermostatic valve as a whole, i.e. as a combination of a radiator regulation valve and a thermostatic head.

2.1.2.1 Impact of the Thermoregulator Location on the Radiator Performance

The location of the thermostatic or electronic head sensor may have a substantial effect on the device operation, varying the measurement of temperature t_c and distorting it in relation to the room temperature t_i. Although the integrated head variant is the most functional in practice, the cheapest and the easiest to make, considering the function it fulfils, it is not the optimal one with respect to the thermoregula-

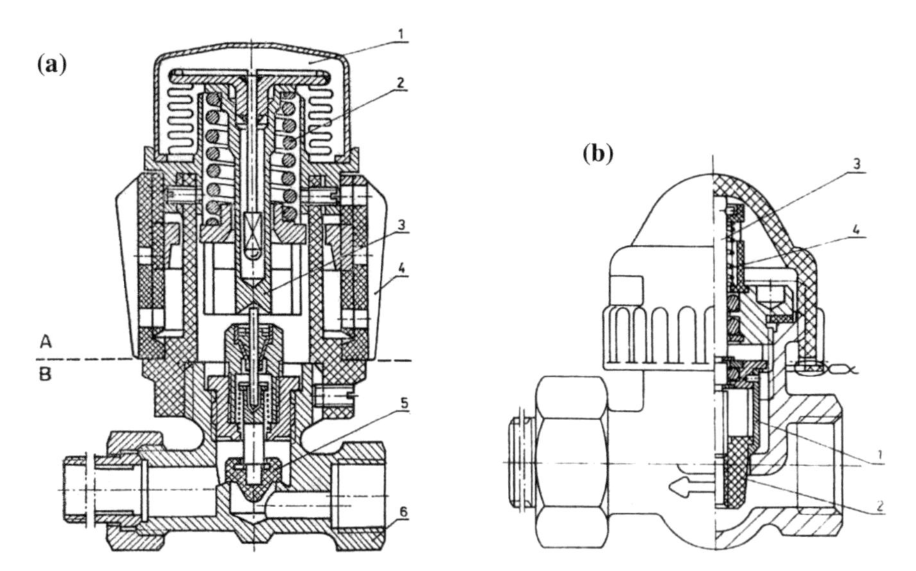

Fig. 2.8 Radiator thermoregulators: **a** *OBD-SPEC* radiator thermoregulator A—thermostatic head, B—regulation valve, 1—sensor, 2—main spring, 3—follower, 4—setting device knob, 5—valve plug (closing element), 6—valve body; **b** radiator *708* regulation valve intended to co-operate with a thermostatic head: 1—closing plug, 2—throttling plug, 3—spindle, 4—spring [16]

tor operating conditions. A sensor incorporated into the head directly connected to the valve is exposed to additional interference that may lead to false temperature measurements, usually inflating the results unduly. It should be remembered that the device task is to regulate the temperature in the room and not to regulate the medium flow through the radiator. Nor is it supposed to regulate the temperature of the radiator itself. Therefore, so that the regulation process should run correctly, the thermoregulator measuring sensor actually has to measure ambient temperature, and this value has to be regulated and stabilized. In the case of an integrated head, the measurement is subject to distortion for the following reasons:

- the thermoregulator is mounted near the radiator, in the zone of air warmer than in the room other areas, and it is additionally exposed to the radiator surface thermal radiation,
- the thermoregulator is mounted on a pipe supplying a medium with a temperature higher than ambient, so there is an additional heat transfer to the sensor from the valve and head elements,
- the thermoregulator is often screened by curtains or other elements inhibiting air circulation in the thermoregulator area and creating zones of warmer air,
- the thermoregulator is close to the windowsill, which favours creation of zones of warmer air.

The effect of the factors mentioned above is that the device "senses" temperature t_c higher than the real room temperature t_i and set temperature t_{zad}, and closes the valve to obtain the value in the thermoregulator area equal to the set temperature. Due to that however, the room temperature beyond the thermoregulator area is lower than actually required. The problem is solved by using thermoregulators with an outer sensor installed beyond the zone of the radiator impact, for example on the room side wall, where the temperature measurement is representative of the room actual temperature. However, the downside of the solution resulting from the need to use a capillary is the usually increased hysteresis zone (due to additional elements causing the sensor substance friction) and the higher thermal inertia (due to the sensor substance increased mass).

Considering the above, a conclusion may be drawn that placing an integrated thermoregulator on the radiator supply pipe, which in fact is the only option currently used, is incorrect because a lower temperature of the radiator surface, of the pipe carrying the working medium and of air, i.e. a temperature value closer to the room temperature, occurs if the device is mounted on the return pipe at the radiator bottom. The advantage of such a location is not only a smaller regulation error, i.e. the difference between the room temperature t_{i} and the value measured and set on the head, t_{c} and t_{zad} respectively, but also smaller variations in the error itself because seasonal oscillations in the heating water temperature are smaller for the return than the supply pipe [33, 34]. Another advantage of this location of the thermoregulator is the fact there is no need to make so big and frequent seasonal changes in the values set on the head as is the case of the device mounted on the supply pipe. Placing the radiator thermoregulator on the supply pipe is popular mainly due to reasons of functionality, i.e. to provide an easier access to the device if the setting needs to be changed.

Figure 2.9 presents results of an experimental verification of the thesis. The testing concerned the difference between temperature t_{i} measured at the height of 1.5 m over the floor in the middle of the room where the tests were conducted and temperature t_{c} in the area of the thermoregulator anticipated mounting on the supply or on the return pipe. It may be assumed that the difference is identical to the difference between temperature t_{i} in the room location as described above and the hypothetical value of set temperature t_{zad}, set on the thermostatic head. This made it possible to examine the impact of the thermoregulator location on the temperature it measured compared to representative temperature. The tests were performed for different temperatures of the outer environment corresponding to different values of temperature of the radiator supply water. Owing to that, it was possible to examine the impact of changes in the supply water temperature on changes in the temperature measured by the thermoregulator compared to representative temperature.

The analysis results confirm the thesis. The difference between measured temperature t_{c} and representative temperature in the room is bigger near the supply pipe compared to the return one. So are the changes in the difference during the heating season. It follows that if an integrated thermoregulator is installed on the radiator supply pipe, in order to obtain a certain value of temperature t_{i}, a bigger correction to the value of temperature t_{zad} set on the head has to be made compared to the theoretical value than if the device is mounted on the return pipe. Additionally, current corrections to the setting necessitated by changes in the temperature of the outer

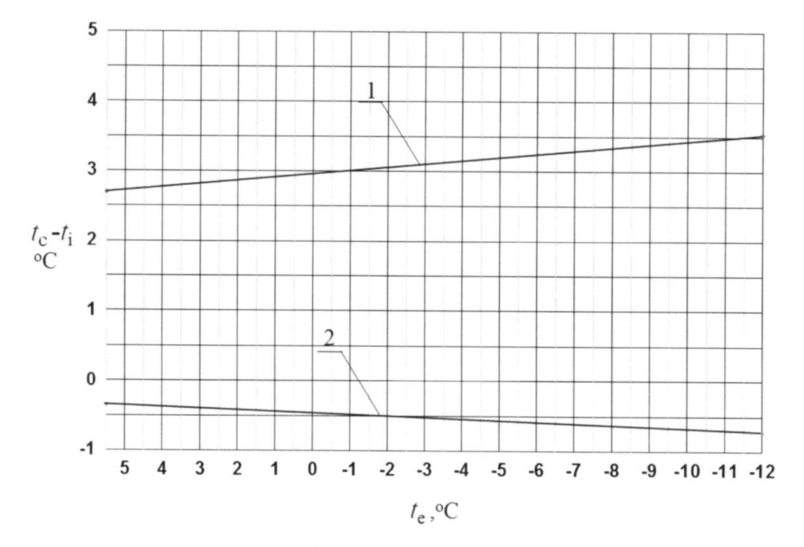

Fig. 2.9 A comparison of the hypothetical error in the regulation of an integrated thermoregulator depending on location: 1—for a thermoregulator mounted on the radiator supply pipe; 2—for a thermoregulator mounted on the radiator return pipe [16]

environment will have to be bigger, too. It can also be seen that if the thermoregulator is installed on the return pipe, the difference under consideration is negative. This means that the thermoregulator detects a temperature lower than the value prevailing in the middle of the room, which is natural because the radiator is installed in the room in a typical way—right over the floor. Circulating in the room, air gradually gives up heat to the surrounding partitions. Under the room ceiling the temperature is the highest, whereas the lowest values prevail near the floor (this is true for the type of heating under consideration; for an underfloor radiator, for example, the situation is different). An intermediate value, higher than the one near the floor, settles halfway through the height. If the radiator is located differently, e.g. higher, on an internal wall, or under the ceiling, the error in regulation may have a positive sign, but absolute values will each time be higher for the case where the integrated thermoregulator is installed on the radiator supply pipe.

Obviously, the presented results give a general picture and the values may differ depending on the room size, the representative temperature measurement location, the room thermal losses, the installation operating temperatures, the valve and the head structure thermal resistance, etc. For example, the lower the radiator operating temperatures, the lower the differences and the less perceivable the phenomenon will become.

Apart from the location of the radiator thermoregulator, or, more precisely—the location of the thermoregulator temperature sensor, the device operation is also affected by the method of assembly on the radiator pipe. The head of an integrated thermostatic or electronic thermoregulator should be installed so that the impact of the air heated by the pipe and flowing up should be minimized. And this means horizontal assembly. If the head was mounted vertically on the supply pipe upwards, the air heated by the pipe and flowing up would inflate the temperature measured by the head sensor. If it was mounted downwards, similar influence, though smaller due to the bigger distance and the medium lower temperature, would be exerted by the radiator return pipe. The same would be the case at vertical upward assembly on the return pipe. The above factors can be minimized by locating the head horizontally.

Due to the fact that the radiator integrated thermoregulator is usually screened by a curtain and is as a rule partially covered by the windowsill, the temperature set on the head has to be a bit higher than the temperature value actually wanted in the room. Opposite situations are very rare. If they do occur, it is usually due to very untight windows or the use of trickle vents. If cold air from the outside flows directly onto the head, temperature t_c it is supposed to measure turns out to be too low compared to set temperature t_{zad}, and the room may become overheated. A situation like this can also occur if the operating parameters selected for the thermoregulator result in a large proportional range.

Figure 2.10 presents different variants of the location and installation of different types of thermostatic thermoregulators indicating correct solutions and those which are not optimal. The examples illustrate different variants of the thermoregulator assembly if it is located on the radiator supply pipe. If it is installed on the return pipe, the situations are the same. The figure presents a few typical instances of the radiator thermoregulator installation. Apart from the most common one, where the radiator is installed under an external window without being covered in any way, there is also the case where the radiator is hidden by a kitchen cabinet. Such location of the radiator is not optimal by principle, but if it does occur, using an integrated thermoregulator is a serious mistake. The regulation error in these conditions will be very high and the device will in fact lose the capacity for automatic regulation and stabilization of the heated room temperature. Therefore in this case, a thermoregulator with an outer sensor should be used, like in the case of the canal underfloor radiator (top, far right).

2.1.2.2 Co-operation of Several Radiator Thermoregulators Within a Single Room

It is generally assumed that each radiator should be equipped with a thermoregulator. This seems natural if there is just one radiator operating in the room. However, an analysis should be conducted of the situation where the room is heated by a few radiators. It may then turn out that the thermoregulators mounted on them will exert an undesirable effect on each other. Let us analyse the case where a single room is heated by two radiators, both of which have their own thermoregulator. Assume that the user wants to lower the room temperature t_i. In order to do so, he/she reduces pre-setting t_{zad} (i.e. turns down the setting knob on one of the heads). Due to that, the flow through the radiator gets smaller and the radiator temperature and, consequently the radiator heat output, drops. As a result, the room temperature

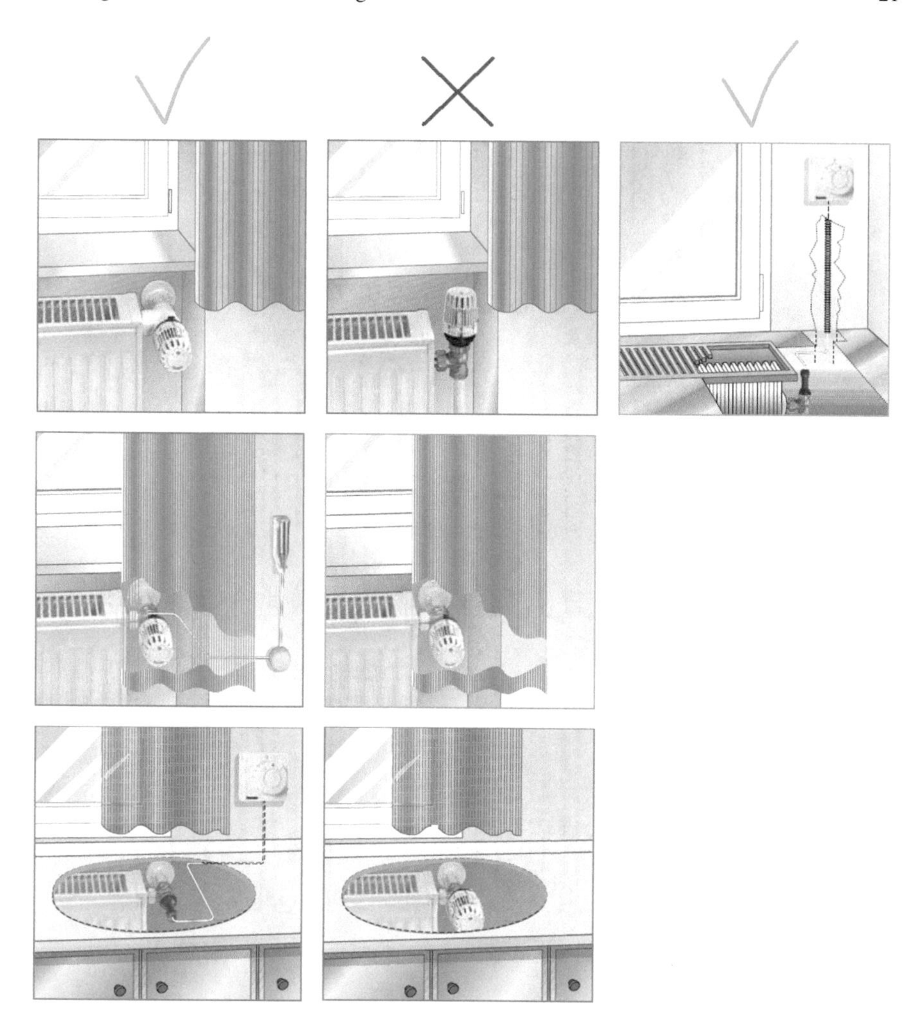

Fig. 2.10 Methods of the radiator thermoregulator installation (based on [6])

starts to decrease. However, the other head is still set to the initial value of set temperature t_{zad}. Detecting a drop in the room temperature below the set value, the head will open the radiator valve to raise the temperature and restore the pre-set level. It may thus turn out that, though the setting of one of the heads is reduced, the decrease in the room temperature will not be achieved. Moreover, such operation may disturb hydraulic order of the installation other circuits due to an unexpected increment in the medium mass flow rate in one of them. This is especially possible in the current practice of the radiator thermoregulator selection, where generally the rule of initial throttling that could prevent the phenomenon is not applied strictly. Consequently, if two or more thermoregulators operate in the room independently, their effect may be contradictory—they will counteract each other and thus fail to

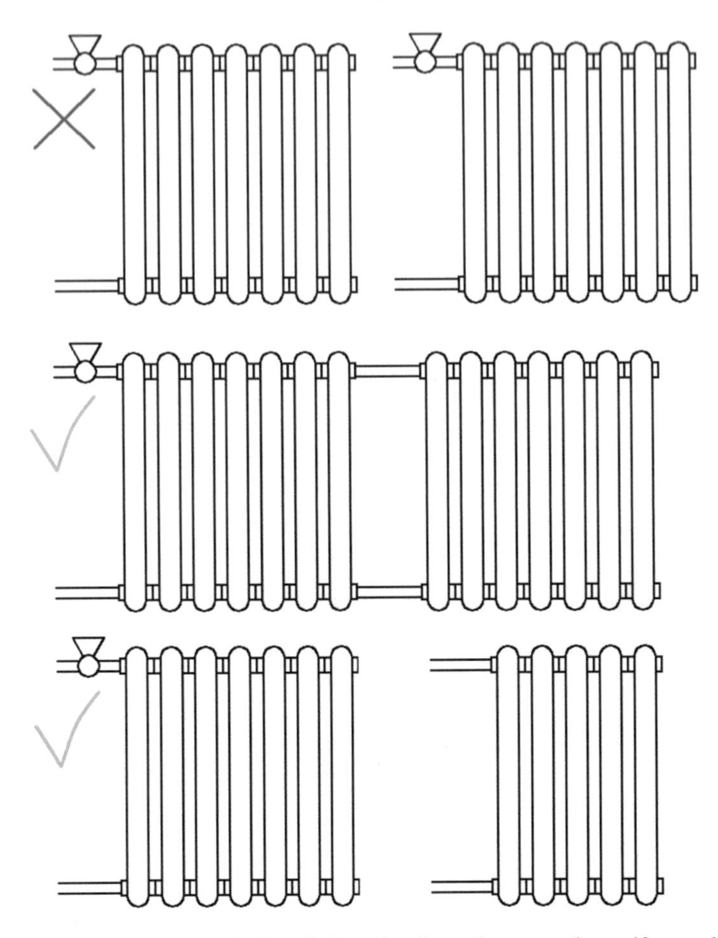

Fig. 2.11 Principles governing the installation of radiator thermoregulators if several radiators operate in a single small room

fulfil their function correctly. The problem can be eliminated either by changing the pre-setting of every head to the same value or leaving only one head to serve all the radiators. In the latter case, the radiators would have to be connected in series. Optionally, if the room is heated by two radiators differing in size and heat output considerably, the thermoregulator should be mounted on the bigger one (the other should then have a manual regulation valve), as presented in Fig. 2.11.

However, the solution is usually not used in practice. Most often this is due to the fact that users are unaware of the described phenomenon. Another reason is that the negative effects are either none or hardly noticeable, which results from one simple thing. A higher number of radiators is usually necessary in large rooms due to high demand for thermal power, impossible to meet by a single radiator, and due to the need for a uniform temperature distribution in the entire room, again impossible to achieve with a single radiator (a convector one—if surface (especially underfloor) radiators are used, the situation is different). The latter reason is especially related to finding justification for the issue under consideration. The radiator has a certain thermal range, i.e. the area it is able to heat uniformly, both due to convection/air circulation and radiation. As the distance from the radiator increases, its thermal impact becomes so small that another radiator has to be used to ensure proper heating of the entire area and a uniform temperature distribution in the whole room. And this is exactly why an independent thermoregulator can be used in such a situation. The individual thermoregulators installed on the two radiators then operate in the room thermally independent zones and do not disturb each other.

The room size and the distance between the radiators for which thermoregulators operate independently cannot be defined easily and accurately because the quantities depend on the room thermal parameters. They are also affected by the number of external walls and rooms with specific temperature values that the room is adjacent to. It may generally be stated that as the room unit heat losses decrease (e.g. in buildings with very effective thermal insulation), the radiator thermal range increases because circulating air is cooled less on warmer internal partitions. In such a situation, co-operation of several thermoregulators may be improper. However, no zones with significantly different temperatures arise in such rooms and therefore there may be no need to install more than one radiator (and, consequently, more than one thermoregulator).

A special case to be considered is a room with long glazed surfaces, such as lecture halls with many windows for example. Installing several in-series radiators with a single thermoregulator may in this situation be unfavourable. Flowing through subsequent radiators, water will get colder and colder and the total cooling may be significant for more distant radiators. Their surface temperature will be low, which means a lower temperature value in their thermal range zones compared to the required level. Located at the beginning, the thermoregulator will "see" a much higher temperature in the thermal range of the first radiator than the value for the last, and the temperature will also be higher than the room mean temperature. In such a situation, raising the thermoregulator head setting will not be the optimal solution. The temperature in the range of the radiators located at the end of the series will rise, but it may then be too high in the first radiator thermal range, which may cause a sensation of thermal discomfort. In such a situation, the classic solution with several radiators each equipped with its own thermoregulator will be more advantageous. However, a good practice in this case would be to use fewer longer radiators rather than more shorter ones. Longer radiators increase the distance between individual thermoregulators from each other and thus reduce their mutual adverse effect.

2.1.2.3 Parameters of the Radiator Thermoregulator Operation

In terms of operation, the radiator manual regulation valve and the radiator thermoregulator differ from each other significantly. In both cases, the direct regulated parameter (but not the target one—this will be the room temperature t_i) is the closing plug position over the valve seat (the valve opening degree). But because the ther-

Fig. 2.12 Room temperature depending on the setting selected on the thermostatic head [6]

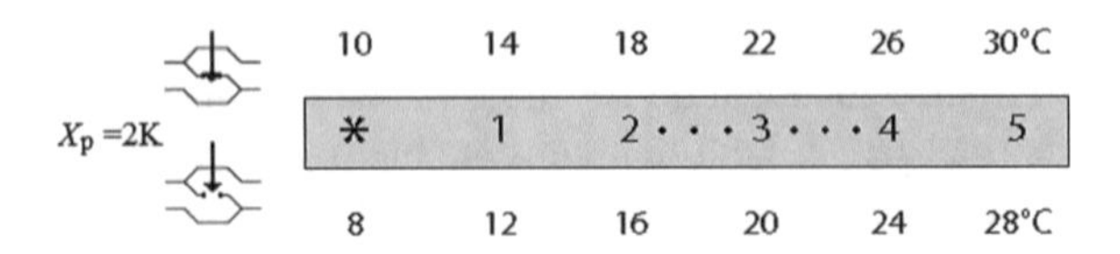

moregulator operates in a closed system of automatic regulation, it is characterized by a much higher number of parameters. For thermostatic and electronic heads the parameters are a bit different, but in the former case they are listed and formally specified. Standard EN 215:2004 [14], concerning the radiator thermoregulators, specifies a number of parameters and characteristics describing the device operation and structure, as well as the testing conditions. The most important quantities and characteristics include as follows:

- set temperature value t_{zad}. This is the room air temperature t_i set by the head setting device, assumed also as equal to the sensor temperature t_c corresponding to proportional range $X_p = 2$ K.
- thermoregulator static characteristic, presented in Fig. 2.13. This is the dependence of the output signal value (the head follower position h) on the input signal value (the regulated quantity—the sensor temperature t_c), determined in steady-state conditions at a change in the input signal from the minimal to the maximum position and back. It is the head characteristic that changes if the head is mounted on the radiator valve in the installation (the impact of the plug differential pressure and of static pressure, the impact of the branch and the radiator temperatures, etc.).
- thermoregulator proportional range X_p. This parameter concerns operation of a regulation valve with a head installed on it. It is not taken into account in the case of the regulation valve only or the thermostatic head only. It expresses the difference in the head sensor temperature t_c at which the head follower and the valve plug are shifted by a certain length of travel from the plug initial position (opening degree) to the valve full closing position. It does not have to be the full available lift of a given valve, but only a partial one. And this is the way in which the range is usually selected. In other words, if proportional range $X_p = 2$ K, the head will close the valve co-operating with it completely as soon as temperature t_c measured by the head exceeds temperature t_{zad} set on it by a value equal to the difference (here—2 K). Relating the above to the typical number scale used on thermostatic heads, the obtained temperature intervals are as presented in Fig. 2.12. The figure also illustrates the dependence of the valve plug position on measured temperature, in relation to the proportional range. Term X_p is also expressed by term S (e.g. in the standard [14] and expression S − 2 K means that the proportional range is 2 K).
- thermostatic head amplification factor k_m. The parameter concerns the thermostatic head. It expresses the head follower shift h per a unit change in the head sensor temperature t_c corresponding to the tangent of the angle between the secant and the static characteristic in the characteristic given point (cf. Fig. 2.13). Therefore, the higher the parameter value, the stronger the head response to variations in the

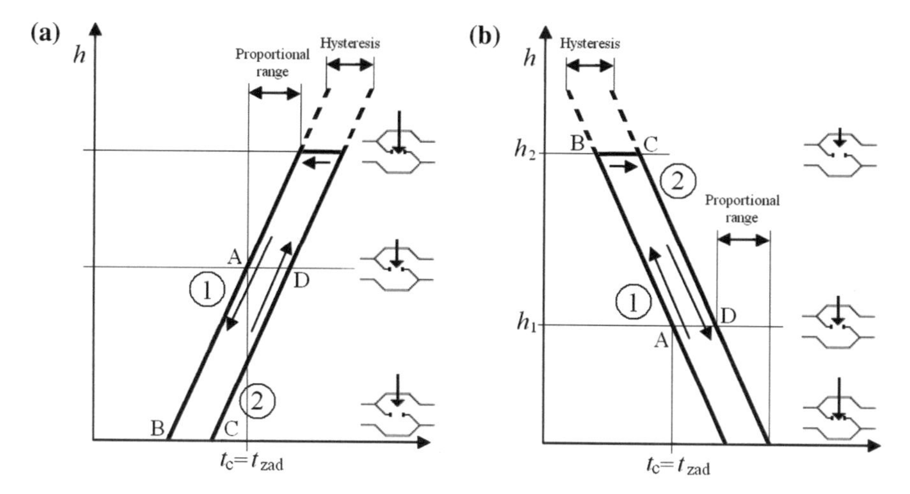

Fig. 2.13 Example static characteristic of a radiator thermostatic head

measured temperature (assuming that the change is bigger than the hysteresis zone, cf. further below), which translates into smaller variations in the room temperature.

Comparing the definitions of proportional range X_p and set temperature t_{zad}, it can be concluded that the thermoregulator valve (the element of the valve second stage of regulation) closes completely due to the head impact when the head sensor temperature t_c exceeds the value set on the head by a specific quantity referred to as proportional range. It is recommended that the proportional range should total 2 K. This is the effect of thermal comfort conditions and permissible oscillations in the room temperature on the one hand, and of the changing stability of the thermoregulator operation on the other. As the proportional range decreases, the stability deteriorates in a natural way because the closing element available range of travel is reduced and the finite accuracy of the making of the elements starts to play an increasingly important role. Balancing circuits hydraulically, it is also possible to design the thermoregulator for a different proportional range value. The range depends on the pressure value that has to be throttled on the valve and on the medium volume flow rate which determines the radiator heat output. Consequently, at a given value of temperature t_i, the proportional range can be changed setting different values of active pressure in the circuit. The higher the pressure, the more the thermoregulator has to close to keep the set flow rate, which results in a decreased value of proportional range X_p and reduced oscillations in room temperature t_i. Such a situation will also occur in a double-regulation valve with two adjustable sections of the medium flow (cf. Sect. 2.2.3) if the rise in active pressure is carried out without a relevant correction to pre-setting and the increment in pressure will have to be throttled on the closing element and not on the throttling one. Nevertheless, both these cases involve an increased demand for pumping power.

The two quantities—proportional range X_p and amplification factor k_m—deserve a few words of comment because they make thermoregulators considerably different from the radiator manual regulation valves. The stipulations concerning radiator thermostatic valves [14, 15, 36] and the guidelines resulting from them with regard to hydraulic and thermal balancing of the central heating installation equipped with the devices (e.g. in Poland it is document [42]) are not unequivocal. The inaccuracy resulting from the above-mentioned standard EN 215:2004 [14] is that it uses the value of proportional range X_p as the target parameter. But the quantity which is essential in terms of throttling and changes in the flow rate throttling depending on changes in the head sensor temperature, i.e. the valve closing element opening degree h (plug lift over the seat), depends also on the co-operating head amplification factor k_m. The factor value may vary depending on the manufacturer and even on the model of the device offered by different companies (nowadays it is usually $k_m \approx 0.2$–0.25 mm/K and $k_m \approx 0.40$–0.55 mm/K for liquid and vapour sensor heads, respectively). A change in the sensor temperature t_c occurs in parallel with a change in the valve plug position h, which is proportional to the temperature change multiplied by amplification factor k_m. The plug position over the valve seat h_x, for a given value of amplification factor k_m, results from the assumed proportional range and can be expressed using the following relation:

$$h_x = X_p \cdot k_m. \tag{2.1}$$

It follows that assuming a given value of X_p, the valve throttling will vary for different values of the amplification factor k_m, i.e. for different heads co-operating with the valve. Alternatively, at a given throttling value determined by opening h_x, different values of proportional range X_p will settle for different values of amplification factor k_m.

Apart from that, the documents mentioned above do not pose the condition of setting the upper limit for the closing element range of operation (regulation second stage, here—thermostatic heads) by assembly pre-setting, which is the case for manual valves. For thermostatic, even double-regulation, valves flow factor k_v is defined for a set value of pre-setting n_i, but not for the maximum opening degree of the second-stage regulation element, where current regulation could proceed downwards only, which is the original idea of and point in using thermostats as devices responding to heat gains (or losses) in the room. Flow factor k_v is specified for a partial opening degree of the valve plug, corresponding to a specific proportional range X_p, e.g. for $X_p = 2$ K. Because the values of the maximum range of travel of valve plugs vary within the limits of $h_{100} \approx 1.3$–2.5 mm, considering relation 2.1 and the practically applied values of the head flow factor k_m, it turns out that values of the opening degree of the valve closing elements for proportional range $X_p = 2$ K are a few times smaller (of the order of 0.4–1 mm) than those which are maximally available. The conclusion is that although a specific value of the valve pre-setting n_i is selected due to the results of the circuit hydraulic balancing, the valve throttling can decrease, which means an increased flow rate. It is indicated in many works [16, 23, 35] that this may be the cause of the installation hydraulic and thermal disorder.

This also raises doubts as to the point in performing hydraulic balance calculations and determining the design (i.e. the maximum) thermal load specified by binding regulations.

Different approaches can be applied to prevent the possibility of the thermoregulator increasing the medium flow rate above the design value. One of them is such selection of pre-setting n_i that the required value of flow factor k_v should be achieved for the closing plug full opening, i.e. for $X_p = \max$, and not for $X_p = 2$ K. In the case of valves with two adjustable sections of the medium flow, where the closing plug travel range is independent of the throttling element pre-setting, this means pre-settings with lower values. However, the problem is that at a given pre-setting manufacturers usually do not specify the flow factor values for $X_p = \max$ but for $X_p = 2$ K. Consequently, the required value of pre-setting n_i is difficult to select. The value of $X_p = \max$ can be calculated using the relations presented herein further below, but this requires the knowledge of additional parameters which are not specified by manufacturers, either. The sought value of flow factor k_v can also be measured by means of the methods presented in this book later on.

The possible high oscillations in the room air temperature t_i create a problem in the design for a large proportional range. This results directly from the parameter definition and the thermostatic thermoregulator principle of operation. In order to close completely, the thermoregulator must detect a change in temperature t_c measured by its sensor by the design proportional range X_p. For this reason, the larger the range, the bigger the increment in the room temperature has to be above set temperature t_{zad} so that the device should close, which is unfavourable from the point of view of the room thermal comfort. However, this can be prevented by selecting a thermostatic head with a possibly high amplification factor k_m to co-operate with the regulation valve. The higher the factor value, the smaller will be the required increments in room temperature t_i to make the thermoregulator close. Using for example a thermostatic head with a vapour sensor, for which the typical value of the amplification factor is $k_m \approx 0.45$ mm/K, and assuming that the typical proportional range $X_p = 2$ K corresponds to the closing plug maximum range of travel (so that the oscillations in the room temperature should not be much higher than 2 K), the maximum range of travel of the closing plug is obtained at the level of $h_{100} \equiv h_{max} = 0.9$ mm. This value is lower compared to most (thermostatic) regulation valves available on the market, which are devices with two adjustable sections of the medium flow where the closing element travel range is constant and independent of the pre-setting value. A larger proportional range will settle in these cases, for example—for $h_{100} \equiv h_{max} \approx 1.3$ mm (e.g. the *Danfoss RTD-15* valve), it will total 2.9 K. But for devices with one adjustable section of the medium flow with a coaxial plug, the range can be regulated and the closing plug maximum travel range can be reduced. In order to satisfy the postulate that the thermoregulator must not increase the medium flow rate, the use of such valves, e.g. the *Herz TS-90-V* valve (cf. Fig. 2.28), in combination with vapour sensor heads is more than justified (at present the *Herz* company is practically the only manufacturer of this type of heads). Moreover, using valves with one adjustable section of the medium flow is much more favourable with regard to the value of what is referred to as the valve inner and total *authority* (cf. Sect. 4.1) and, thereby, to the impact on the valve regulation characteristic and, ultimately, the regulation quality, compared to valves with two adjustable sections of the medium flow.

- hysteresis (zone). The hysteresis zone may concern the thermostatic head itself as well as the head combination with a thermostatic valve. It is the latter case that is essential in practice because the head always co-operates with a valve. The hysteresis zone will then increase. According to the standard definition, the term defines the temperature difference between points of identical flow rates on the thermoregulator closing and opening characteristics, i.e. the interval of changes in the input signal t_c within which the static (and static and flow) characteristic is ambiguous (cf. Figs. 2.13 and 2.15). In other words, it is the difference in the position of the head follower and, consequently of the valve plug lift over the seat (h), and in the working medium mass/volume flow for a given temperature t_{zad} set on the head, for two alternative processes: opening and closing because at the same value of set temperature t_{zad}, the head follower position (h) is different depending on whether the thermoregulator closes or opens. The characteristic that illustrates this phenomenon is referred to as the radiator thermostatic head static characteristic. It is presented in Fig. 2.13. The figure presents a fragment of the thermostatic head static characteristic for which the curves illustrating changes for the opening and the closing process are approximately linear. For clarity, the hysteresis zone is presented as wider than it is in practice. The figure shows an example static characteristic plotted in two versions. The chart in Fig. 2.13a is the most common in practice and it is also utilized in standard EN 215:2004 [14]. It presents the follower temperature-dependent position h in the form of its protrusion from the sensor. As the h value rises, the opening degree of the valve co-operating with the head does not rise but diminishes. This fact has to be borne in mind to interpret the chart correctly. Figure 2.13b presents the same static characteristic, but as a reversed one, where parameter h denotes the follower position depending on the sensor temperature t_c as its degree of insertion into the sensor. In this case, this equates with the valve opening degree. This type of characteristic is less common but it offers a clearer illustration and relates directly to the thermoregulator operation principle. Therefore, it is a good idea to refer to it and use the characteristic in the analysis of the thermoregulator operation.

An example proportional range X_p, as defined above, is marked in the figure. It corresponds here to less than half of the valve plug entire available travel h_{100}, which is usually the case in practice when the nominal proportional range is $X_p = 2$ K. Still, the value can be different and it may comprise a different range of the line and of the plug travel in the figures presented above. This was already discussed above.

Assume that the initial design head follower position h_1 ($h_1 > 0$) corresponds to a certain room temperature equal to the working medium temperature t_c and to the temperature set on the head, e.g. $t_i = t_c = t_{zad} = 20$ °C (point A in Fig. 2.13). If the room temperature t_i and, thereby, temperature t_c measured by the head sensor, starts to fall, the head will start opening the valve according to Line 1 in Fig. 2.13. The lower temperature t_c measured by the head sensor, the more the valve will open.

The moment temperature t_c measured by the sensor has stabilized, the valve opening process will stop and a new working point will be established for it, characterized by a bigger opening degree h_2 (point B in the figure). If now the room temperature t_i starts to rise, temperature t_c measured by the sensor will rise, too. The head should start closing the valve, reducing its opening degree h. The process, being identical to the previous one except for the opposite sense, should run along the same line that the valve opening ran along. But this will not be the case. The head will not respond to start closing the valve till the increment in the sensor temperature t_c above the value for which the opening process was previously stopped is big enough to overcome the resistance of static friction in the device elements (point C in Fig. 2.13). This surplus in temperature is referred to as the hysteresis zone or the ambiguity zone. The zone is delimited in the figure by lines 1 and 2, i.e. by points B and C, and D and A. If a sufficient surplus in the sensor temperature t_c occurs, the head will start the valve closing process, according to Line 2 in the figure.

Due to hysteresis, for the head follower identical position h, the values of the head sensor temperature t_c and of the room air temperature t_i for the closing and the opening process must be different. This means that in order to maintain a given value of the room air temperature t_i, and—consequently—the valve opening degree h, the values of temperature t_c measured by the sensor for the two processes must be different. So that the thermoregulator should reach the same opening degree h_1 as at the previous opening process and ensure keeping the same room temperature t_i, temperature t_c measured by the sensor must be higher than the opening process temperature by the hysteresis value. In practice this means that different temperatures may settle in the room for a given value of temperature t_{zad} set on the head depending on whether additional heat gains occurred in the room, making the thermoregulator close, or heat losses arose, making the thermoregulator open. Whether it will happen or not and how strong the phenomenon will manifest itself also depends on other factors that affect the thermoregulator performance, such as the impact of water pressure triggering a force on the valve plug and follower, the impact of changes in the water temperature, etc. For one process, e.g. the thermoregulator opening, the adverse effects may accumulate, whereas for the other—they may eliminate each other.

The effect of the hysteresis phenomenon is that the thermoregulator may not respond to variations in the temperature of the surroundings if values of the variations do not go beyond the hysteresis range. For this reason, the hysteresis zone should be as small as possible. According to standard EN 215:2004 [14], it should not exceed 1 K.

Figure 2.14 illustrates static characteristics of real thermostatic heads plotted conventionally and as reversed ones.

Figure 2.14a, b relate to the *Danfoss* complete integrated RAVL valve. Figure 2.14a presents the "reversed" static characteristic for the set value of t_{zad} = 22 °C, whereas Fig. 2.14b—for several different set values. As indicated by Fig. 2.14b, the device maximum proportional range is $X_{p,max}$ = 6 K. However, if the value gets higher than about X_p = 3 K, the static characteristic gets curved to become horizontal. Due to that, if the differences between the sensor temperature

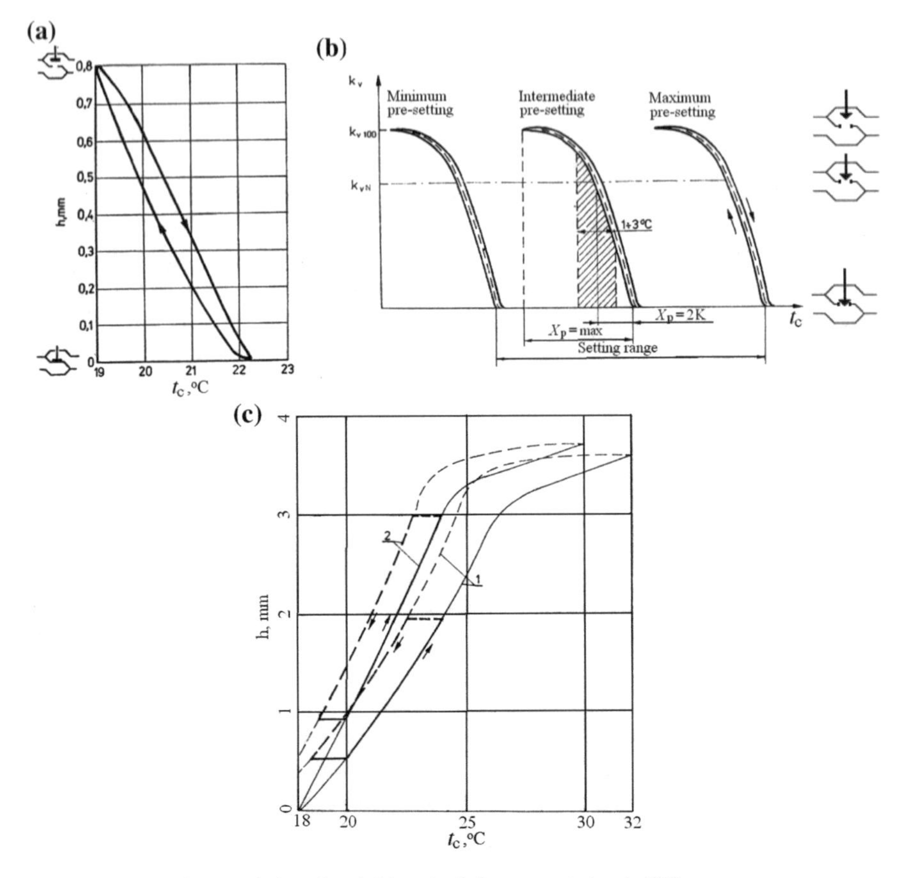

Fig. 2.14 Static characteristics of real (historical) thermostatic heads [18]

t_c and the set value of t_{zad} exceed 3 K, the valve opening degree gets only slightly bigger, which helps to prevent unexpected and unwanted increments in the medium flow rate that might cause the installation hydraulic and thermal disorder.

- characteristic flow rate (q_{ms}). This is the flow established for $Xp = 2$ K and differential pressure of 10 kPa at any value of the head setting.
- nominal flow rate (q_{mN}). This the characteristic flow for the intermediate position of the thermostatic head selector (20–24 °C). In the case of a valve with the possibility of pre-setting, this is the mass flow through a valve with no pre-setting.
- maximum flow rate ($q_{m\ max}$). This is the maximum mass flow of water (fully open valve) at differential pressure of 10 kPa.
- thermoregulator valve flow characteristic. This describes the dependence of the flow factor value on the thermoregulator valve closing element position (equivalent to the closing characteristic, second-stage regulation).

- valve (regulated section) authority a. This is a dimensionless indicator of the share of a variable, regulated hydraulic resistance generated by the valve closing element in the valve total resistance (cf. Fig. 2.16):

$$a = \frac{\Delta p_1 - \Delta p_2}{\Delta p_1} = 1 - \frac{\Delta p_2}{\Delta p_1}. \tag{2.2}$$

If quantity a is defined in this way, it cannot be treated as the valve inner authority (as stated in [18, 38, 39] for example), but as the authority of the regulated section only, which will be explained and proved in Sect. 4.1 herein. From the above formula, the valve measuring conditions (extremely slight pressure losses on the pipework other elements, i.e. the valve outer authority $a_z \approx 1$) and the comparison with the chart in Fig. 2.16 it follows that term $\Delta p_1 - \Delta p_2$ denotes the pressure loss arising in the valve closing element, which is related to the valve total pressure losses Δp_1. Using the formula defining the flow factor and the chart in Fig. 2.16, the relation can be transformed to the following form:

$$a = \frac{\Delta p_1 - \Delta p_2}{\Delta p_1} = \frac{\frac{\dot{V}_N^2}{k_{vN}^2} - \frac{\dot{V}_N^2}{k_{vs}^2}}{\frac{\dot{V}_N^2}{k_{vN}^2}} = 1 - \left(\frac{k_{vN}}{k_{vs}}\right)^2 = 1 - \left(\frac{q_{m\,N}}{q_{m\,max}}\right)^2. \tag{2.3}$$

Subscript N means that the value, according to the standard, relates to the nominal opening degree, corresponding to proportional range $X_p = 2$ K ($S - 2$ K).

- thermoregulator static and flow characteristic. This illustrates the dependence of the value of the thermoregulator valve flow factor k_v on the value of the regulated quantity (the head sensor temperature t_c) at a constant difference in pressure and unchanged set value, measured in steady-state conditions. The characteristic is obtained from the thermoregulator measurements or by combining the characteristics of the head and the valve. Figure 2.15a presents the static and flow characteristic of an example thermoregulator obtained by means of the graphical method, and the curves obtained taking account of the factors that start to exert their influence after the device is mounted in the installation (cf. Fig. 2.15b). The terms *closing characteristic* and *opening characteristic* are not identical with but equivalent to the characteristic of the regulation second stage, i.e. to the term *opening characteristic* or *closing characteristic* in the case of manual regulation valves. In the testing of these elements, the hysteresis zone is not determined and for this reason there is only one curve in question.
- $\Delta p = f(q_m)$ characteristic. This is a chart illustrating pressure losses depending on the medium flow rate. Standard EN 215:2004 [14] requires at least three curves: for $X_p = 1$ K, $X_p = 2$ K and for the fully open valve ($X_p = \max$), where the selector is in every case in the intermediate position. If the valve pre-setting is possible, the measurement should be performed for all available values of pre-setting n_i. The characteristics for a typical thermoregulator (valve + head) with no possibility of pre-setting are presented in Fig. 2.16.

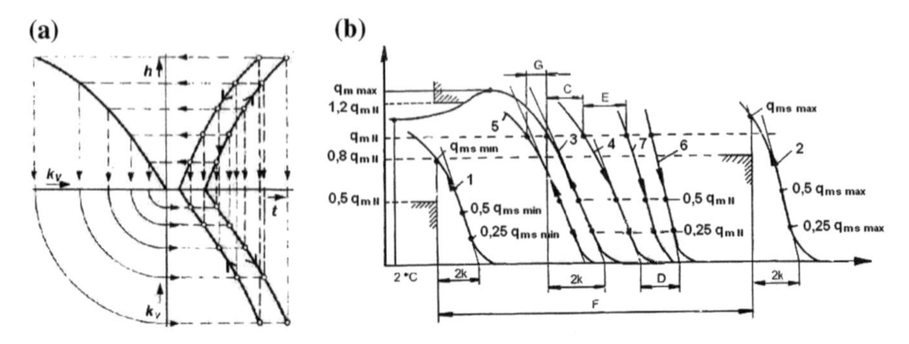

Fig. 2.15 Example thermoregulator static and flow characteristics $t(k_V)$; 1, 2—opening characteristic for the lowest and the highest pre-setting, 3, 4—opening and closing characteristic at the selector intermediate position, 5—opening characteristic of a thermoregulator with transmitters (remote sensor) at the selector intermediate position, 6—closing characteristic at the selector intermediate position and the pressure difference higher than 10 kPa, 7—closing characteristic at the selector intermediate position and static pressure of 1000 kPa [14, 16]

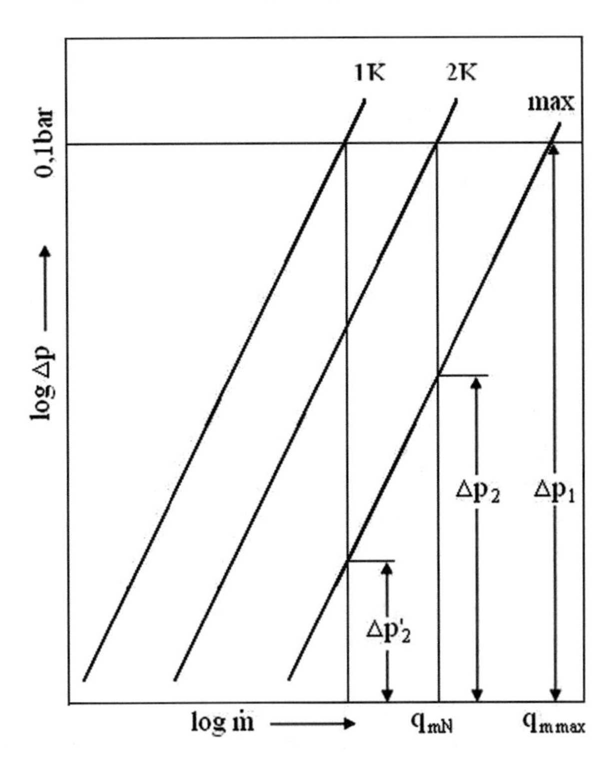

Fig. 2.16 $\Delta p = f(q_m)$ characteristics of a typical thermoregulator

- differential pressure influence of the working medium. The pressure difference arising on the valve plug triggers a longitudinal force carried over by the head follower and acting on the sensor. The impact is illustrated by the temperature difference (shift) between closing characteristics 4 and 6 in Fig. 2.15 at the same nominal flow rate. As stipulated by standard EN 215:2004 [14], the impact should not exceed 1 K.

- influence of static pressure. The effect is similar to the impact of the pressure difference. It is illustrated by the temperature difference (shift) between closing characteristics 4 and 7 in Fig. 2.15 at the same nominal flow rate. As stipulated by standard EN 215:2004 [14], the impact should not exceed 1 K.
- influence of ambient temperature on thermostatic valves (thermoregulators) with transmission elements (capillary tubes to sensors and/or remote setting devices). As stipulated by standard EN 215:2004 [14], the impact should not exceed 1.5 K. It is illustrated by the temperature difference (shift) between closing characteristics 3 and 5 in Fig. 2.15 at the nominal flow rate.
- water temperature effect. The influence manifests itself in distorting the ambient temperature measured by the sensor due to the heat transfer from the radiator pipe to the sensor through the thermoregulator structure. According to standard EN 215:2004 [14], it is the difference in the sensor temperature t_c to which corresponds a change in the water flow rate caused by a change in the temperature of the heating medium flowing through the thermoregulator valve by 30 K. As stipulated by standard EN 215:2004 [14], the impact should not exceed:

- 1.5 K for integrated thermoregulators (with a built-in sensor),
- 0.75 K for thermoregulators with transmitters (outer/remote sensor).

- response time. This is the time after which, at a jump rise in the air temperature by 3 K, a change in the water flow rate will occur that corresponds to a rise in the sensor temperature by 1.5 K. As stipulated by standard EN 215:2004 [14], the time should not be longer than 40 min. Qualitatively, the response time is an equivalent of the time constant.

From the point of view of automatics, a thermoregulator with an electronic head operates differently from one with a thermostatic head. The electronic regulator controller sends signals to the executive element—an electric motor usually, which drives the follower with a constant speed. The sequences and the duration of the signals depend on the difference between measured temperature t_c and set temperature t_{zad}. The thermostatic head regulator is able to respond only when the value of measured temperature t_c differs from set temperature t_{zad} by at least the hysteresis value. Only after this difference is exceeded, does the head respond and the follower position h is shifted to set a new value of t_c, still however differing additionally from the value of t_{zad} due to the proportionality of action. An electronic regulator uses the information about the difference between the temperature set on the regulator and measured in the room (t_{zad} and t_c, respectively) to adjust its response to the current situation, determining not only the state of action (closing, no response or opening, i.e. conventional states -1, 0 and +1), but also the intensity/length of individual states (close more/less). The bigger the difference between measured temperature t_c and set temperature t_{zad}, the more the follower position h will change. At set intervals, the regulator sensor performs subsequent temperature measurements, thus detecting the difference between the set and the measured temperatures t_{zad} and t_c.

The collected data make it possible to determine the required state of action and respond adequately. For example, if measured temperature t_c is much lower than set temperature t_{zad}, the regulator will respond sending a relatively long signal opening the valve (conventionally—+1). The next measurement of temperature t_c will show that the difference between the measured value and t_{zad} is smaller. The regulator will then respond again, but the signal opening the valve will be shorter, etc. If it turned out that after another cycle the t_c value is higher than $t_{zad,}$ the regulator will send a signal closing the valve (conventionally—−1) If there is no difference between the set and measured values of t_{zad} and t_c, respectively, or if the difference is smaller than the insensitivity (hysteresis) range, there will be no response on part of the device. Such a controller can therefore demonstrate oscillations depending on the sampling frequency, i.e. the intervals between subsequent measurements of temperature t_c, and on values of the amplification factor. The higher the amplification of the controller proportional term (P) and the shorter the time of the integration term (I), the closer the system is to oscillations and instability, even though the device operates faster and with better accuracy. In contrast to the proportional thermostatic thermoregulator (P controller), this controller type makes it possible to achieve the required level of the stabilized quantity mean value. Regulators operating based on this principle are referred to as step or quasi-continuous controllers (the control signal is discrete and not continuous), or three-state controllers (three states of action: −1, 0, +1), and the effects of their operation are similar to those obtained by means of continuous proportional-integral (PI) controllers.

Electronic radiator heads have many additional advantages compared to self-actuated thermostatic regulators. They make it possible to program the time of operation and set periods of reduced temperature in the room or periods when the room heating is completely switched off. They also enable selection of the regulation type (P control, PI control or, sometimes, PID control) and determination of the insensitivity/hysteresis range, they can detect an open window, etc. The possibility of defining the insensitivity range can be especially useful. If the range is increased, the number of switching between states +1 and −1 gets smaller, or no switching occurs at all once the range exceeds the difference between the measured and the set values of temperatures t_c and t_{zad} established by means of subsequent steps of readjustment (cf. Fig. 2.18c). Such devices also create an opportunity for remote and mobile control by means of central home controllers and/or computer equipment, smartphones, etc. using an Internet connection.

Figures 2.17 and 2.18 present charts illustrating time-dependent changes in the output and input quantities for the two types of temperature stabilizers for one variant of the control process—heating. In both cases, example time-dependent (dynamic) changes in the room response are presented (changes in the room temperature t_i over time τ). Real characteristics have similar shapes and they are described by exponential equations, with curvatures similar to logarithmic ones (cf. [33, 34]). The two figures present characteristics of the following:

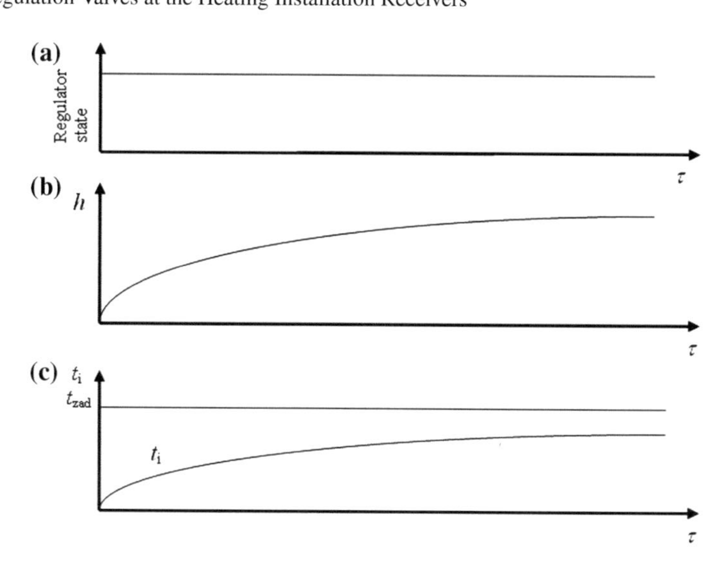

Fig. 2.17 Dynamic (time) characteristics for regulation by means of a thermostatic thermoregulator

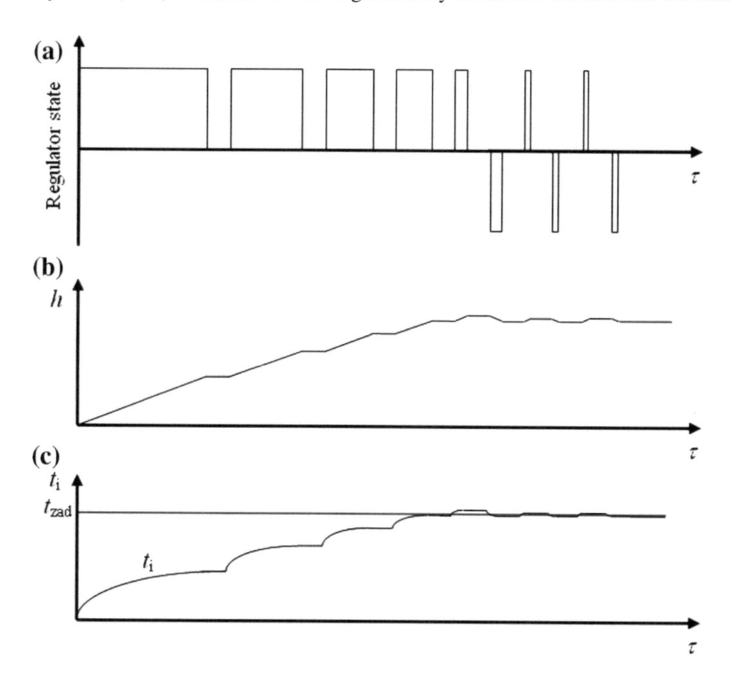

Fig. 2.18 Dynamic (time) characteristics for regulation by means of an electronic thermoregulator

(a) the device state—informing whether it exerts any effect on the valve. In the case of a thermostatic thermoregulator, this is the temperature sensor deformation and transmission thereof onto the follower; for an electronic thermoregulator this is the signal driving the motor and opening the valve (+1), or lack thereof (0), or a signal driving the motor and closing the valve (−1).
(b) the head follower position h (represented by the follower insertion into the head) and, thereby, the valve plug lift over the seat, as an output signal.
(c) the change in room temperature t_i. In order to illustrate the analysis more clearly, individual states in time during the changes in the room temperature t_i are put for the same time points as changes in the thermoregulators output signals. In reality, the room response to each reaction of the thermoregulator will be delayed in time and the line illustrating changes in temperature t_i will be shifted to the right in relation to the thermoregulator output signal line even though it naturally corresponds to the thermoregulator individual output signals. The non-zero time of reaching the set state by the regulated object, here—the set point of the room temperature t_i, after every change in the thermoregulator output signal is related to thermal inertia of the object (and generally—of the entire regulation system). The non-zero time of the beginning of the reaction to the change is referred to as the transportation lag, being the effect of the time needed by the thermoregulator output signal to reach the regulated object, i.e. the room.

As already mentioned, radiator thermostatic heads are now the most common and widely used devices for regulation and stabilization of the room temperature and for the reduction in energy consumption for heating purposes, which helps to rationalize the installation operating costs. For this reason, the quality and the energy-related properties of the elements are so important.

The ever increasing European requirements concerning energy consumption of buildings, together with the consumers' growing awareness of the necessity to save energy and their will to optimize household budgets, create an urgent need to gain access to clear and reliable information before they decide to purchase a given product. Most household goods, audio and video devices and other electric appliances, such as bulbs, vacuum cleaners, fridges, water pumps, etc., have already been assigned an energy efficiency rating. In the case of thermoregulators, this type of information is given by the European Valve Manufacturers Association, which issues relevant certificates based on the Thermostatic Efficiency Labelling (TELL) product classification system. The information comprises a number of key parameters which are essential in terms of energy consumption rationalization, such as:

- Hysteresis value—marked as C
- Influence of water temperature—marked as W
- Influence of differential pressure—marked as D
- Response time—marked as Z.

All these parameters, together with their permissible values and determination methods, are described in standard EN 215:2004 [14]. The method of the energy

Table 2.1 Energy efficiency classes of radiator thermostatic thermoregulators

Energy efficiency class	VI	V	IV	III	II	I
EEI	≤1.0	≤0.9	≤0.8	≤0.7	≤0.6	≤0.5

efficiency class determination by means of the parameters is specified in the TELL recommendations. It is based on calculating the Energy Efficiency Indicator (EEI), to which an appropriate energy efficiency class is assigned in the scale from I—the highest one (formerly: Class A) to VI—the lowest (formerly: Class F). The classification indication should be added to the list of the radiator thermoregulator parameters. The EEI is calculated using the following relation:

$$EEI = \frac{C + W/X + D + Z/40}{4} \tag{2.4}$$

where $X = 1.5$ for a thermoregulator with an integrated sensor or $X = 0.75$ for one with a remote sensor. The calculated value is compared to the intervals specified in Table 2.1 and the device is assigned an appropriate energy efficiency class.

2.1.2.4 Impact of the Radiator Thermoregulator Operation on the Room Temperature

Obviously, changes in the setting of the radiator regulator, whether it is in the form of a manual or an automatic (e.g. thermostatic) valve, affect the room temperature because they vary the radiator heat output. However, when it comes to the impact of the thermoregulator operation on the room temperature, this should be understood as the impact of automatic regulation because manual regulation is not the case for thermoregulators.

From this perspective, manual valves have no effect on the indoor temperature because no matter whether additional heat gains or losses exceeding the design values arise in the room, such a valve will not react and the room temperature will vary only depending on the said heat gains or losses.

In the case of the radiator automatic thermoregulator, the device itself will also affect the quantity because the thermoregulator will always make an effort to restore and maintain the initial pre-set temperature. For a number of reasons, this is impossible for self-actuated thermostatic thermoregulators and there will always be some deviations.

The first reason is the hysteresis phenomenon, already described above. The thermoregulator will react only if the changes in the room thermal power are so big that the temperature oscillations resulting therefrom exceed the hysteresis range. Even then, however, the thermostatic head will set the valve plug position and, thereby, the room heat output, with an error imposed by the range.

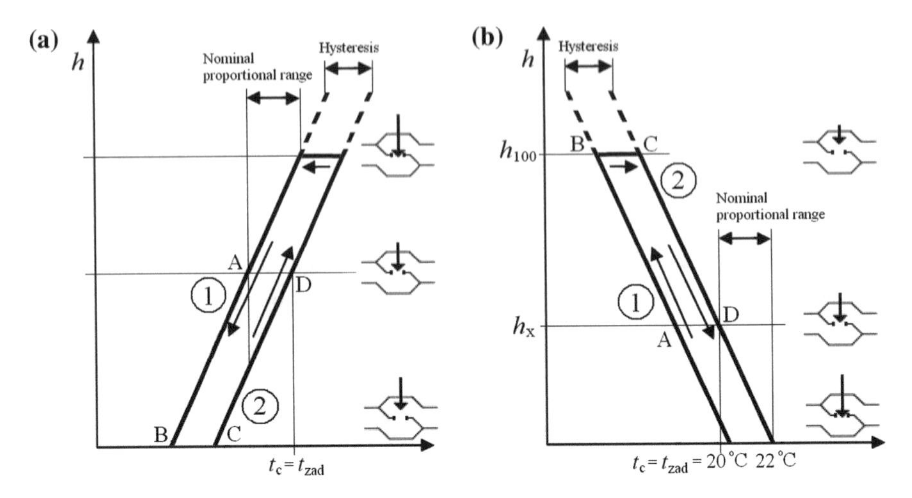

Fig. 2.19 Example static characteristic of the radiator thermoregulator

Another thing is the ratio between the room heat gains and the heat output of the radiator (and of the pipes, if any). If, for example, heat gains appear in the room at the level of 50% of the radiator momentary heat output, the thermoregulator will try to close the regulation valve so that the new value of thermal power supplied by the radiator should total 50% only, to ensure with the other 50% resulting from the heat gains the initial, required room temperature. If, however, the gains exceed 100% of the radiator heat output, the temperature in the room will rise above the required value even if the valve closes completely.

Yet another and in fact the most essential factor is the proportional character of operation of a thermoregulator with a thermostatic head. As a rule, proportional regulators are unable to maintain a constant value of a parameter, here—the room temperature, if disturbance occurs. For the device to react, it is absolutely necessary that the difference between measured and set temperatures t_c and t_{zad} at a certain level exists. The issue can be described using the thermoregulator static characteristics presented in Fig. 2.13 with example temperature values as shown in Fig. 2.19.

Assume that the thermoregulator initial working point is {D}, for which set temperature t_{zad} is equal to temperature t_c measured by the sensor of 20 °C. If heat gains arise in the room, the room temperature t_i and the sensor temperature t_c will rise. Due to that, the thermoregulator will start to close, trying to prevent a rise in the room temperature t_i. It should be noted, however, that it is the constant surplus $t_{zad} - t_c$ that is required for the head to set the valve plug new position h_x and reduce the medium flow rate, which results in a reduction in the radiator heat output. The higher the difference, i.e. the bigger the regulation error, the more the valve will be closed. If the surplus is equal to the design (nominal) proportional range that the thermoregulator is selected for, e.g. $X_p = 2$ K ($S - 2$ K), the valve will close completely. Nevertheless, the surplus $t_{zad} - t_c$ will still be the case necessary for the thermoregulator to maintain the valve new, smaller opening degree h. This is a serious disadvantage of thermoregulators with thermostatic heads and other proportional (P) controllers. They will not react unless a constant difference occurs between temperature t_{zad} set on the head and the room temperature t_i, which by principle is something the user wants to avoid. This weakness does not affect e.g. electronic thermoregulators, which are usually integral, proportional-integral or proportional-integral-derivative controllers (marked P, PI and PID, respectively).

The figure indicates that the smaller the valve initial, design opening degree h_x/h_{100}, the smaller the increments in the room temperature t_i will be. Similarly, the bigger the characteristic slope, i.e. the higher the value of the head amplification factor k_m, the smaller the oscillations. Reducing the design opening degree is unfavourable because at a reduction in the valve plug range of travel, regulation becomes less accurate due to the growing impact of the inaccuracy of the valve elements making and fitting, as described earlier. Therefore, it may again be concluded that a high value of the head amplification factor k_m is favourable.

Another element that conditions possible oscillations (and, to be more exact, increments) in the room temperature is the head setting. In practice, the radiator thermoregulator users select the maximum setting, i.e. 5, which for most thermostatic heads corresponds to the temperature of (at least) 28 °C, as presented in Fig. 2.12. Such temperature levels are usually not achieved due to the fact that radiators are selected to ensure temperatures of the order of 20–24 °C; their heat output is then naturally too small. Still, if such a setting is selected, the head will open the valve closing element completely (within the boundaries of the possible pre-setting, should it limit the element travel range—cf. Sects. 2.2.2 and 2.2.3), thus establishing the maximum available proportional range, which, due to technical and operating parameters of typical valves and thermostatic heads is often a few times higher than the permissible value of 2 K (cf. Sect. 2.1.2.3). If in such a situation heat gains arise in the room, e.g. a few people enter it, the temperature will start to rise. If for example the proportional range maximum value, at the head setting 5, totals $X_p = 8$ K, the head will not close the valve until the head sensor temperature, being the equivalent of the room temperature, reaches 28 + 8=36 °C. Before the level of 28 °C is reached, the head will not react at all. If the thermoregulator is operated in such an incorrect manner, the device basic function is practically eliminated. Each time then, the selector setting should correspond to the design temperature and be correlated with the calculated mass/volume flow of the medium supplied to the radiator to ensure the required temperature level. For example, if the room temperature is to total 20 °C, setting 3 should be selected; for 22 °C the correct setting is 3.5, for 24 °C—4, etc. (this, naturally is a simplification; in practice additional factors should be taken into account, as described e.g. in Sect. 2.1.2.1).

From the point of view of temperature regulation, each room has a certain static thermal characteristic illustrating changes in its temperature depending on the supplied thermal power. If it is considered as a straight line, it can be described by means of the amplification factor and proportional range, like in the case of the radiator thermoregulator. The proportional range will then be the available range of changes in thermal power and the amplification factor, like previously, will be the tangent of the angle of the characteristic inclination in a given point.

The characteristic will of course differ for two rooms of the same size if they are characterized by a different design demand for thermal power (the maximum demand for a specific minimum outdoor temperature and a given indoor temperature). The difference in the parameter value may result for example from different values of the partitions heat transfer coefficient. A room with better thermal insulation will need less thermal power. It is obvious that at the same time it will be more sensitive to changes in the supplied thermal power.

In order to estimate the changes in the room temperature at a change in the supplied thermal power in the steady state, it is necessary to know the room static thermal characteristic. Knowing it, it is also possible to determine changes in the room temperature taking account of the operation of a thermoregulator with a known static characteristic. For the purposes of the analysis, the room static thermal characteristic can be described using a straight line. In reality, though, its shape is a bit different and non-linear because usually the temperature changes are not linearly dependent on changes in thermal power—this would be the case if all partitions exchanged heat with the surroundings with identical temperature values, for example if all the partitions were external walls. But partitions can be external walls as well as internal ones adjacent to other rooms or floors on the ground. Consequently, the temperatures on their external sides can be different in relation to the space under analysis. Determined changes in indoor temperature thus do not usually result in identical changes in the difference between the outdoor and indoor temperatures for individual partitions. This means that changes in thermal power are also different. Moreover, one of the elements of total heat losses and, thereby, of the room static thermal characteristic, are ventilation heat losses, which in the case of natural (gravitational) ventilation rise non-linearly (faster than the temperature difference would suggest) with an increase in the difference between the outdoor and indoor temperatures due to the rise in the pressure difference and, thereby, in the intensity of natural draught and the incoming air flow. The non-linearity of the building facility total heat losses can also be observed analysing what is referred to as heating curves programmed in heat sources. They usually illustrate the installation supply temperature depending on the temperature of the environment. Most often, they are in the form of curves rather than straight lines. However, the analysis of the thermoregulator co-operation with the room is as a rule conducted on a small fragment of the room static thermal characteristic because the temperature changes are slight—the temperature value is stabilized by the thermoregulator operation. For this reason, no big errors arise if this small fragment of the curve is approximated by means of a straight line.

Knowing the indoor and outdoor temperature, the heat transfer coefficients of all partitions and their surface area, and taking account of the heat transfer between adjacent rooms, it is possible to determine the thermal characteristic shape, i.e. the inclination and the starting and the end points. Although such calculations are not performed to analyse the room thermal response from the point of view of temperature regulation, they are part of the standard EN 12831:2003 [13] intended for finding the design thermal load, i.e. the required thermal power needed to keep the design (required) indoor temperature values. The standard includes determination of the heat loss coefficient (H, expressed in W/K). The coefficient informs by how much the power supplied to the room must change to change the room temperature by 1 K. Using the above simplification concerning the linear shape, an example room thermal characteristic plotted in this way is presented in Fig. 2.20a. The bigger the room sensitivity to changes in the supplied thermal power, the steeper the characteristic. By transposing the chart appropriately as presented in Fig. 2.20b, the curve can be used to create a characteristic describing the temperature changeability depending on the supplied thermal power. For comparison, Fig. 2.20c additionally presents typical heating curves.

In order to create this chart, it is necessary to calculate the angle of the line inclination to the horizontal axis, as well as the starting and the end point. The characteristic slope factor is the inverse of the coefficient of heat losses H and, thereby, the room amplification factor. It may therefore be written that:

$$k_{\mathrm{s}} = \frac{1}{H}. \tag{2.5}$$

As indicated by the function mathematical description included in the chart presented in Fig. 2.20b, the example room is characterized by the amplification factor value at the level of k_s=0.04 K/W (°C/W), which means that each 1 W increment in the supplied thermal power involves a rise in temperature by 0.04 K (°C). The amplification factor can also be related to the design benchmark value of thermal power (here: $X_{\mathrm{p,s}} = 100\%$ of the proportional range). It can also be written in percentages, as an increment in temperature at a rise in thermal power by 1 percent, according to the following formula:

$$k_{\mathrm{s},\%} = k_{\mathrm{s}} \cdot 1\% \cdot \dot{Q}_{\mathrm{i}}. \tag{2.6}$$

1% of 1000 W is 10 W, which gives $k_{s,\%}$=0.4 K/% (°C/%).

The change in the room temperature occurring at a change in supplied thermal power can be found using the following formula:

$$\Delta t_{\mathrm{i}} = k_{\mathrm{s}} \cdot \Delta \dot{Q}_{\mathrm{i}}. \tag{2.7}$$

The final temperature value can be calculated by means of the following relation:

$$t_{\mathrm{i,k}} = t_{\mathrm{i}} + \Delta t_{\mathrm{i}}. \tag{2.8}$$

The line in Fig. 2.20b illustrates the change in the room demand for thermal power depending on outdoor temperature for the required indoor temperature of $t_{\mathrm{i}} = 20$ °C. The chart starting point denotes what is referred to as design conditions. It is described by design values of the outdoor temperature (in winter), which depends on the building climatic zone and the design thermal load resulting therefrom. It is assumed here that the demand for thermal power totals 1000 W. Any higher value of the outdoor temperature means lower demand for thermal power. Zero demand is achieved if the outdoor temperature is equal to the required design room temperature

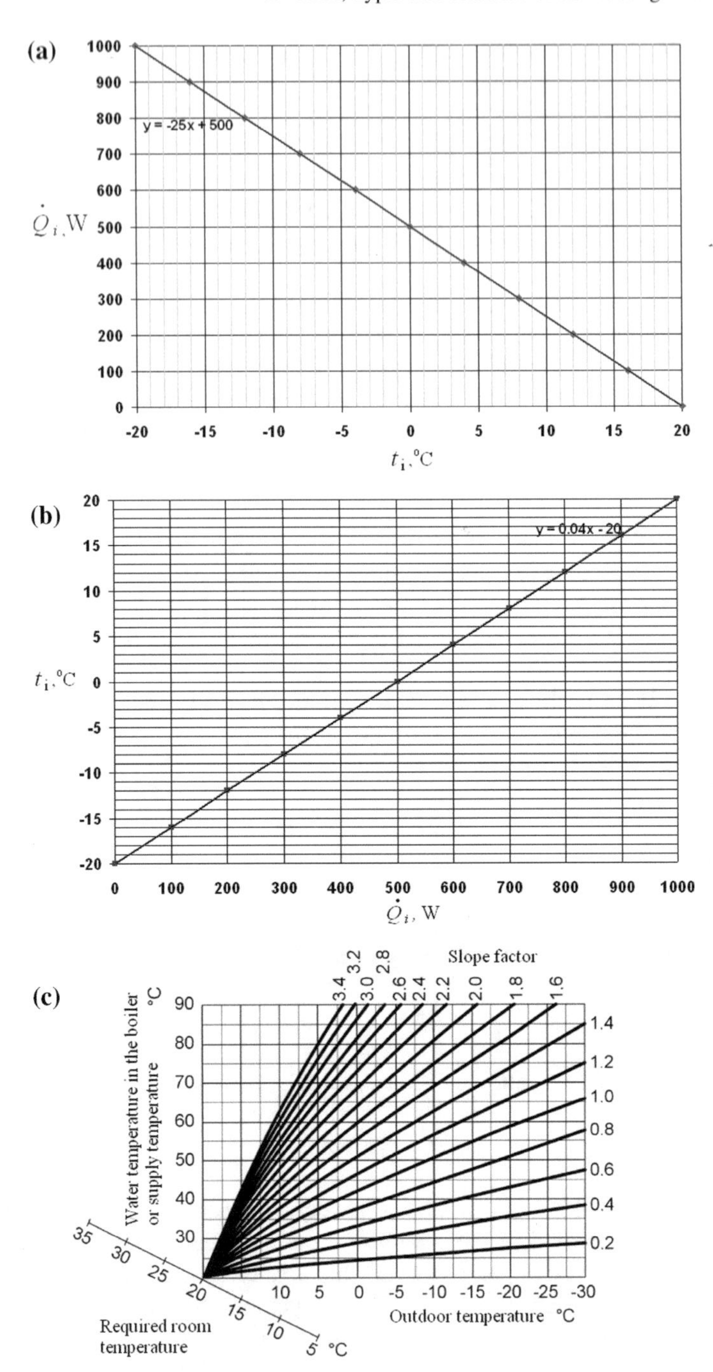

Fig. 2.20 Example static characteristics of a room (**a**, **b**) and the heating curve (**c**)

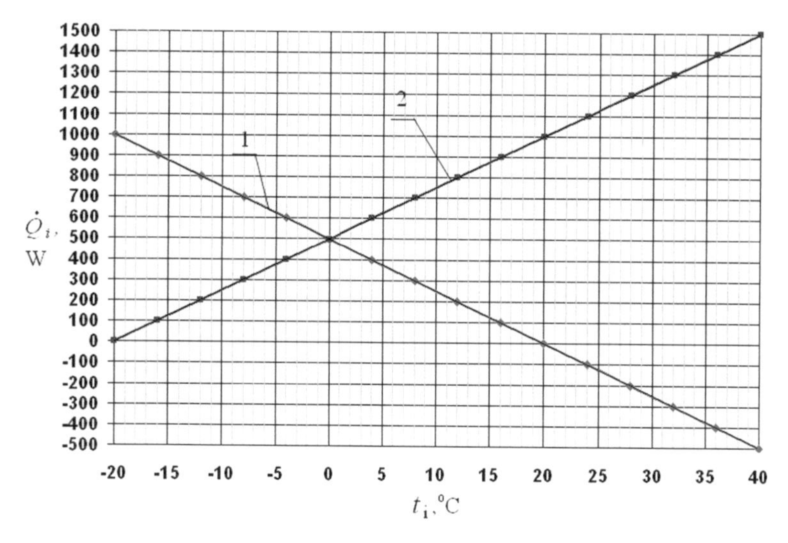

Fig. 2.21 Example static characteristics of a room; 1—required thermal power depending on indoor temperature; 2—indoor temperature depending on supplied thermal power

resulting from the room type and purpose. This is obvious because if no thermal power is supplied to the room, the indoor temperature will settle at the level of the outdoor value. In other words, if the room required temperature is 20 °C and outdoor temperature is at this very level, there is no need to supply thermal power to the room. For simplification, it is assumed here that no additional heat gains (from e.g. humans, equipment or insolation) arise in the room that would raise the indoor temperature above the required value. This is illustrated by the further part of the line in Fig. 2.21, which results in negative values of the demand for thermal power, which means that this heat has to be carried out of the room to maintain the initial temperature in it. The figure presents both lines for a wider range of data in a single system of coordinates. Therefore, the situations should be clearly distinguished from each other and the system axes should be treated as the x- or the y-axis depending on which of the lines in the chart is taken into consideration.

The change in the temperature of a room where a radiator thermoregulator is operating with a known static characteristic results from a shift in the working point—the point of intersection of static characteristics of the thermoregulator and of the objects supplying thermal power to the room. The objects are radiators (and the pipes carrying the medium, if any), as the basic elements of the initial heat balance, the other elements being those generating gains in the heat flux and thus disturbing the original equilibrium and creating a new one. Considering that changes in the room temperature are proportional to changes in supplied thermal power (as described earlier), the characteristic of a thermoregulator co-operating with heat sources in the room, i.e. the medium (relative) flow rate/heat output characteristic can be put directly on the medium (relative) regulated flow rate/room temperature one.

The radiator static thermal characteristic illustrates the dependence of the radiator heat output on the medium flow rate (for a given supply temperature). Depending on the radiator type, its thermal load and operating conditions, its shape may vary, but it is always similar to a logarithmic curve. This is discussed in detail in [33, 34]. The static thermal characteristic {1} of a typical radiator is presented in Fig. 2.22a. The figure also shows the curve illustrating water cooling in the radiator {2}. The analysis of the co-operation of the radiator with the thermoregulator is based on the assumption that the medium nominal mass flow $\dot{m}_N$ ($\dot{m}_N = 0.01184$ kg/s) and the nominal thermal power $\dot{Q}_N$ ($\dot{Q}_N = 1000$ W) are taken for water cooling at the level of $\Delta t_w = 20$ °C, which is the most common case in practice in Poland. Therefore, the nominal (design) proportional range at the level of $X_p = 2$ K is taken for the point on the radiator static characteristic for which water cooling reaches the value mentioned above. This particular point of the curves intersection {A} is the initial working point, for the initial state of balance. Figure 2.22b presents the said characteristic together with some additional ones. The horizontal line {3} illustrates extra heat gains at the level of 20% (200 W) of the radiator heat output. The last line {4} illustrates the resultant static thermal characteristic of the radiator and other additional heat sources. It is a sum of two individual characteristics. Allowing for some simplifications made for the purpose of the analysis,[2] the radiator original curve does not change because the radiator operating conditions (the medium mass flow and temperature) do not change, either. It is only increased by a constant value in the form of the considered gains in thermal power. Due to that, for each value of the medium mass/volume flow, the total power is higher by a constant value of heat gains, which produces a new, shifted curve. Point {B} is the system new working point. The room temperature t_i resulting from the value of supplied thermal power is additionally put on the vertical axis. The analysis is performed for a room characterized by the parameters mentioned above, i.e. the design thermal load at the level of $\dot{Q}_i = 1000$ W at the design outdoor temperature $t_e = -20$ °C, indoor temperature $t_i = 20$ °C and the amplification factor value at the level of $k_s = 0.04$ K/W. Consequently, as follows from Formula (2.7), the supply of extra 200 W, if there is no thermoregulator, means an increment in the room temperature by $\Delta t_i = k_s \cdot \Delta \dot{Q}_i = 0.04 \cdot 200 = 8$ K.

Figure 2.23a presents the same characteristics (the representative fragment) but in a "reversed" form to enable direct putting thereon the radiator thermoregulator static characteristic as presented in Fig. 2.19. Figure 2.23b presents cumulative characteristics of the radiator and the thermoregulator.

It can be observed that a rise in thermal power supplied to the room involves a rise in the room temperature t_i despite the fact that there is a thermoregulator operating in the system. The rise results directly from the selected proportional range X_p. In this case the range is $X_p = 2$ K. The bigger proportional range X_p, the higher the possible temperature increments, as results from the comparison between the intersection points of lines 4 and 3 and 4′ and 3. As described ear-

[2] In reality, if the room temperature rises, the radiator heat output decreases due to the smaller difference between the temperatures of the radiator surface and the room, which is the heat transfer driving force. This is explained in detail in Chap. 6.

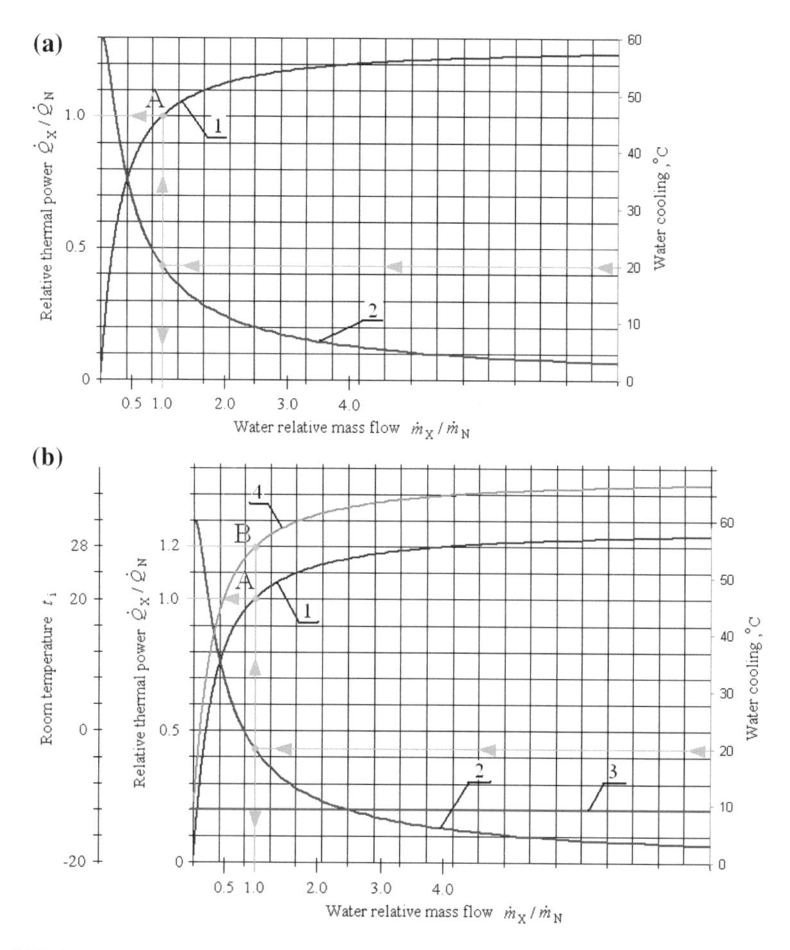

Fig. 2.22 Example static thermal characteristics of heat sources in the room: 1—static thermal characteristic of the radiator; 2—water cooling in the radiator; 3—thermal characteristic of extra heat sources; 4—resultant thermal characteristic of heat sources in the room

lier, proportional range X_p, at the valve given opening degree (conditioning the medium mass flow and the radiator heat output), results from the value of the thermostatic head amplification factor k_m expressed in mm/K. The higher it is, the more vertical the thermoregulator static characteristic and, consequently, the smaller the oscillations in the room temperature t_i. However, the characteristic in Fig. 2.23 is presented not in relation to the valve opening degree, but to the medium mass flow. The quantities are related to each other directly, but the relation is not proportional. Defining the relation between the two parameters requires a more comprehensive analysis, which is presented in Sect. 3.3 and in Chap. 4. But in the case of a qualitative analysis (considering not the precise values, but only the

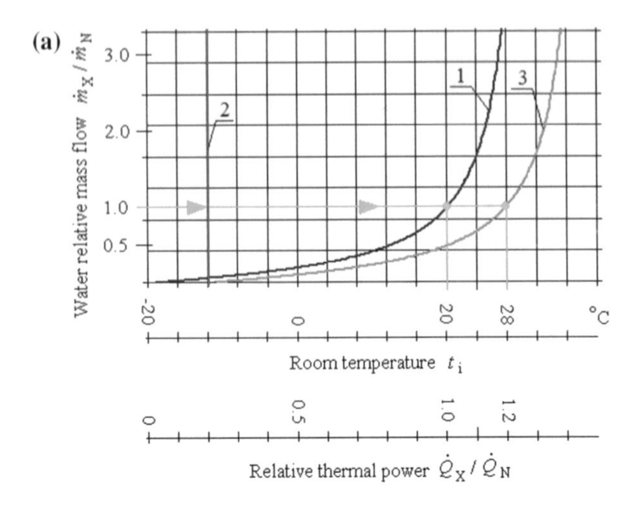

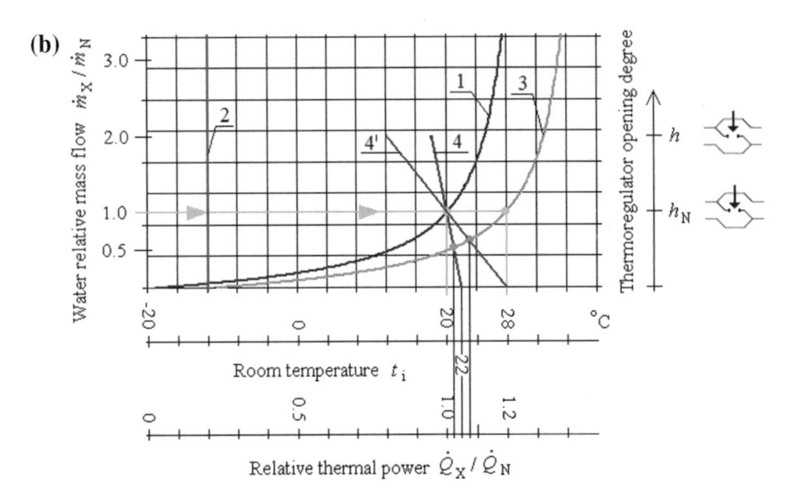

Fig. 2.23 **a** example "reversed" static thermal characteristics of heat sources in the room, **b** co-operation of the radiator thermostatic thermoregulator with the radiator: 1—static thermal characteristic of the radiator; 2—thermal characteristic of extra heat sources; 3—resultant thermal characteristic of heat sources in the room; 4, 4′—static characteristic of the radiator proportional thermoregulator for proportional range $X_p = 2$ K and $X_p = 8$ K

trend and sign thereof) as presented above, it is possible to use the value of the thermoregulator amplification factor expressed in percentages, like for the room amplification factor. It is then not necessary to know specific values of the thermoregulator opening degree h_x/h_{100} and the head amplification factor k_m expressed in mm/K. As described earlier, proportional range X_p is (approximately) directly proportional to the valve opening degree h_x/h_{100}. Therefore, changes in the thermoregulator sensor temperature t_c cause proportional changes in the valve opening degree. The amplification factor can be expressed in percentages using the following formula:

$$k_{m,\%} = 100\% \cdot 1/X_p. \tag{2.9}$$

Consequently, if the design proportional range (also referred to as control range) totals $X_p = 2$ K, a change in the sensor temperature e.g. by 1 K means a 50% difference. And this means that the percentage amplification factor totals $k_{m,\%} = 50\%/\text{K}$.

The rise in the room temperature that takes account of the operation of a proportional thermoregulator, i.e. one with a thermostatic head for example, can be calculated using the following relation:

$$\Delta t_i = \frac{1}{1 + k_{s,\%} \cdot k_{m,\%}} \cdot k_{s,\%} \cdot \Delta \dot{Q}_{i,\%}. \tag{2.10}$$

The amplification factors and the change in thermal power in the formula should be expressed in percentages.

2.1.3 *Return Water Temperature Limiters*

Return temperature limiters, also referred to as RTL valves, are used to obtain a required, appropriately low, temperature of water leaving the radiator by limiting the water flow rate. The aim of this action is in fact not to regulate (limit) the medium temperature at the radiator outlet, but to do so with regard to the water mean temperature. The device is used almost exclusively for underfloor radiators co-operating with convector radiators within a single installation. Elaborate systems of control of the operation of underfloor radiators in rooms are a prerequisite for keeping the thermal comfort parameters if the installation includes a great number of such radiators and its temperature parameters are selected as typical for the operation of the other convector radiators [33, 34]. This results from the fact that, compared to convector radiators, underfloor radiators require different (lower) operating temperature parameters and different values of the working medium pressure in the system (higher—due to high hydraulic resistance of coils). For this reason, mixing/distribution systems and individual pumps are used for them. The former are responsible for the radiator operation temperature parameters, whereas the latter ensure appropriate pressure. However, a situation should be analysed where the installation is mainly composed of convector radiators that the operating parameters are selected for, with very few surface radiators operating as auxiliary devices. It will then turn out that there is no economical justification for implementation of an overall pump mixing/distribution system with devices intended for the underfloor radiator surface temperature regulation due to both investment expenditures (purchase-related costs) and operating costs (energy needed to drive the pump). The underfloor radiators are then incorporated directly

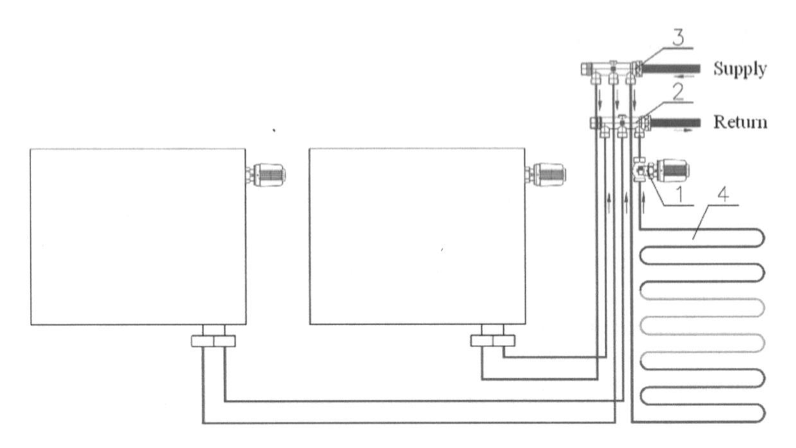

Fig. 2.24 Direct incorporation of a surface radiator into a high-temperature installation using a return temperature limiter [33, 34] 1—thermostatic return temperature limiter (RTL), 2—working medium collector bar, 3—working medium distributor bar, 4—underfloor radiator

into the installation that the convector radiators are connected to. Obviously, such a connection makes it necessary to apply a system reducing the underfloor radiator surface temperature so that permissible values thereof should not be exceeded [33, 34]. It is done indirectly by means of return temperature limiters. As for their operation principle, RTL's are identical to the radiator control valves with thermostatic heads. Their external structure is also similar. The main difference is that the sensor placed in the head measures the temperature of the medium flowing through the valve and not the temperature of the surroundings. This is possible owing to the valve special internal structure which carries the temperature from the valve plug to the head very well. A diagram of the structure of a system with such an element is shown in Fig. 2.24.

The return temperature limiter, like the radiator thermostatic valve, keeps the set temperature of the working medium—here: water flowing through the valve—in the required proportional range, taking the hysteresis value into account. The heating medium temperature that the regulator is to keep is set on the head. If the working medium flow temperature is equal to the value set on the head (t_{zad}), the valve opening degree h_x/h_{100} controlled by the head is equal to the nominal (design) value, i.e. to the value for the selected proportional range X_p. A rise in the working medium temperature is accompanied by a rise in the head sensor temperature t_c. As a result, the thermally expandable substance expands and the valve starts to close. Due to that, the medium mass flow is decreased. The decrease in the mass flow involves greater cooling of the working medium. This means the medium lower mean temperature and, consequently, a drop in the mean temperature of the radiator surface. If the temperature measured by the sensor decreases, the opposite situation occurs. It should be remembered that the set temperature value must not be lower than the temperature of the return temperature limiter surroundings because the device will then remain closed all the time (the installation place has to be taken into consideration). The return temperature limiter may also remain closed if it finds itself close to another heat source, for example if it is installed directly in a closed box of the underfloor heating system distributor (manifold box).

The description of the principle of operation may suggest that such a device should be equally effective if it is installed on the radiator supply. The only necessary correction would then be an appropriate increase in the maximum temperature setting, like in the case of radiator thermoregulators, because the temperature at the radiator inlet is higher than at the outlet. It might even seem that such a location will be more beneficial because, considering the fact that the device task is to limit the maximum temperature of the radiator surface, installing it at the radiator beginning creates a bigger chance that the condition will be satisfied as the valve will not allow water with too high a temperature to flow into the radiator. But for a simple reason this is not so. The drop in the temperature of water flowing in the pipe with a given rate results from the thermal power the liquid loses by giving up heat to the surroundings. The loss arises in the pipe irrespective of whether or not it is built into the underfloor radiator. If it is within the radiator structure, the phenomenon is even stronger because in this case the surface area of heat exchange with the surroundings is increased (the floor, not just the pipe).

The longer the distance covered by water in the pipe, the bigger the liquid loss of power and the larger the drop in its temperature. This means that, even at a small flow rate, the drop is slight at the pipe beginning. Only if the water flow through the pipe is stopped completely, i.e. the pipe is not supplied with hot water, can the medium cool down "completely"—to the temperature of the surroundings. However, even in this situation the water temperature in the area of the pipe connection to the distribution system, where water is still flowing, is higher due to the heat transfer from the connector through the immovable volume of the liquid in the pipe and due to heat conduction in the pipe itself. Even if the device were placed at the pipe beginning, upstream the radiator, it would still "see" the temperature close to the temperature in the distribution duct and it would keep trying to cut off the flow completely to reduce the temperature to the required value of t_{zad}. For this reason, RTL valves should be installed on the return.

It should also be borne in mind that the RTL valve is unable to realize current automatic regulation of the room temperature, but only of the mean temperature of the radiator surface. The return temperature should be set at a level that ensures that the mean temperature of the radiator (floor) surface should not exceed the permissible value. The method of calculating the floor mean temperature and the requirements concerning the temperature values are described in [33, 34].

2.2 Realization of the Principle of Single and Double Regulation in Modern Regulation Valves

Considering current legal regulations in various countries manual radiator regulation valves have lost their significance. For example in Poland, in specific situations the regulations impose the use of radiator control devices that ensure automatic regulation and stabilization of the room temperature [42]. But some of the manual regulation valves solutions were applied in thermostatic radiator valves. This explains the great number of similarities and the need to make reference to manual valves whenever the radiator thermostatic valves are under discussion.

The now common principle of double regulation in valves intended for heating installations is realized by:

2.2.1 *Single-Plug Valves (One Adjustable Section of the Medium Flow)*

The initial hydraulic resistance (regulation first stage) is determined by establishing the plug specific lift over the valve seat, which was the case e.g. in the Polish *M-3176* manual valve. The second-stage regulation is thus due to the plug range of travel limited by the pre-setting. Because throttling and closing are realized using the same element, the throttling and the closing characteristics are identical. The solution is still present in the offer of many companies, especially when it comes to manual valves. Such valves can differ in the manner of setting the medium flow cross-section underneath the regulation valve. The plug can move inside the valve seat or over it. Figure 2.25 presents examples of valves where the closing plug moves inside the valve seat and over it, whereas Fig. 2.26—a valve where the plug moves only over the seat.

Generally, such solutions are no longer found in the case of radiator thermostatic valves. Structures similar to the one described above—with a single plug—are available from most manufacturers but they do not offer the opportunity to limit the closing element travel range or select the pre-setting. This results from the fact that they are intended for single-pipe or/and gravitational installations that should be characterized by small hydraulic flow resistances practically excluding the pre-throttling principle. They are structures with an increased flow capacity (compared to typical radiator thermostatic valves) and they are fully open in their neutral state. In such a valve, a change in the opening degree occurs only due to the action of a current regulation device, e.g. a thermostatic head. This type of valve is presented in Fig. 2.27.

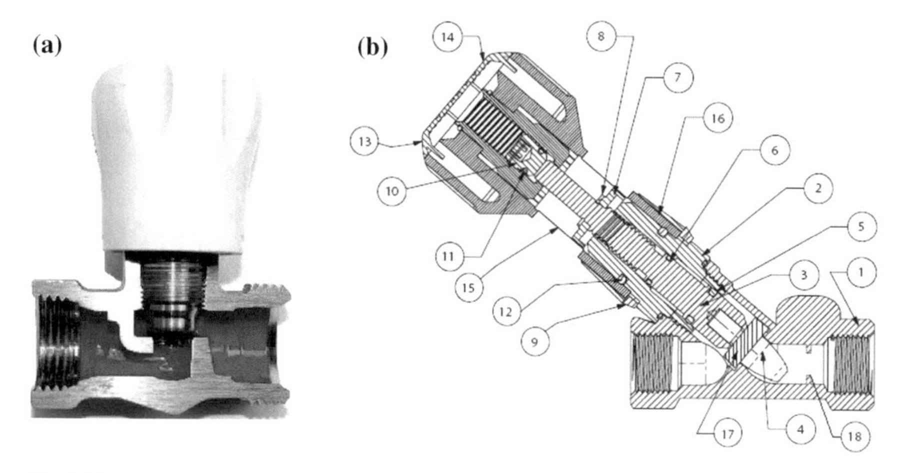

Fig. 2.25 Manual regulation valves with a plug moving inside the valve seat: **a** the *Herz GP* valve (own materials); **b** the *Danfoss MSV-C* valve [5] 1—valve body, 2—casing, 3—pre-setting spindle, 4—valve plug, 5, 6—O-ring (EPDM), 7—pre-setting indicator, 8—locknut, 9—casing, 10—pre-setting locking screw, 11—knob, 12—locking bolt, 13—knob cover, 14—information label, 15—pre-setting scale (whole numbers), 16—pre-setting scale (fractions), 17—plug seal, 18—measuring orifice

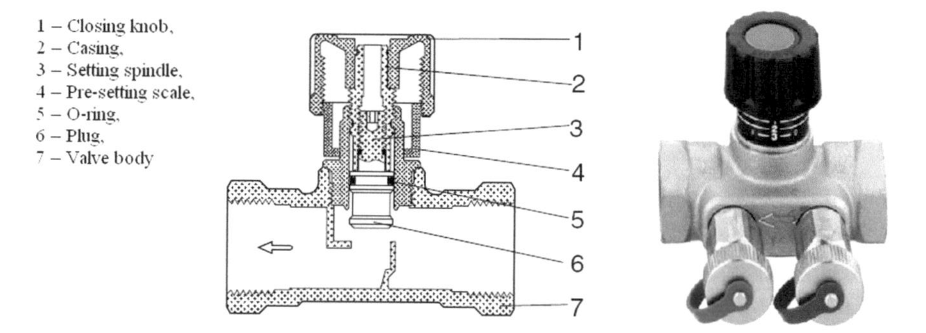

Fig. 2.26 *Danfoss MSV-I* manual radiator regulation valve with a closing plug moving over the valve seat [5]

2.2.2 *Valves with a Double Coaxial Plug (One Adjustable Section of the Medium Flow)*

In terms of the regulation principle, they do not differ much from the devices described above. By setting a specific lift over the valve seat, the external plug (regulation first stage) determines the initial hydraulic resistance and throttling values, whereas the internal plug (regulation second stage) serves the purpose of current regulation controlled by a thermostatic head for example. The consequence of the described mutual position of the plugs is that at a flat seat the element of the

Fig. 2.27 *Herz TS-E* radiator thermostatic regulation valve with increased flow capacity (own materials)

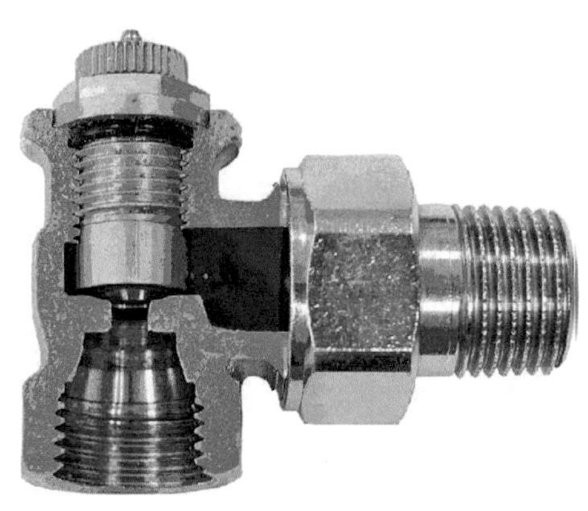

Fig. 2.28 *Herz TS-90-V* radiator thermostatic regulation valve (own materials)

second stage of regulation can only act within the stroke imposed by the lift of the first-stage regulation element. The solution is not optimal with regard to the value of the maximum proportional range $X_{p,max}$, which in this situation cannot be kept constant as a function of the pre-setting. The problem becomes less serious in the case of small proportional ranges that heating installations are usually designed for. The advantage of the solution is that the closing plug can be shaped freely to obtain the assumed closing characteristic. Another good point is that the solution favours achieving high values of what is referred to as inner *authority* (cf. Chap. 4) because hydraulic resistance is generated in one adjustable section of the medium flow. In the past, a solution could also be found where the mutual position of plugs was opposite to the discussed one. Proportional range X_p is then constant, but the closing plug cannot be shaped freely and has an unfavourable, flat profile, which was the case for the Polish *708* type valve. Radiator thermostatic valves with a double plug are now made by few manufacturers. The leader is the *Herz* company, whose valve realizes the principle of a closing plug embedded inside the throttling one. The solution is presented in Fig. 2.28.

2.2.3 Valves with Cylindrical Elements with a Coaxial Internal Closing Plug (Two Adjustable Sections of the Medium Flow)

This is the most popular solution. The cylindrical shutters constituting the regulation first stage feature appropriately shaped slots or holes in the side surface. The cylinder rotation in relation to the valve body outlet channel, or a change in the position of the inlet hole in relation to outlet holes in the cylinder, causes a change in the slot covering or uncovers an opening with a different diameter, which changes the outflow surface area and, thereby, hydraulic resistance. The liquid usually flows into the cylinder. Unfortunately, because the first-stage regulation element is located beyond the channel through which the liquid flows out of the element of the regulation second stage, the value of the element inner authority a_w decreases, which means deformation of the closing characteristic and deterioration of the valve regulation capacity. The solution is similar to those in common use many years ago, also in Poland—valves with an integrated throttling orifice. The difference is that here it is not necessary to replace the valve or the orifice mounted in it to change initial throttling; a change in the value of pre-setting n_i will suffice. The advantage is the possibility of achieving a constant value of proportional range X_p. Examples of this type of solutions are shown in Fig. 2.29.

Figures 2.30 and 2.31 present the typical method of integration of the two regulation stages in thermostatic valves for the two types of double-regulation valves, i.e. for valves with one and with two adjustable sections of the medium flow. Figure 2.30 presents the *Herz TS-90-V* valve of a linear figure for selection of the minimum value of pre-setting n_i (a and b), the pre-setting maximum value (c and d), for the internal closing plug full lift (a and c) and for the regulating plug coming out maximally (b and d). As previously described, limitation of the available range of travel can be observed of the internal regulating plug being the element of the current second-stage regulation.

Figure 2.31 presents the throttling-closing mechanism of the *Herz TS-FV* valve. Holes with different diameters can be seen on the cylinder side surface (a few are visible in the photograph but they are placed on the entire perimeter). The medium flows out of the cylinder, from underneath the plug. Each hole diameter corresponds to a specific value of pre-setting n_i. In the photograph, the internal cylinder with a hole is set so that the medium flows through the cylinder outlet hole and then through the external cylinder biggest hole, whose diameter corresponds to the highest value of pre-setting n_i ensuring the smallest throttling.

There are also thermostatic valves with an integrated stabilizer of differential pressure which is built into the valve body. The device makes it possible to minimize pressure oscillations on the valve regulating element during the regulation process, which ensures minimization of variations in outer authority a_z and an increase in the value thereof to almost one. Its operation also results in minimal deformation of the valve final (or operating) regulation characteristic (cf. Sect. 3.3). Moreover, owing to pressure stabilization, there is a reduction in the phenomenon of changes in

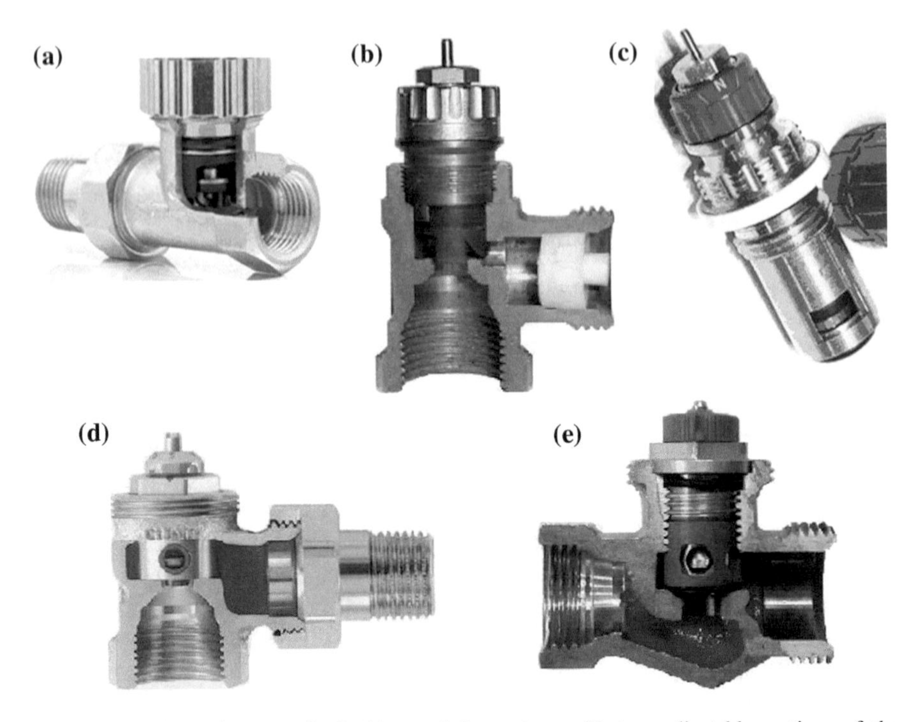

Fig. 2.29 Radiator thermostatic double-regulation valves with two adjustable sections of the medium flow: **a** *Vector* valve made by *Valvex*; **b** *Danfoss RTD-N* valve; **c** *RA-N* valve insert made by *Danfoss*; **d** *V-exact* valve made by *Heimeier*; **e** *Herz TS-FV* valve [5–7, 11]

the medium mass/volume flow through the valve at variable pressure in the system resulting from e.g. changes in the pump impeller rotational speed or redistribution of pressure in the pipework at a change in throttling on the other regulation valves and/or thermoregulators. Consequently, selection of a given value of pre-setting n_i results in a specific maximum value of the medium flow rate, which is something that traditional radiator thermostatic valves are unable to ensure.

The solution minimizes many problems related to the operation of the radiator thermoregulators, just like the use of separate differential pressure stabilizers in the installation, as discussed further below, but the effects are even better. Because of their cost, separate pressure stabilizers are usually mounted only on rather large parts of installations with at least a few circuits (radiators). Due to that, stabilization concerns parameters of the separate part and not each radiator valve individually. The radiator valves on this part may thus be affected by phenomena similar to those occurring in the entire installation without pressure stabilizers, i.e. pressure redistribution between the valves, oscillations in flow rates, overflows, room over/underheating, noise on the valves, etc. The problems can be reduced if each thermoregulator has its own pressure stabilization. A similar effect can naturally be produced by using a separate differential pressure stabilizer in every circuit and for each radiator valve. But invest-

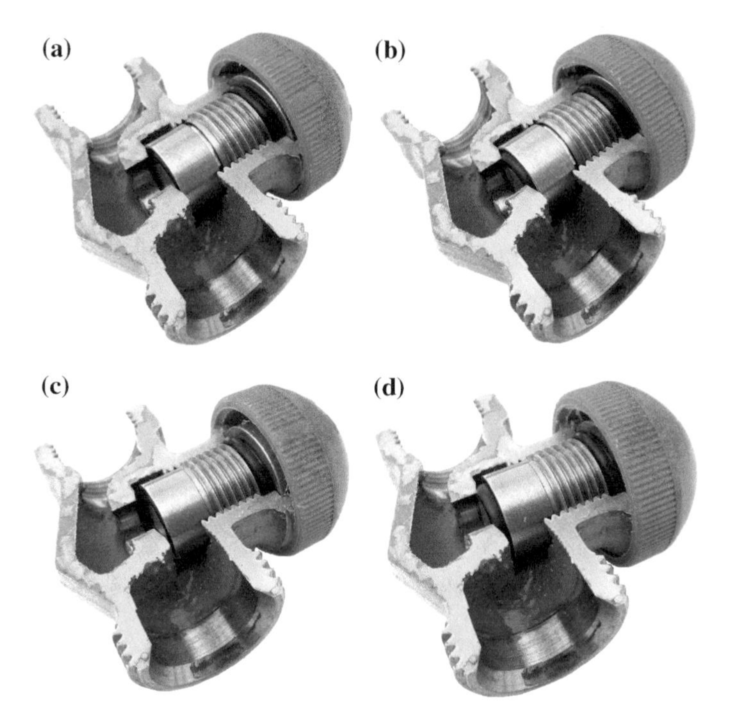

Fig. 2.30 Combination of the throttling and closing element in the *Herz TS-90-V* valve (own materials)

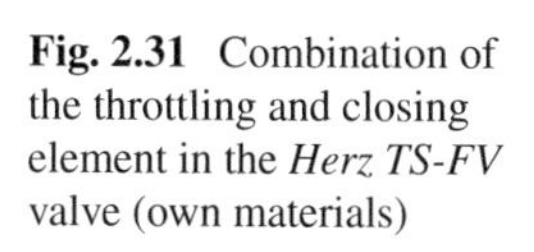

Fig. 2.31 Combination of the throttling and closing element in the *Herz TS-FV* valve (own materials)

ment costs then rise considerably. A radiator thermostatic valve with an integrated differential pressure stabilizer is a cheaper solution than two separate devices used in a given circuit, but it is noticeably more expensive than the radiator thermostatic valve only. It should therefore be considered carefully which solution will turn out to be more profitable in a given circuit, i.e. will the reduction in operating costs exceed the investment expenditures. Sometimes using a single overall differential pressure stabilizer in combination with the installation "ordinary" radiator thermostatic valves could be a better option. Considering the radiator valves operating conditions, the

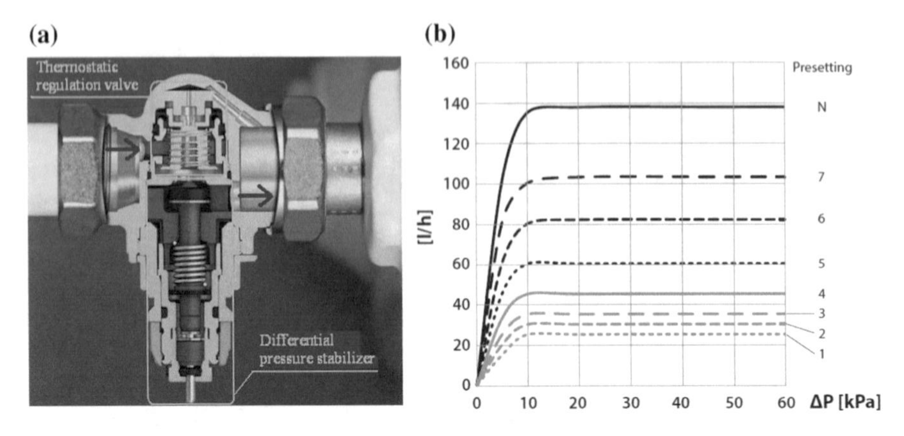

Fig. 2.32 *Danfoss* radiator thermostatic *Dynamic Valve RA-DV* with an integrated differential pressure stabilizer (**a**) and the valve hydraulic characteristics (**b**) [5]

regulation quality and stabilization of the required thermal and hydraulic parameters of a given facility, the solution is certainly the most profitable one.

The pressure signals reaching the stabilizer are supplied inside the valve body. The principle of the differential pressure stabilizer operation is discussed in Sect. 2.3.5. Figure 2.32 presents the view of a thermostatic valve with an integrated differential pressure stabilizer and the device hydraulic characteristics. Comparing the characteristics with those of a typical thermoregulator (like the one shown in Fig. 2.51 for example), it is possible to notice the described phenomenon of the medium flow rate stabilization for a given pre-setting of the valve as a function of the pressure difference arising on the valve. The chart itself is a simplification because in reality it is impossible to keep a constant value of the medium flow rate if the valve differential pressure varies. The self-actuated differential pressure stabilizer is in fact a proportional controller and as a rule absolute stabilization of the regulated quantity value is impossible (cf. Sect. 2.1.2.3 on radiator thermostatic thermoregulators). A rise in the difference in the system pressure will to a certain extent involve a rise in the medium mass/volume flow (the lines in Fig. 2.32 will not be horizontal).

The "standard" radiator thermostatic valve and the solution under consideration are shown for comparison in Fig. 2.33.

According to the data provided by the manufacturer, the minimum value of differential pressure arising on the valve to ensure the device proper operation is 10 kPa. This means that this is the value of the stabilized difference in pressure, and the heating circuit equipped with this kind of valve should be designed for at least this level.

Radiator double-regulation valves are also made as valve inserts, as presented in Fig. 2.29c. These elements, with no typical body, are mounted directly on the radiator, whose supply and return pipes are then most often connected at the bottom. This type of integration is fairly common. An example of such a solution is presented in Fig. 2.34.

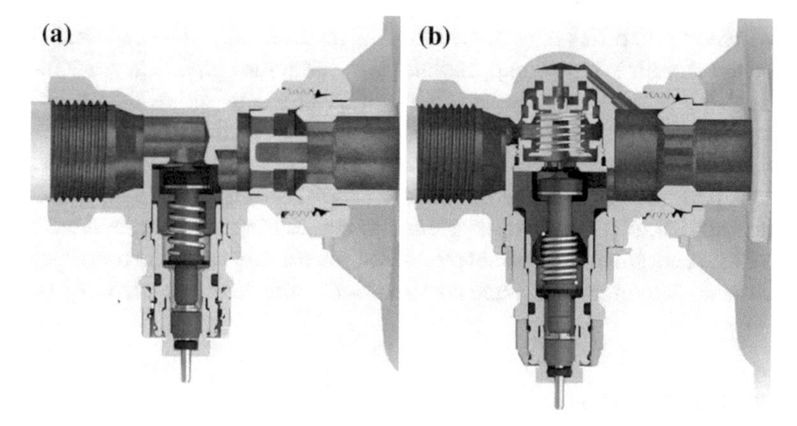

Fig. 2.33 *Danfoss* classic radiator thermostatic *RA-N* valve (**a**); *Danfoss* radiator thermostatic *Dynamic RA-DV* valve with an integrated differential pressure stabilizer (**b**) [5]

Fig. 2.34 Radiator valve insert and method of assembly on the radiator [6]

Considering all the structural options of the radiator regulation valve discussed herein, it seems that the most favourable solution for the radiator valve intended for co-operation with a thermostatic head is one that integrates the elements of the regulation first and second stage in one adjustable section of the medium flow, ensures independence of the regulation second-stage element range of travel as a function of the opening degree of the first-stage element and does not limit the possibility of shaping the geometry of the regulation second-stage element to an unfavourable flat profile only. Among all the valve discussed, the *Goeke* historical design of the *Optimal* valve is the closest to meet these requirements.

The market also offers regulation and regulation/cut-off valves (also called *lock-shield valves*) with a pre-setting, mounted on the return pipe. They are the same as manual regulation valves, but smaller and usually deprived of the possibility of current regulation. The valves are intended for pre-throttling if for any reason the main regulation valve at the receiver is unable to throttle the required pressure value or cannot perform the function of setting initial throttling. However, it should be remembered that, if added to the system and considering the extra hydraulic resistance which is not subject to current regulation, such a valve causes a drop in the main regulation valve total authority and deterioration of the device regulation properties.

2.3 Balancing Valves

Generally, each regulation valve is in fact a balancing one as it is intended for compensating for pressure losses in the circuit, i.e. it ensures the circuit hydraulic balancing. But it is commonly agreed that valves are categorized in terms of their location in the installation into regulation valves at heat receivers and the other ones, which are used, among other things, to ensure appropriate working conditions for the former. The task of the devices included in the group of balancing valves is to keep the following parameters in the section subject to regulation:

- the set value of pressure for static operating conditions and/or in installations with a constant flow rate. In less extensive installations, the function is performed by manual balancing valves, most often—straight-through ones.
- the set value of pressure for static and/or dynamic operating conditions and/or in installations with a changeable flow rate. In more extensive installations, the function is performed by balancing valves equipped with electronically controlled actuators.
- the set value of temperature or flow rate if more than one regulated sections are connected in the installation. This task is performed by mixing or distribution valves. They most often operate with electronically or electrically controlled actuators or they are equipped with thermostatic elements.
- the value of pressure not exceeding the required value. The function is performed by automatic pressure relief valves. They usually operate without any auxiliary elements, as self-actuated devices. In larger installations, the valves are not self-actuated and they are equipped with appropriate actuators.
- the set value of pressure for dynamic operating conditions and/or in installations with a changeable flow rate. The task is performed by automatic differential pressure stabilizers. Like in the previous case, they may be self-actuated or driven by actuators.
- the set value of the flow rate in a given section of the installation. The task is fulfilled by flow stabilizers and automatic flow limiters. Like in the previous case, they may be self-actuated or driven by actuators.

Fig. 2.35 *Herz Stromax M* manual balancing valve (own materials)

Apart from that, one device often combines at least two of the functions described above. In practice, these are most often combined differential pressure and flow stabilizers and differential pressure stabilizers combined with flow limiters.

2.3.1 Manual Balancing Valves

Manual balancing valves are used for pressure throttling and setting a specific value of the medium mass/volume flow for conditions of operation at a constant flow rate. They are usually mounted on the installation separated parts where there is a group of operating heat receivers or where considerable corrections have to be made to the supplied pressure value. A typical location is the beginning of the heating riser, upstream the first heat receiver, or the radiator manifold/distributor. This makes it possible to reduce pressure in a given riser to the required value. The installation active pressure value is calculated for the most unfavourable (critical) circuit, i.e. one that is characterized by the biggest pressure losses (in practice—the one which is the most distant and/or which requires the highest thermal power and, consequently, the medium biggest mass/volume flow). Therefore, it will be too high for the other circuits. For static conditions, with a constant flow rate of the medium, the required pressure level can be maintained sufficiently by installing a straight-through valve. The valve is characterized by a pre-set hydraulic resistance value, which for a set value of the medium mass/volume flow throttles the required value of pressure, thus maintaining a constant pressure value in the regulated section. The device example structure is shown in Fig. 2.35.

2.3.2 Balancing Valves with Actuators

Manual straight-through balancing valves are the most popular solutions in installations with a constant flow rate, in home heating installations and also in less extensive installations in multi-family buildings and small blocks of flats. However, in large installations the efficiency of the solution is slight. Extensive installations usually

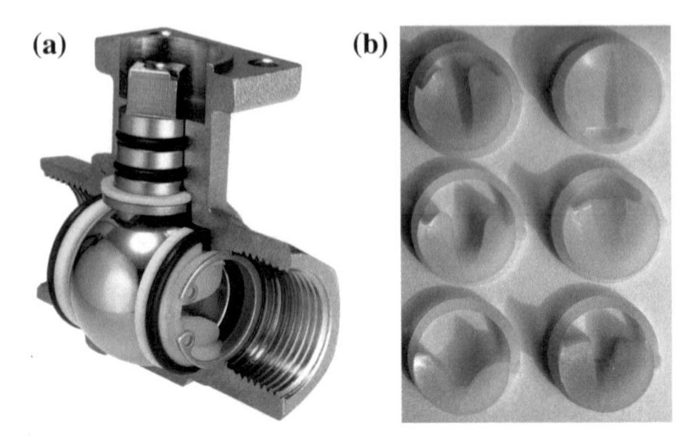

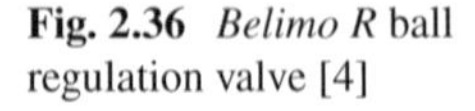
Fig. 2.36 *Belimo R* ball regulation valve [4]

require continuous monitoring of and corrections to operating parameters. They are also characterized by considerable flow rates and high pressure values of the medium. On-line corrections to the flow rate and pressure cannot be made by means of a manual valve. Due to high values of the medium flow rate and/or the medium pressure, big forces arise on the valve plug along its axis, which makes it difficult to change the valve settings by hand. In order to solve the problem, straight-through regulation valves with electric actuators are used or hydraulic ones which are most often controlled electronically. The actuator is sent a signal from the control system based on current and pre-set values of operating parameters. In this way, corrections can be introduced continuously.

In most cases, straight-through balancing valves are structures identical with manual balancing valves—they are plug valves. The difference is that the reciprocating motion is not the effect of a bolt rotation on a threaded spindle, but of the bolt insertion (due to pressure exerted on it) or pulling out. But there are also solutions in the form of ball regulation valves used as balancing ones. One of them is presented in Fig. 2.36a (product made by the *Belimo* company, actuator not installed).

In terms of structure, this is a typical cut-off ball valve. However, it has an appropriately shaped orifice. On one side, the orifice adheres to the ball surface tightly, which creates one adjustable section of the medium flow. Appropriate geometry of the hole makes it possible to obtain the required regulation characteristic of the valve. As specified by the manufacturer, original regulation curves are equal-percentage ones.

The orifices presented in Fig. 2.36b are factory-mounted permanently and they come in different variants and shapes. This makes it possible to obtain different values of flow factor k_{vs} and, possibly, different regulation characteristics, as may be required. The solution is similar to the one described in Sect. 2.1.1—the historical Polish version of a regulation plug valve with a replaceable orifice. Its advantage, however, is that it does not create an additional section of the medium flow and as a result—does not reduce the valve inner authority. Original regulation characteristics of the *Belimo* regulation ball valves are similar to equal-percentage ones, which means that the devices are a favourable solution in the case of regulation of heating installations.

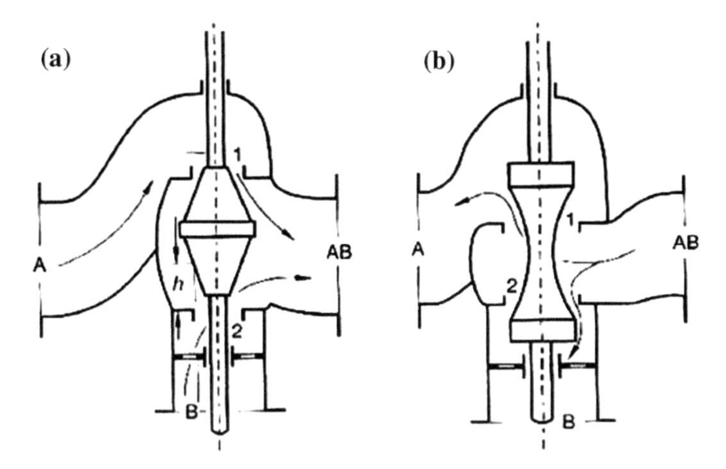

Fig. 2.37 Two-way valve operating as a mixing (**a**) and distribution (**b**) device [40, 41]

At present, regulation ball valves are rather seldom in Europe, whereas they are fairly popular in the USA. Regulation ball balancing valves are more and more often selected to replace classical plug valves due to the lower cost of the valves themselves and of the controllers/actuators co-operating with them.

2.3.3 Mixing and Distribution Valves

The valves are used to regulate and stabilize the set value of the medium temperature or flow rate in the installation given section if more than one regulated sections are connected in it. For this reason, they find application in installations with underfloor radiators, or as elements preventing too big a drop in the temperature of the water return to the heat source/boiler. They are made as two-way valves with three or four ports, in which case in practice they are commonly referred to as three- or four-way valves. It should be mentioned that this does not reflect the actual principle of operation because in a valve with three ports the medium can flow out through two of them, the third must be used by the medium to flow in, which means that there are maximally two ways of flow. In practice, however, the name "two-way valve" is extremely rare and even experienced designers and fitters do not use it.

The function of mixing or distributing the medium flow is carried out inside the valve. If the medium flows in through two ports, the individual streams are joined on the valve and flow out through the third, i.e. mixing occurs. The valve operates as a mixing one. If the medium flows in through one port, it is separated into two streams flowing to the other ports. The valve operates as a distribution one. This operation principle is illustrated in Fig. 2.37.

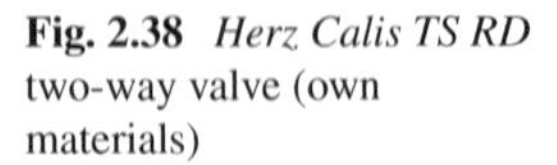

Fig. 2.38 *Herz Calis TS RD* two-way valve (own materials)

If a medium with different temperature and mass flow values flows into a two-way valve through two ports, it leaves through the third port with an intermediate temperature value and summarized mass flow value, as the effect of the mixing inside the valve. In a four-port valve, the medium can flow into and out of the valve through two ports simultaneously. Whether a given device operates as a mixing or a distribution valve depends on whether the collective port AB (cf. Fig. 2.37) is connected to the pump suction or pumping side. Figure 2.38 presents a cross-section of the *Herz Calis TS RD* two-way valve intended for co-operation with thermostatic heads. Due to that, it can be easily used in a system with quantitative temperature regulation (by varying the medium flow rate) of the receiver, e.g. an underfloor radiator, in a mixed heating installation. The device is recommended by the manufacturer as a distribution valve.

Generally, two-way mixing and distribution valves can be used interchangeably, but most often they differ in structure because, like in the case of one-way control valves, it is advantageous if the working medium flows in under the plug (cf. Fig. 2.37), thus avoiding a force arising on the plug and tending to close the valve. Nevertheless, valves intended for mixing are offered in a wider range of flow capacity (flow factors k_v) and they may operate at bigger pressure differences. Moreover, such valves are often made in a manner that ensures a certain minimum flow of the working medium through one port even if the other is fully closed.

The aim of the two-way valve application is mainly to maintain a set constant value of the medium temperature in the installation regulated section at a specific flow rate or, alternatively, to keep the medium flow rate constant. Two-way valves operating as mixing ones are most often used to regulate the temperature of water supplying the entire installation or a part thereof, e.g. in cases where underfloor radiators requiring a lower supply temperature are used with typical convector radiators. In such a situation, the medium mass flow in the circuit through the heat receiver where the pump is installed is constant (oscillations may arise due to the pump throttling characteristic). The quantity that is regulated (by the degree of mixing in the

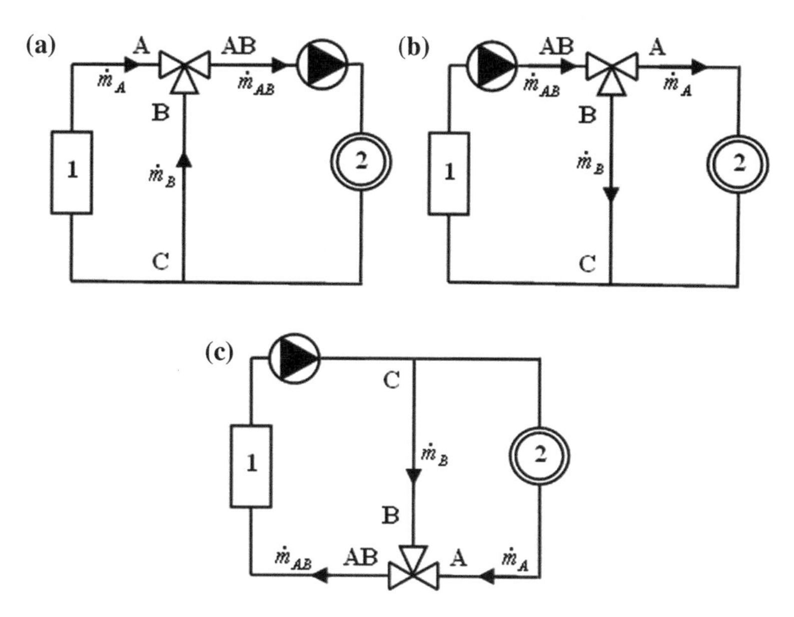

Fig. 2.39 Two-way valves: **a**, **c** mixing system, **b** distributing system; 1—heat source, 2—heat receiver

valve) is the temperature of the medium outflow. The medium mass flow in the heat source part varies from 0 to 100%.

Valves operating as distribution ones are useful in installations with central qualitative operating regulation, i.e. at changeable flow rates and constant temperatures of the medium flowing out of the valve. The heat source circuit then operates at a constant value of the medium flow rate, whereas in the receiver circuit the value varies from 0 to 100%. Example diagrams of solutions incorporating the devices under consideration are presented in Fig. 2.39.

Diagrams {b} and {c} in Fig. 2.39 are similar to each other. In both configurations the valves can perform a similar function within the installation, i.e. the receiver quantitative regulation, with the possibility of keeping a constant flow rate in the heat source. They can be used interchangeably, for example configuration {c} can be used if the valve is structurally not intended for the flow distribution (compare Figs. 2.27 and 2.39) but only for mixing, which means that the device could not operate correctly in configuration {b}. The two-way valve is also typically used to ensure appropriate temperature of water returning to the boiler to avoid excessive cooling of flue gases and water vapour condensation from them, which might give rise to sulphuric acid formation (in cases where the boiler is fired with coal, which often contains large amounts of sulphur). The valve can then operate either as a distributing or a mixing element. During the installation start-up, when the heat receivers (radiators) are cold, water supplied to them is cooled considerably and returns to the boiler with a temperature much lower than assumed for design conditions. In order to prevent that, some of the water leaving the boiler is returned by means of the valve directly to

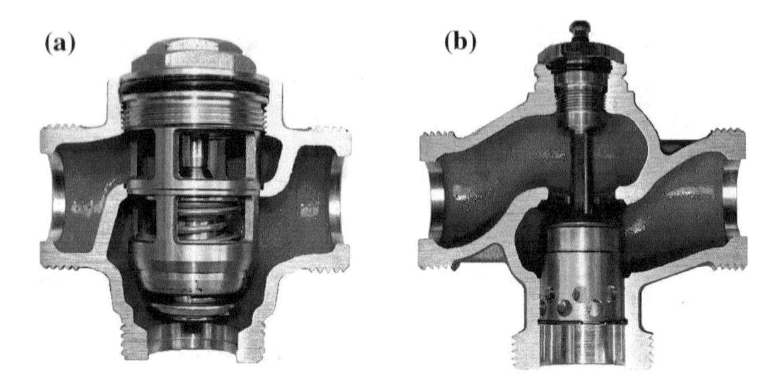

Fig. 2.40 *Herz Teplomix* two-way valves: **a** mixing valve, **b** mixing-distribution valve (own materials)

the return pipe, where it is mixed with cooled water returning from the installation. The two streams get mixed either in the pipe connection point {C} (in the case of the distribution valve, variant {b} in Fig. 2.39) or inside the valve (in the case of the mixing valve, variant {c} in Fig. 2.39), thus raising the overall temperature of return water. Naturally, this involves a partial reduction in the amount of the medium and in the thermal power transferred from the supply pipe to the installation at the start-up stage, which practically lengthens the process. Minimal mixing occurs in the initial phase of the start-up, i.e. the valve directs a very small amount of the installation return water to the boiler return pipe and, consequently, a very big amount directly from the boiler supply pipe. Ports {A} and {B} are then characterized by the medium minimal and maximum flow rates, respectively. With time, as the temperature of the heat receivers gets higher and higher and the water cooling in them gets smaller, the valve directs more and more water returning from the installation to the boiler return pipe. As a result, the amount of water returned directly from the supply pipe is reduced. There comes a moment when the highest degree of mixing is obtained and the two streams—the one returned directly from the boiler and the one from the installation—are equal. Further on, the mixing degree gets lower because the valve decreases the amount of water returned from the boiler and increases the flow returning from the installation until the water supply from the boiler is cut-off (if the valve structure permits it, a small leakage usually occurs).

Figure 2.40a presents the *Herz Teplomix* two-way mixing valve intended for ensuring appropriate temperature of water returning to the boiler. It has a built-in thermostatic mechanism activated at the temperature of 55–61 °C, depending on the selected variant. Figure 2.40b illustrates the same manufacturer's two-way valve which can operate as a mixing or a distribution valve driven by an actuator.

Moreover, performing similar functions, such devices are used in installations with solar collectors and in municipal systems of hot water preparation. As mentioned above, the element can be driven using thermostatic devices, manually or by means of electric actuators of the boiler automatic control system. In the first case, the operation principle is similar to radiator thermoregulators. The main difference is that

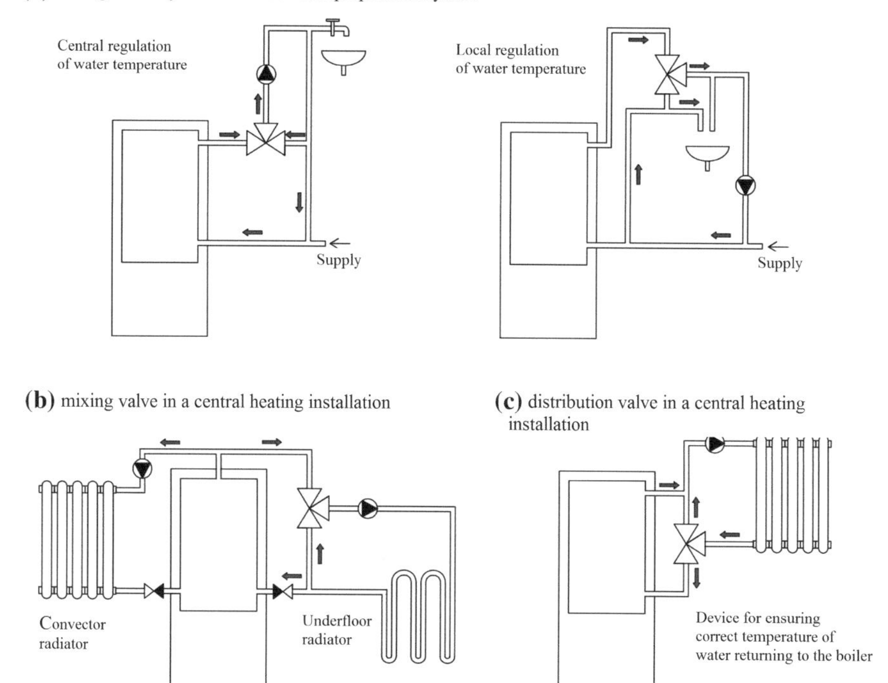

Fig. 2.41 Examples of practical applications of two-way valves

here the water temperature is measured and regulated in the connector pipe to which a given installation part is connected, e.g. the boiler return pipe. The measurement can be performed using a thermostatic head with an outer sensor mounted in the target location or inside the valve. The plug and the bolt structures then ensure heat conduction to a specially designed head. Alternatively, the thermostatic element is mounted inside the valve, like in the *Herz Teplomix* option. Practical realizations of installations with two-way valves are illustrated in Fig. 2.41.

Four-port two-way valves are most often used to ensure appropriate temperature of water returning to the boiler and, at the same time, enable distribution of hydraulic resistances of the boiler and the radiator circuits such that a change in parameters of the streams mixing should not necessarily be related to changes in the flow rate values in either of the circuits. A diagram of a system incorporating such a valve is shown in Fig. 2.42. An element which is also used for similar purposes, which is not a valve though, is a *fluid coupling* [24–32, 40, 41, 43, 45]. This is a favourable solution, especially in extensive multi-boiler installations with large capacities and considerable flow rates.

Like in the case of one-way straight-through regulation valves, two-way mixing and distribution valves can come as plug or ball devices. Which valve type is selected is often determined by its price—two-way ball valves and their actuators are usually

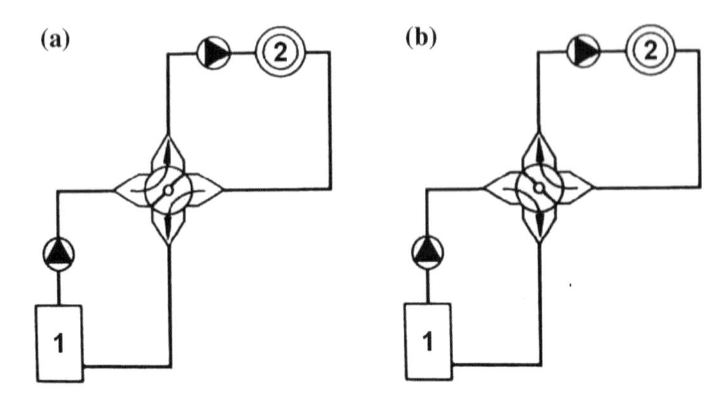

Fig. 2.42 System with a four-port valve: **a** under full load; **b** under no load; 1—heat source, 2—heat receiver

cheaper compared to plug valves. However, if the valve is to co-operate with a thermostatic head, the two types cannot be used interchangeably—a typical head operates using the reciprocating motion, whereas a ball valve is controlled by rotation.

There are also combined double two-way valves (colloquially referred to as six-way valves). They find application in combined heating-cooling devices, e.g. fancoils (commonly used in large-sized stores for example), to which pipes are connected from the heating and the cooling system separately. In such cases, individual two-way valves, as well as individual actuators, are used for each pipework. Combining two valves into one instead of using two separate two-way valves is usually a more expensive solution, but it makes it possible to reduce costs related to actuators because in this case it can be just one element.

Such a valve can also be extended to incorporate an ultrasonic flow meter unit (like in the *EPIV* valve, cf. Fig. 2.65), which makes it possible to gain the advantage of current regulation and stabilization of the flow in circuits of heat and cold using a single measuring-executive unit. Owing to the solution, heating and cooling devices such as ceiling heating bars, fancoils or climaconvectors can be supplied with two kinds of the medium economically.

Figures 2.43a, b present ball valves made by the *Belimo* company: a two-way valve and a double, combined, two-way valve. Figure 2.43c shows a double, combined, two-way valve supplemented with a flow meter unit.

2.3.4 Automatic Pressure Relief Valves

Automatic pressure relief valves are intended to prevent exceeding the set pressure value in the installation or a part thereof. Moreover, if needed, they make it possible to maintain a certain minimal level of the medium circulation or its constant value in the heat source, as for most boilers constant-flow operation is favourable in terms

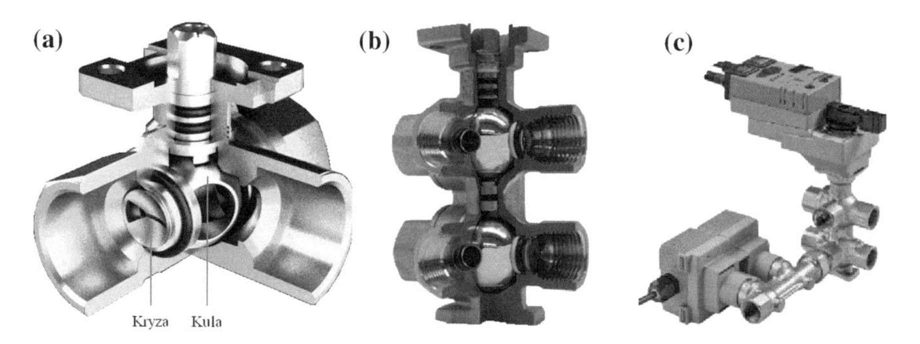

Fig. 2.43 *Belimo* ball valves: a two-way valve (**a**), a double, combined, two-way valve (**b**), a double, combined, two-way valve with a flow meter [4]

of efficiency and life. In this light, they are technically pressure reducing valves. Like radiator thermoregulators, they can be self-actuated or driven by actuators, e.g. electric ones. Oscillations in the installation active pressure, which also determine changes in flow rates, may result from the pump throttling characteristic. They also arise additionally in individual sections of the pipework due to pressure losses in common branches (sections through which the medium reaches points of branching into individual circuits).

Basically, depending on their location in the system and the principle of their operation, they can be divided into pressure reducing valves and relief valves. The former are arranged in series. In such a configuration they generate an extra hydraulic resistance thus throttling the required pressure value and limiting the flow in the secondary section, downstream the valve. The higher the pressure in the primary section, upstream the valve, the more the valve closes, reducing pressure downstream, in the secondary section. This means that the valve is an element of proportional control, with no supply of auxiliary energy (unless the device is equipped with an extra drive). So that the function of automatic pressure reduction should be fulfilled, the valve must have the pressure measuring element—a sensor. Usually, this is an elastic membrane connected to the plug. A diagram of a typical relief valve, in two versions of functionality, is presented in Fig. 2.44.

The valve in Fig. 2.44a responds to pressure upstream the plug. Owing to the built-in setting spring, in the neutral state the valve is open. After the medium upstream-valve pressure p_E is supplied to the "top" part of the membrane with surface area A_M through what is referred to as the impulse tube, force $F_M = p_E \cdot A_M$ arises, opposed by force F_F, being the effect of the spring tension (the pressure on the membrane other side is equal to atmospheric pressure). Once the pressure value rises enough for the first force to exceed the value of the second, the plug will be shifted downwards until the new value of the spring tension is obtained to balance the force arising on the membrane in the plug new position. This means a reduction of pressure downstream the valve, on the valve secondary side. The pressure required for the valve to start to close is obtained by appropriate tension of the spring.

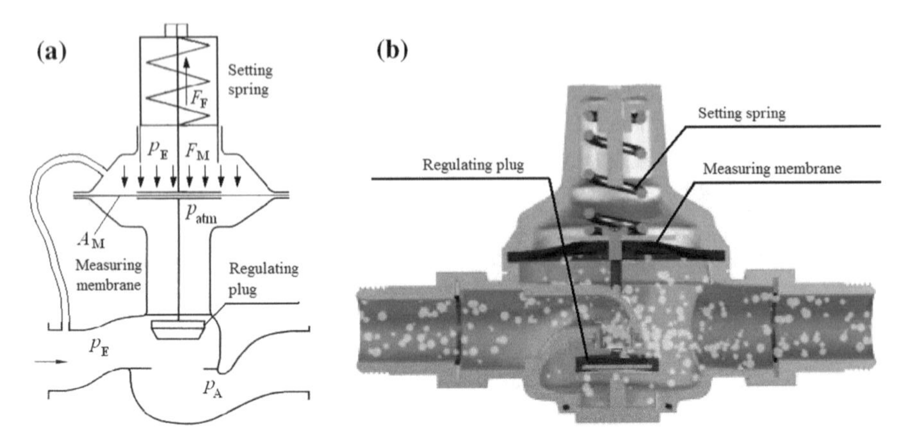

Fig. 2.44 Pressure reducing valve with pressure measured upstream the plug (**a**) and downstream the plug (*Socla BIS* valve) (**b**) F_F—spring tension force, p_E—primary side pressure, p_A—secondary side pressure, F_M—force on the valve membrane and plug, A_M—membrane active surface area (based on [10, 40, 41])

The valve in Fig. 2.44b responds to pressure downstream the plug. The principle of operation is the same as above, except that the supply of the force onto the measuring element, the membrane, and the motion of the executive element, the plug, are opposite. Once the pressure value downstream the plug rises, the force acting on the membrane is increased, the spring is squeezed, the membrane gets deformed upwards and the plug is shifted upwards, towards the valve seat, closing the valve.

Pressure reducing valves perform their task by closing (which involves a rise in their hydraulic resistance) and preventing a given section of the installation, such as the receiving (secondary) network, against too high a pressure value carried over from the primary network. In this case, the receiving (secondary) side can be for example the building internal heating installation, and the primary side—the district heating network, or a part of the internal installation equipped with a pump. Moreover, it is possible to consider the cold water supply and distribution network as the primary side, which is connected to the building internal water supply installation, being the secondary (receiving) side. Therefore, the valves should be mounted on the supply pipe. Moreover, they can be used as elements making sure that the installation is continuously filled up to the set level, which is desired in open-vented heating systems (cf. Fig. 2.46). They should then be mounted on the return pipe.

Relief valves differ from pressure reducing valves in that pressure is supplied to the membrane opposite ("bottom") side and the spring tension is opposite (or the spring is squeezed or placed on the other side of the valve plug). This means that in the neutral state the valve remains closed and it is opened due to a rise in pressure. Relief valves can be made with springs or springs and membranes. Figure 2.45 presents diagrams illustrating the device structure.

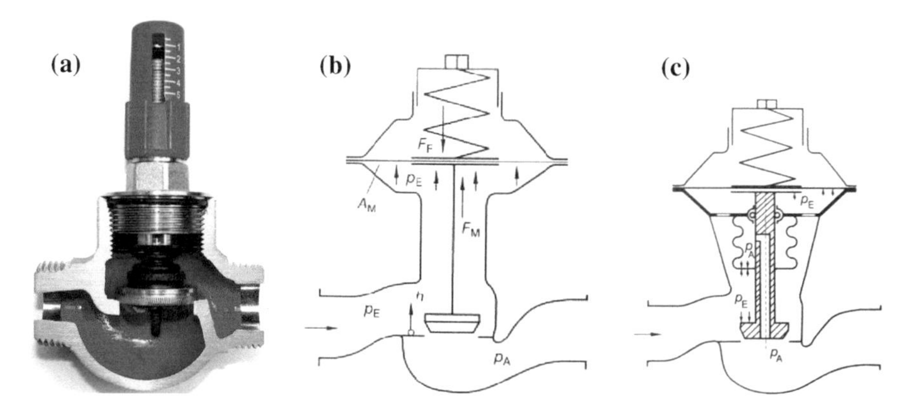

Fig. 2.45 Cross-sections of relief valves: *Herz 4004* spring valve (**a**) (own materials); membrane valve with and without pressure relief (**b**, **c**). F_F— spring tension force, p_E—primary side pressure, p_A—secondary side pressure, F_M—force on the valve membrane and plug, A_M—membrane active surface area (based on [40, 41])

At present self-actuated relief valves are usually made with a spring, as presented in Fig. 2.45a. The medium flows in under the plug so the pressure beneath the plug is higher than over it. A force directed upwards is created as a result, trying to open the valve. It is balanced by the spring tension force. Once the value set on the spring is exceeded, the valve starts to open. In this case therefore, the plug is not only a flow throttling element, but also a sensor measuring the value of the arising force.

If the relief valve has an additional membrane (cf. Fig. 2.45b), the membrane is the measuring element. It is subjected to pressure that the spring is to balance, like in the case of the pressure reducing valve. However, because the medium is throttled on the plug, the difference in pressure and the force arising thereon disturb the measurement performed by the membrane. This effect can be eliminated, e.g. by means of an additional pressure capsule, into whose inside the medium with a pressure value of the regulated secondary side is supplied through the regulating plug hollow (cf. Fig. 2.45c). This produces a force opposite to and balancing the one acting on the plug bottom part. The solution is referred to as pressure relief and is used in the same way in pressure reducing valves. Currently, heating installation pressure relief valves hardly ever have an additional membrane so as not to complicate the valve structure unnecessarily (and keep the price low, too).

Pressure relief valves are mounted in the system in parallel, connecting the supply and the return pipes. In practice, the solution is referred to as a *bypass*. As a rule, the device is used so that, when opening, it should increase the medium total flow rate upstream, in the primary circuit, and, thereby, reduce pressure (e.g. produced by a pump) in this branch. In other words, its task is to reduce pressure on the secondary side connected to it in parallel. But it will be effective only if the pump (or another pressure source) is characterized by a distinct pressure drop as a function of the medium flow rate. If the pressure source remains unresponsive to changes in the medium flow rate, the system with a pressure relief valve will fail to serve its

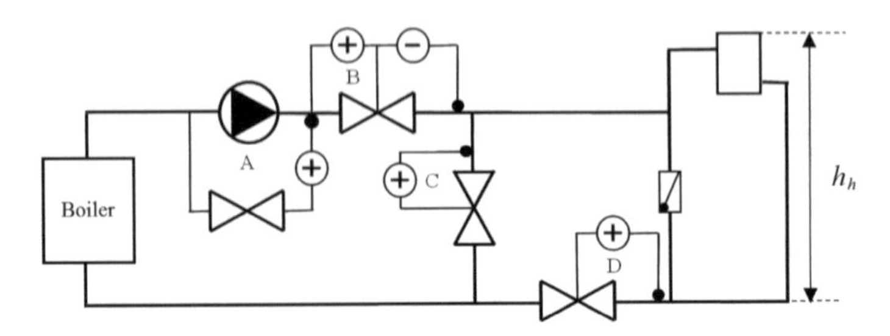

Fig. 2.46 Selected typical locations of the pressure reducing and the pressure relief valve in the installation (based on [40, 41])

function because pressure on the primary side will cause the valve to open and a rise in the medium total flow rate only, without a reduction in generated pressure. Considering that the valve is connected in parallel to the secondary side, where the thermoregulator is located, the same pressure will arise on it. To some extent, the pressure drop from the common point (i.e. the valve) to the thermoregulator will be due only to hydraulic resistances of the pipes in this section, rising with a rise in the flow rate and, in this way, causing a slight reduction in the phenomenon of a pressure increase on the thermoregulator.

In terms of this mechanism, in-series pressure reducing valves have an advantage over pressure relief valves—when they close at a pressure rise on the primary side, they each time reduce the pressure value on the secondary side. Moreover, their action involves a reduction in the medium flow rate on the primary side, and not an increase therein.

Figure 2.46 shows typical locations of pressure reducing and pressure relief valves operating in an installation. Analysing the diagram, it can be noticed that an in-series mounted valve, operating as a pressure reducing device, should open if upstream pressure rises ("+" sign) or downstream pressure decreases ("−" sign). In the case of a parallel configuration, operating as a pressure relief valve, the device should open if upstream pressure rises ("+" sign).

In cases {A} and {C} the valve fulfils the same role, operating as a pressure relief device. The only difference is that in case {A} water with supply temperature is returned directly onto the pump suction side and not into the return pipe, like in {C}. Due to that, the temperature of water in the pipe running to the heat source is not raised. This is required e.g. in the case of condensing boilers, which should operate at low return temperature values to enable water vapour condensation from flue gases. In case {B} the valve operates as a classic in-series pressure reducing valve. In case {D}, the valve is installed in a manner that is to prevent a rise in the medium level above the liquid column height h_h in the protected tank. A liquid column with height h_h produces the following hydrostatic pressure upstream the valve:

$$p_\mathrm{h} = h_\mathrm{h} \cdot \rho \cdot g. \tag{2.11}$$

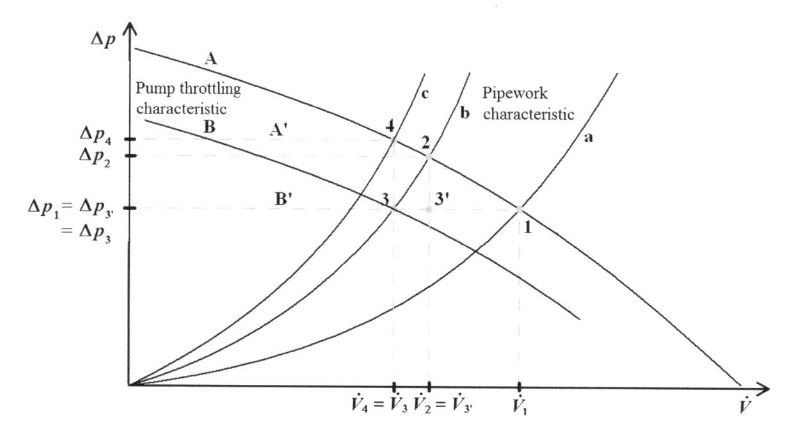

Fig. 2.47 Change in the installation working point location—quantitative regulation

The valve is set to this particular pressure value. If water flowing into the tank through the upper pipe fills the tank to a level exceeding the value of h_h, the valve will start to open, causing water to flow into the return pipe and reducing the value of parameter h_h.

Figure 2.47 illustrates the sense in using pressure reducing and pressure relief valves in a heating installation with changeable flow rates. It analyses the change in the installation working point location in the case of quantitative regulation, i.e. throttling on valves, or changes in active pressure.

Assume that the medium given values of pressure and volume flow (Δp_1 and $\dot{V}_1$, respectively) correspond to the initial working point marked as 1 (the point of intersection of the pump initial characteristic {A} and the pipework initial characteristic {a}). Assume also that pressure Δp_1 is the maximum permissible value, e.g. due to the condition of noiseless operation of thermoregulators. Any excess of pressure has to be eliminated. If the installation thermoregulators start to close, a rise will occur in hydraulic resistance and the pipework hydraulic characteristic will get steeper in relation to curve {b}. Consequently, the medium flow rate will be reduced due to the shifting of the working point to the left, along the pump throttling characteristic {A}, with a parallel rise in pressure, as results from the shape of the curve. A new working point will settle, marked as 2 and characterized by parameters Δp_2 and $\dot{V}_2$. Compared to the initial value of pressure Δp_1, an excess of pressure with the value of $\Delta p_2 - \Delta p_1 = \Delta p_2 - \Delta p_3$ will arise in the pipework. This might suggest that this is the pressure value to be throttled by an extra pressure reducing valve, but it is not. If point 1 is projected horizontally to the left, keeping the condition of pressure constancy (the pressure value should not be exceeded), the new pipework characteristic will be intersected in point {3} and not in point 3'. For point 3 the excess of the pump pressure (projection carried out vertically upwards till intersection with the pump characteristic) totals $\Delta p_4 - \Delta p_3 = \Delta p_4 - \Delta p_1$ and not $\Delta p_2 - \Delta p_{3'}$. Knowing the location of the required point {4} on the pump characteristic {A}, the required shape of curve

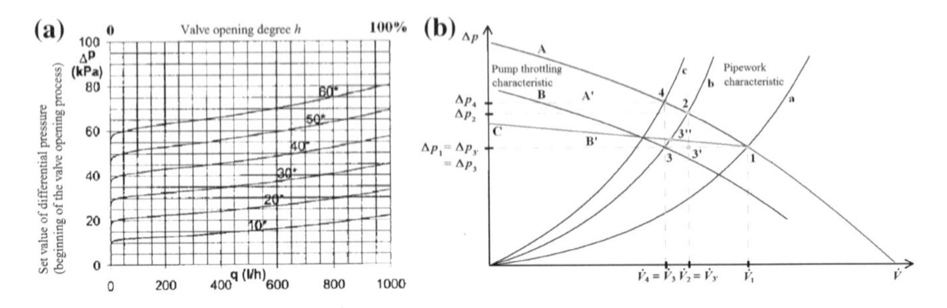

Fig. 2.48 Hydraulic characteristics of the BPV DN 15 *IMI Hydronic Engineering* pressure relief valve [8]; **a**, **b**, **c** pipework characteristic at a rise in hydraulic resistance (closing of thermoregulators); (A, B) pump throttling characteristic at the pump impeller different rpm values

{c} of the pipework with the pressure reducing valve can be determined. It can be seen that it is steeper, which is natural because the additional valve means additional hydraulic resistance.

Pressure on the pipework secondary side can also be reduced by means of a relief pressure valve installed in parallel. The valve task will be to reduce the pump pressure by shifting the working point from {2} to {1}, where the maximum permissible pressure value occurs. For this purpose, it will have to increase the medium flow rate on the primary side appropriately. Conducting an analysis similar to the case with the pressure reducing valve, it may be stated that the valve will have to cause an additional flow of the medium through its own branch (through the valve itself and the bypass pipes it is installed on). The additional volume flow will total $\dot{V}_3 - \dot{V}_1$.

For both types of the valve, the target action in such a situation will be to reduce the secondary side pressure to a value not higher than the set one. If for example the set value is Δp_1, the aim will be to ensure that the curve illustrating pressure on the secondary side runs horizontally, according to line {B′} till the intersection with line {A} or line {B} (depending on whether the initial working point was point {1} or {3}), and then downwards along the lines. In practice, due to the fact that such valves are proportional controllers, they are unable to ensure constancy of the regulated quantity (secondary side pressure) and a certain impact occurs of changeable primary side pressure on the value of pressure stabilized on the secondary side. Figure 2.48a presents hydraulic characteristics of a typical relief valve which illustrate the phenomenon. The lines in Fig. 2.48a relate to set values of the valve opening pressure. It can be seen that an increment in the valve opening degree and a rise in the volume of the medium flowing through it, being the conditions of the pump pressure reduction, cannot be achieved without a rise in this pressure. Figure 2.48b is an example illustration of how the secondary side pressure, described by line {C}, is affected. The figure relates to the analysis presented in Fig. 2.47.

If point {1} is assumed as the initial working point with curve {a} and the pressure value on the valve is set at the level of Δp_1, the thermoregulator closing, e.g. to obtain curve {b}, will result in working point {3″} with pressure $\Delta p_{3''}$, higher than pressure Δp_3 in the case of pressure constancy. The pressure relief valve has to be selected so that the permissible value of differential pressure should not be exceeded in the entire operating range of thermoregulators. The pressure value will therefore be lower than the expected one by a value that it rises by due to the operation of a given valve, when, due to the closing of thermoregulators, the medium flow rate decreases from the design value to zero. In practice, this also means that for the medium design flow rate through the valve, the assumed pressure loss on the valve (point {1}) must be lower than the permissible value. Whether a given valve will be able to reduce the pump pressure to the required value is conditioned by the relation between the valve flow capacity and the required flow rate of the medium that has to be returned on the valve if the flow rate in the secondary network is decreased. The flow capacity must be higher than the value of the said flow rate. For this reason, the valve has to be selected based on the value of the unwanted pressure rise occurring despite the valve application on the one hand, and the required value of the device flow capacity on the other.

Alternatively, a situation may be considered where the opening degree of thermoregulators remains unchanged (no variations in hydraulic resistance and no change in the shape of the pipework characteristic), but the pump pressure varies. The initial case, for example, will be working point {3} on the pump throttling characteristic {B} characterized by parameters Δp_3 and $\dot{V}_3$. A rise in the pump pressure, i.e. an upward shift in the pump throttling characteristic position to curve {A}, will involve shifting the working point on the pipework to point {2}, which gives a new and higher value of the medium volume flow $\dot{V}_2$. In order to restore the initial value of the medium volume flow $\dot{V}_3$ in the pipework, it is also necessary to restore the initial value of pressure Δp_3 in it. Like previously, projecting point 3 vertically upwards, along a constant volume flow value, onto the pump new characteristic {A}, intersection point {4} is obtained with pressure Δp_4. Consequently, the excess pressure value is Δp_4–Δp_3. In the case of a pressure relief valve, the excess of the medium volume flow is $\dot{V}_3 - \dot{V}_1$.

Both solutions involve energy losses in the heating installation. In the case of a pressure reducing valve, the medium flow rate is reduced but an additional excess of pressure arises upstream the valve, in the primary network, which is lost on the valve and converted primarily to thermal energy due to the medium friction. A pressure relief valve reduces produced pressure, but an excess of the medium volume flow arises in the primary network. Flowing through the valve and the bypass pipe, the excess of the medium also results in conversion to mainly thermal energy. The additional pressure (in the case of a pressure reducing valve) or the additional volume flow of the medium (in the case of a pressure relief valve) arising on the primary side are generally not carried over onto the secondary side and, due to that, they do not participate in forcing circulation in this part. In terms of the installation total energy balance, they are losses. But these can be different for each of the two valve types even if the devices ensure identical operating parameters, i.e. the medium pressure and volume flow values on the secondary side. The power needed to drive the pump is the product of the generated pressure and the medium volume flow values, taking account of the pump changeable efficiency depending on the device working point (cf. Formula 6.3). Therefore, it is impossible to decide which of the solutions is more advantageous in terms of energy consumption unless calculations are performed for a specific case, taking account of the pump throttling characteristic, the shift in the working point location on it for the two variants, the efficiency characteristic and, finally, finding the electric energy needed to drive the pump (cf. computational example 11 in Chap. 6). Still, it is usually the solution with pressure reducing valves. The reason for that is simple—in the case of an in-series pressure reducing valve, the total volume flow of the medium decreases gradually as thermoregulators (and, consequently, the valve itself) close, to reach zero at full closing. Not having to ensure the medium circulation, the pump then takes approximately only the idling power value. For a pressure relief valve, when the thermoregulator closes, the volume flow of the pumped medium remains more or less the same (within the hydraulic characteristic of the valve operation) because a pressure relief valve opens causing a rise in the medium flow rate in its own branch. Therefore, even if thermoregulators close completely and the flow in the secondary network is stopped, the pump still takes a similar amount of power. However, it should be remembered that in practice the above advantage of the pressure reducing valve is not that essential because a typical heating system pump and a heating boiler (in particular) should not operate at a zero flow rate of the medium. Some minimal circulation in the system has to be maintained.

The most favourable method of regulation and adjustment of the pressure generated on the primary side to the required value is to change the pressure of the pump. There are then no other devices reducing the excess of energy. A change in the pump pressure is achieved by varying the impeller rotational speed, usually by means of frequency converters or inverters. As the number of impeller revolutions drops, the throttling characteristic is shifted downwards. Figure 2.49 presents the principle of lowering the pump pressure depending on the current demand for pressure in the pipework. Older pumps intended for heating installations are most often equipped with a three-stage rpm control switch changing the value of generated pressure in steps as presented in Fig. 2.49. New pumps (fulfilling new European energy use requirements regarding to these devices) and pumps in larger and more complex installations are frequently regulated using infinitely variable adjustment.

Parameter n in the figure denotes rotational speed. In terms of the consumption of electric power needed to drive the pump (P_{el}), it is essential to know the relation between rotational speed and the pumped medium generated pressure and flow rate. It may be written that:

$$\Delta\dot{V} \sim \Delta n \tag{2.12}$$

$$\Delta\Delta p \sim \Delta n^2 \tag{2.13}$$

$$\Delta P_{\text{el}} \sim \Delta n^3 \tag{2.14}$$

The change in the volume flow is in a natural way proportional to the change in rotational speed. Due to the fact that the change in pressure is proportional to the square of the change in the medium volume flow, it is also proportional to the square of the change in rotational speed. Considering that the power needed to drive the

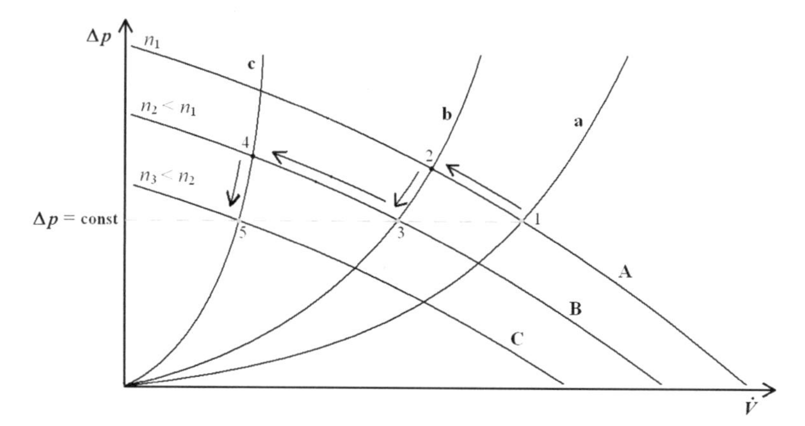

Fig. 2.49 Adjustment of the pipework working point by means of changes in the pump rotational speed

pump is the product of the medium volume flow and generated pressure, changes in power are proportional to changes in the product of the two parameters, which means they are proportional to the cube of changes in rotational speed. It follows [cf. Eq. (2.14)] that reducing pressure by half for example, the required value of rotational speed has to be decreased by about 1.41 times (which is also the value by which the volume flow of the medium will change), and the necessary driving power (equation: $\Delta P_{el} \sim \Delta n^3$), will then be about 2.82 times smaller. A reduction in the medium volume flow by half will involve a fourfold and an eightfold drop in the required pressure and power, respectively. Therefore, appropriate selection of the pump operating parameters or fitting the pump with automatic rpm controllers may bring about a substantial reduction in operating costs compared to using a pump with too high a pressure value or installing additional throttling devices. It should be remembered that a change in the power needed for pumping must not be equated with the change in the demand for power needed to drive the pump because the efficiency of a pump or a motor depends on the impeller rotational speed. It drops with a drop in the number of revolutions per minute. The throttling and efficiency (as well as electricity consumption) characteristics that illustrate the phenomenon are presented in Fig. 2.50.

The above equations, which are common in literature, are right if a square relation is assumed of the pressure drop as a function of the medium flow rate, i.e. assuming constancy of hydraulic resistance of the pipework the pump co-operates with. From the practical point of view, the assumption is right, but it should be remembered that the hydraulic resistance value is not absolutely constant, as discussed in detail in Sect. 3.2.

The changes in the medium volume flow through a heating installation or its given part, which are the reason for the application of pressure relief valves and differential pressure stabilizers described below, are the effect of the specificity of operation of the radiator thermoregulators. As previously described, the elements control the water

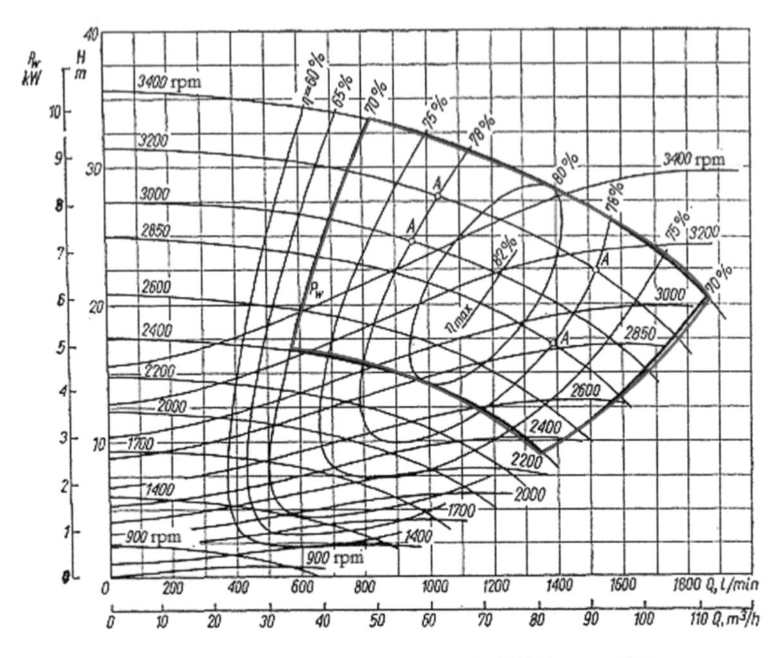

Fig. 2.50 Collective operating characteristics of the *KRZ 1-80/140* pump [47]

flow through the radiator automatically, responding to factors causing a change in the heated room temperature (e.g. heat gains from insolation or humans, or an unexpected rise in the supply temperature or an increase in active pressure resulting in a bigger flow rate and, thereby, the radiator higher heat output). This means that if thermoregulators are installed in the system, the installation becomes characterized by changeable flow rates of the medium. This in turn means that traditional manual regulation straight-through valves are unable to ensure pressure stabilization in the regulated section. As a result, hydraulic and thermal disorder may arise in the installation, which may also become unstable because oscillations in pressure cause oscillations in the amounts of the medium flowing through the radiators. This unfavourable phenomenon occurs even if thermoregulators are installed to stabilize the room temperature because the devices, characterized by a certain time constant and thermal inertia, always respond with a delay which may last up to several dozen minutes. If it is additionally taken into account that as a rule each thermoregulator in the installation operates independently, which means that their time-dependent throttling characteristics do not coincide, it can be seen that the process of pressure redistribution in the installation and the prevalence of unsteady-state operating conditions can be a continuous phenomenon. For this reason, balancing elements which introduce constant throttling are unable in this case to ensure differential pressure constancy.

Moreover, the "downward" correction is each time realized by the thermoregulator through a rise in throttling and, thereby, an increase in the medium pressure. But for every valve there is a certain combination of the throttled pressure and flow rate values which, if exceeded, make generated noises increasingly noticeable and this has a negative impact on the room comfort. Moreover, the closing of a thermoregulator or a group of them, i.e. a reduction in the total volume flow of the pumped medium, involves a rise in the pressure generated by the pump. Due to that, excess pressure arises on the other thermoregulators, increasing the flow rate in the initial phase (before the head detects the change in temperature and responds thereto) and creating the need to increase their closing degree further on. In order to avoid that, the task of excess pressure control and throttling should be transferred to other elements, firstly characterized by higher thresholds of noiseless operation and secondly—located beyond residential areas. Pressure relief valves belong to the group of such elements.

It should be borne in mind that the noises caused by the medium flowing through the valve are not related to throttled pressure only. They depend on the combination of the throttled pressure and the medium flow rate values. The pressure loss arising on the valve rises as the valve closes, reaching its maximum at the full closing position, but this is not accompanied by a monotonic rise in the loudness of the noises. The situation is illustrated by Fig. 2.51. It presents curves of what is referred to as the A-weighted noise pressure level generated if the medium is throttled in a typical radiator valve, for a few values of pre-setting n_i, as a function of the medium volume flow and the throttled pressure values.

In practice, however, the criterion for the device noiseless operation is the maximum pressure throttled by the valve. It is generally assumed that it must not exceed 20–35 kPa. Using this condition as the starting point, it is possible to conclude that active pressure in the installation regulated section should not be higher than the value mentioned above. As subsequent thermoregulators in this part of the installation start to close, active pressure will approach this level. Setting a higher pressure value would involve noise generation. It is practically agreed that a separate regulated section of the installation is at least the installation riser, but the smaller the regulated part where differential pressure is controlled (i.e. the smaller the pressure losses on the sections between the stabilization point and the radiator thermoregulator), the better the device operating parameters. A common situation in small installations (e.g. in single-family houses) is that the maximum pressure generated by the selected pump is smaller than the value conditioning the thermoregulators noiseless operation. In such a case, application of pressure relief valves to co-operate with radiator thermoregulators may not find economic justification. Moreover, the device may sometimes generate undesirable effects in some installation types, e.g. in a system with a gas-fired condensing boiler. Then, the pressure relief valve should not be mounted as a bypass close to such heat sources because they should operate at low return temperatures to ensure water vapour condensation from flue gases. A pressure relief valve would return the medium from the supply pipe directly to the return one, raising the temperature of the medium in the pipe.

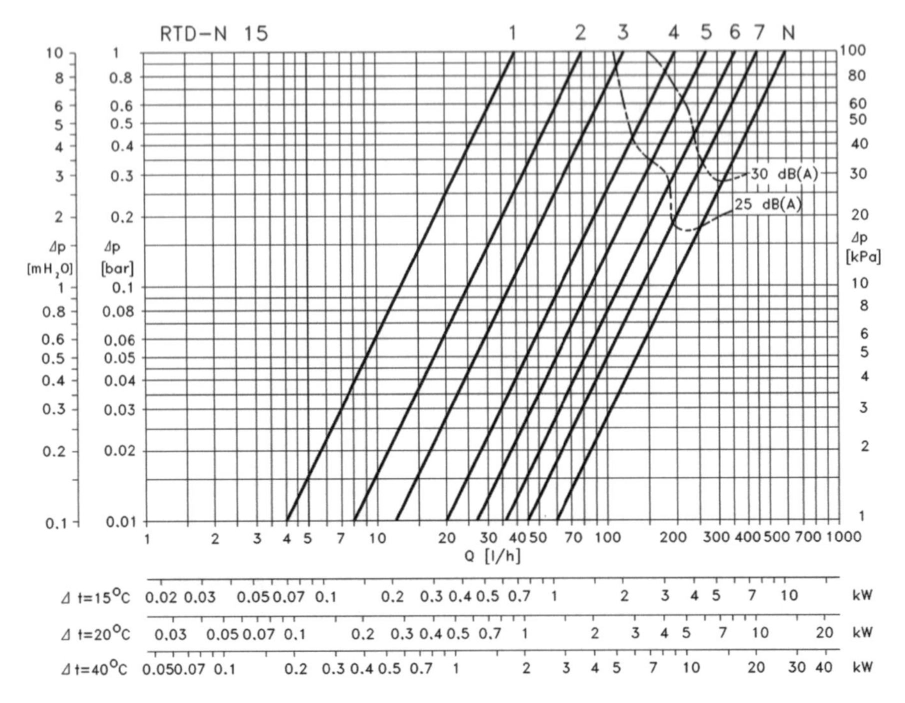

Fig. 2.51 Hydraulic characteristics of the *Danfoss RTD-N 15* radiator regulation valve with curves illustrating changes in the noise pressure level [5]

2.3.5 Automatic Differential Pressure Stabilizers

Automatic differential pressure stabilizers operate on the same principle as pressure relief valves and from the technical point of view it is possible to use them for this purpose. The main difference is that here pressure measurement and stabilization are carried out based on two signals which may come from two different points in the installation as they are delivered to the measuring membrane both sides. This means that the force arising on the membrane and compared to the spring tension force results from the difference in pressure of the two pressure signals and not from the pressure upstream/downstream the valve only, which is the case for pressure reducing and pressure relief valves. As a result, the pressure value between the two measurement points will be stabilized. Therefore, in contrast to pressure reducing and pressure relief valves, stabilizers are able to take account of changes in pressure/hydraulic resistance on the pipework secondary side, i.e. in the opposite pipe, from where the second pressure signal is most often taken. However, this fact does not substantially differentiate one type of the valves from the other in every case. Nor does it absolutely condition which of them should be selected for a given task because pressure changes in any point of the circuit are the effect of changes in the circuit previous parts and they are proportional to them. If the pump pressure rises, which

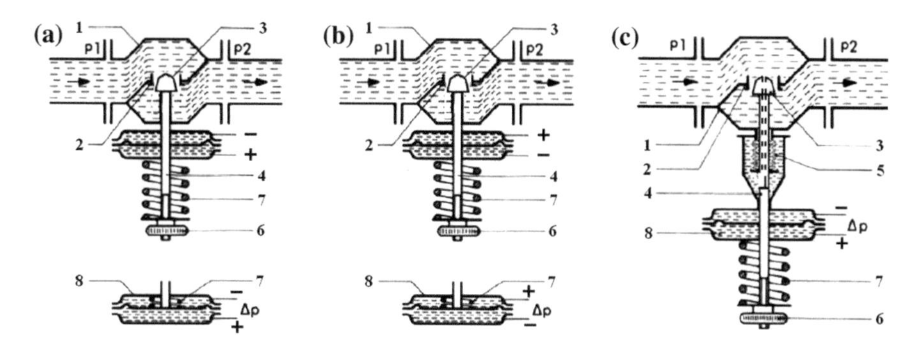

Fig. 2.52 Diagrams of differential pressure stabilizers: **a**, **c** valve operating as a closing element; **b** valve operating as an opening element; **c** valve with the plug pressure relief; 1—valve body; 2—valve seat; 3—valve plug; 4—valve pin; 5—metal pressure capsule of the plug relief mechanism; 6—setting device; 7—setting spring; 8—membrane measuring mechanism [17]

also involves a rise in pressure in the pipework primary side, a proportional rise in pressure also occurs in the secondary part, on the opposite pipe. A difference arises only in the pressure value resulting from the medium pressure drop occurring on the flow path from one part to the other. Still, the difference is proportional to pressure changes in the primary part. The conclusion is that pressure reducing and pressure relief valves can also be used as differential pressure stabilizers. Stabilization of the pipework primary side pressure will also mean stabilization of pressure on the secondary side. The main difference between the two types of devices is the pressure value to be set on them. In the case of a pressure reducing valve, this will be the value at the beginning of the secondary pipework. In the case of a differential pressure stabilizer—the pressure value at the beginning of the secondary pipework reduced by the medium pressure losses arising during the flow through this part. This, however, will be right only in the case of systems with constant flow rates, where hydraulic resistance values and pressure losses in the secondary pipework are constant. In the case of installations characterized by changeable flow rates, e.g. ones equipped with radiator thermoregulators, variable hydraulic resistances in the pipework secondary part do occur and may cause changes in the medium flow parameters. If hydraulic resistances are to be kept constant, it is necessary to use a differential pressure stabilizer. The pressure changeability has an impact on the degree of deformation of regulation characteristics and on the local regulation quality, as described in detail in Chap. 4.

Like in the valves described previously, this valve type also makes use of the plug pressure relief. Figure 2.52 presents diagrams of the differential pressure stabilizer.

Pressure signals can be delivered onto the membrane either by means of additional impulse tubes connected to the target point in the installation (cf. Fig. 2.52) or inside the valve, through the hollow of the regulating plug. The medium is then supplied to one side of the membrane by the plug, from underneath, and to the other—through the impulse tube. Reference is then sometimes made to devices with internal and external differential pressure measurement, respectively.

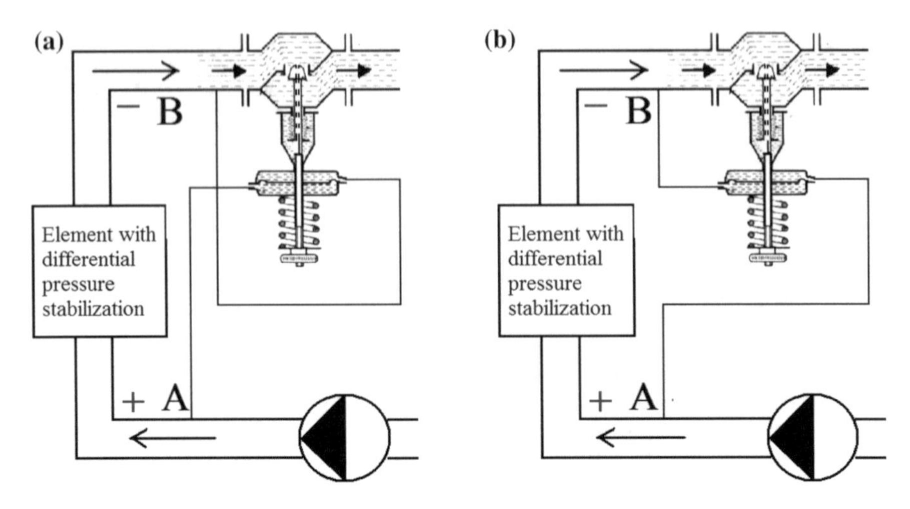

Fig. 2.53 Location of a differential pressure stabilizer with external differential pressure measurement installed in the pipework in series: **a**, **b** differential pressure stabilizer with external differential pressure measurement operating, respectively, as a closing or an opening element at a rise in the pressure of the pump (or another pressure source) or at a pressure drop downstream the stabilized object

The differential pressure stabilizer in the pipework can be connected in series or in parallel. Figure 2.53 presents a system with an in-series configuration for a stabilizer with external differential pressure measurement. Depending on the configuration of impulse tubes connections, it can operate as a closing one or an opening one at a rise in pressure generated by the pump (or another pressure source), or at a drop in pressure (due to a rise in hydraulic resistance) downstream the stabilized object.

Figure 2.54 presents a cross-section and diagram of a differential pressure stabilizer with internal differential pressure measurement. One pressure signal is delivered onto the membrane through a hollow in the plug and a system of channels (black lines with arrows) and the other—using an impulse tube.

It should be remembered that the location of such a valve in the system determines whether at a rise in the pump pressure it operates as a closing or an opening element. For a given location it is impossible to select the principle of operation, which can be done for a valve with external differential pressure measurement, where external impulse tubes can be connected in any configuration in the system. So that an in-series valve with internal differential pressure measurement in the configuration of the spring, membrane and plug location in a valve as presented in Fig. 2.54 should operate as a closing element at a rise in pressure in the primary network (e.g. at a rise in the pump pressure) or at a drop in pressure in the secondary network, downstream the stabilized object (e.g. due to the closing of thermoregulators), the pressure supplied to the membrane "top" must be higher than the pressure supplied to its "bottom". If one pressure signal is already delivered under the membrane and comes from the front of the plug (from underneath the plug), the other signal cannot come from

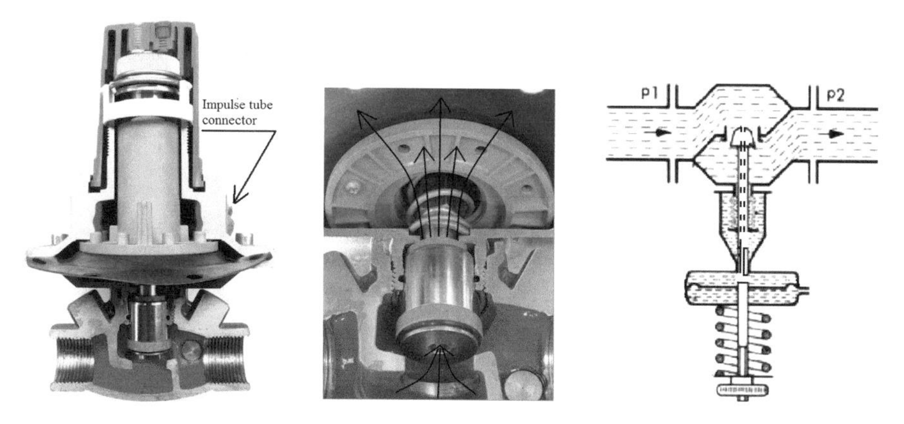

Fig. 2.54 *Herz 4007* differential pressure stabilizer with internal differential pressure measurement (own materials) [17]

behind the plug, looking in the medium flow direction, because the pressure in that region is higher. If that was the case, at a rise in the pump pressure the pressure value upstream the plug would all the time be higher than downstream and the valve would keep opening. The external signal must then come from a point characterized by a higher pressure value, i.e. from an earlier point of the installation. Therefore, this type of valve should be installed on the return pipe, and the external signal of pressure should come from the supply pipe, from upstream the point defining the installation secondary part to be covered by stabilization, as presented in Fig. 2.55a. Installed on the supply pipe, the valve will operate as an opening element if the pump pressure rises upstream the device or if the pressure level drops in the secondary network, downstream the stabilized object. Moreover, the effect of installing the valve on the supply pipe is that stabilization concerns not only the required part of the pipework (the stabilized object) but also a part of the valve itself, including the valve seat-plug interface. As shown in Fig. 2.55b, the pressure signal is taken upstream the valve plug and not from one of the points marked as {A} of the stabilized part of the pipework, located downstream the valve. The pressure in this part of the pipework, downstream the valve, is equal to the pressure value upstream the device reduced by the pressure drop arising on it. Due to the fact that the pressure drop arising on the valve is a function of the valve opening degree (plug lift over the valve seat), pressure at the other end of the pipework stabilized part will be a function of not only changes in the pump pressure and in pressure losses of the pipework given section or the stabilized object, but also of changes in the valve opening degree. In such a situation, the valve does not measure the pressure drop on the stabilized object {A–B} directly. The problem does not affect valves installed on the return pipe, as presented in Fig. 2.55a. If the internal pressure signal was delivered to the valve onto the “top” side of the membrane, the situation for the two variants would be the opposite.

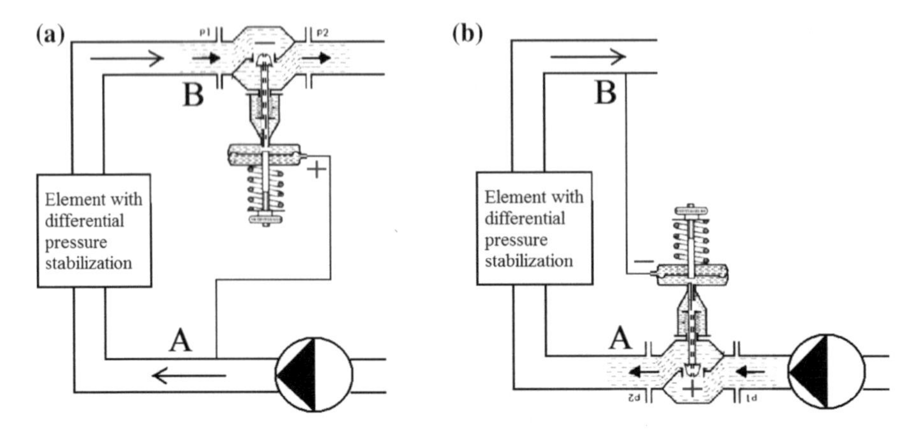

Fig. 2.55 Location of a differential pressure stabilizer with internal differential pressure measurement installed in the pipework in series: **a** with a valve on the installation return pipe; **b** with a valve on the installation supply pipe

It should be noted that, if installed on the return pipe downstream the object of pressure stabilization, the valve does not reduce the pressure level on it but only the pressure difference arising up- and downstream the device. Pressure on the object can only be reduced by installing a valve upstream the object, on the supply pipe.

Considering life and reliability, it is more advantageous to locate differential pressure stabilizers on the return pipe, where they operate at the medium lower temperature and pressure, which slows down the rate of the measuring membrane wear. Whether the valve operates as a closing or an opening element is less important because the device main task is differential pressure stabilization. Therefore, the solution presented in Fig. 2.55a is (or as a rule should be) used in practice.

If connected in the system in parallel, a differential pressure stabilizer can operate both as an opening and a closing element, just like a pressure relief valve. Such a choice, however, is only possible for a valve with external differential pressure measurement, where the two pressure signals may come from any installation point, regardless of the location of the device itself, as illustrated in Fig. 2.56.

In the case of a valve with internal differential pressure measurement installed in parallel (like in Fig. 2.57), the operation mechanism is imposed.

If the internal pressure signal is delivered onto the "bottom" side of the membrane from the pipework primary side, characterized by higher pressure, the other signal is taken from the pipework secondary side, with a lower pressure value, and the valve operates as an opening element at a rise in pressure of the pump (or another pressure source) or at a pressure drop downstream the stabilized object. If on the other hand the impulse tube delivering the signal is not connected, the membrane top side will be under constant atmospheric pressure and the differential pressure stabilizer will operate as a pressure relief valve, responding only to pressure increments on the primary side and not to the pressure difference arising in the pipework given section. This means that a differential pressure stabilizer can be easily transformed into a

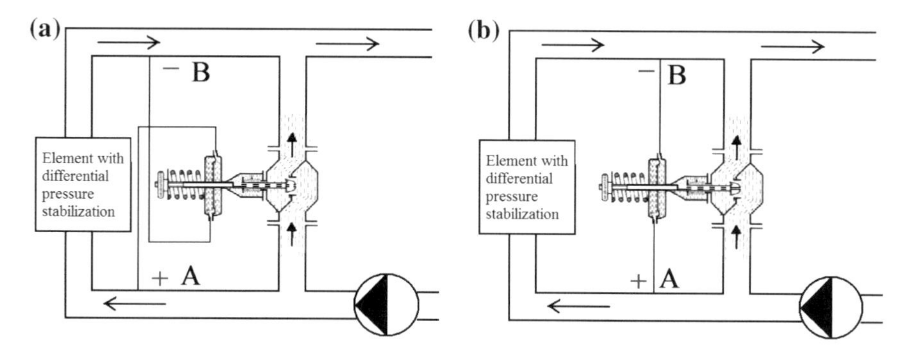

Fig. 2.56 Location of a differential pressure stabilizer with external differential pressure measurement installed in the pipework in parallel: **a**, **b** differential pressure stabilizer with external differential pressure measurement operating, respectively, as an opening or a closing element at a rise in the pressure of the pump (or another pressure source) or at a pressure drop downstream the stabilized object

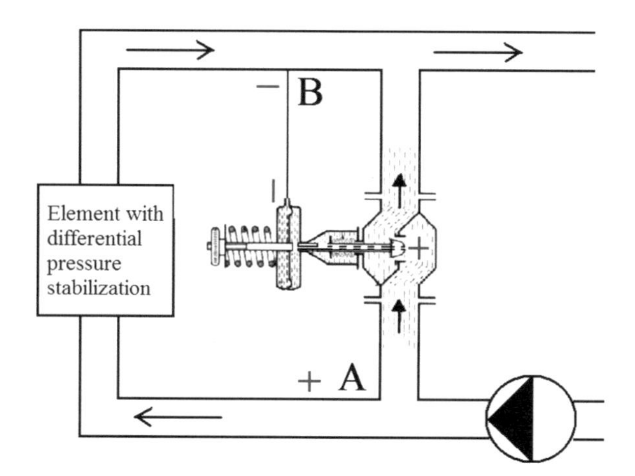

Fig. 2.57 Differential pressure stabilizer with internal differential pressure measurement installed in parallel

pressure relief or a pressure reducing valve. Like previously, if the internal pressure signal is delivered onto the membrane "top" side, the valve operation will be the opposite.

Differential pressure stabilizers can also be used to maintain a constant flow rate in the pipework regulated section by connecting the stabilizer to a measuring orifice or a regulation valve installed in series, at a short distance on a single pipe. Such a case is illustrated in Fig. 2.58.

If the measuring orifice or the additional regulation valve is set to a specific hydraulic resistance value, it gives a specific pressure drop at the medium given flow rate value. A change in the flow rate involves a change in the pressure drop. If therefore the device is included in the pressure stabilization loop, the differential pressure stabilizer will in fact also stabilize the medium volume/mass flow trying to keep the set value of the pressure difference. The value of the stabilized flow rate

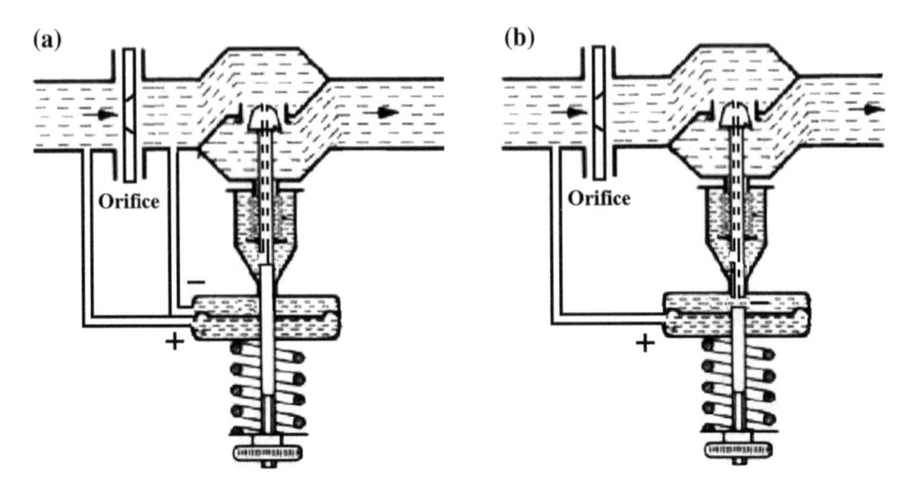

Fig. 2.58 Flow rate stabilization using a differential pressure stabilizer and a measuring orifice; **a**, **b** a system with external and internal differential pressure measurement, respectively (based on [17])

of the medium can be regulated in two ways: either through changes in the setting of the manual regulation valve/orifice replacement or by changing the setting of the differential pressure stabilizer. The smaller the hydraulic resistance set on the regulation valve/orifice (i.e. the more the valve opens), the higher the value of the medium stabilized flow rate. Similarly, the higher the difference in pressure values set on the differential pressure stabilizer, the higher the value of the medium stabilized flow rate. The operating principle will be the same if the pressure measuring tubes are incorporated into the pipe, without an additional valve. However, due to the pipe relatively small hydraulic resistance, the system will not operate efficiently.

Figure 2.58a, b illustrate the system operation using a stabilizer with external and internal differential pressure measurement, respectively. In both cases, the valve plug pressure relief system is applied. As described above, the orifice can be replaced with a manual regulation valve. The essential thing is the configuration in which the impulse tubes delivering pressure signals are connected. If the goal is the flow rate stabilization, the valve has to close at a rise in the system pressure. Therefore, the signal with a higher value of pressure (+) upstream the throttling element, looking in the flow direction, must be delivered onto the membrane "top" side, whereas the one with lower pressure (−)—to the membrane "bottom" side.

Basic diagrams with differential pressure controllers are presented in Fig. 2.59. Moreover, the devices can be installed in the manner presented in Fig. 2.46, serving similar functions.

There are also solutions of the differential pressure stabilizer where both pressure signals are taken inside the valve. In this situation, the device is in fact a flow rate stabilizer, which means that it can play the role of a differential pressure stabilizer if it is installed in parallel and not in series. Consequently, it has the same limitations

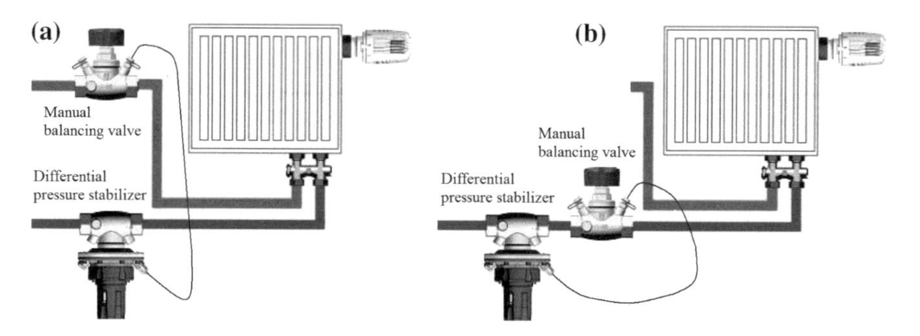

Fig. 2.59 Differential pressure (**a**) and flow rate (**b**) stabilization on the installation regulated section (based on [7])

as relief pressure valves, i.e. devices with pressure measurements performed on one side only.

Like the pressure relief valves discussed above, differential pressure stabilizers are proportional devices, and likewise they are unable to keep the set differential pressure value at a constant level. In this case also the stabilized differential pressure value is selected taking account of the required value of the medium flow rate. However, considering the valve function, which is differential pressure stabilization, from this differential pressure level the valve should have a chance to both open and close, in contrast to the pressure relief valve, which opens only. The working point should then be established not at the beginning or end of the proportional range for a given pre-setting (the valve plug range of travel), but in between the two points. Locating it at either of the proportional range ends will mean that the device will operate as either a closing, reducing pressure element or as an opening, increasing the secondary side pressure element, and not as a differential pressure stabilizer, i.e. a closing/opening device. Figure 2.60 presents hydraulic characteristics of a typical differential pressure stabilizer.

The proportional range X_p marked in the chart hardly depends on the valve pre-setting (the set value of differential pressure to be stabilized). However, for each pre-setting value the range of variability in the medium volume flow rate is different. This results from the fact that the flow rate in the pipework given section depends on the pressure in it, and its value decreases with a drop in pressure. The characteristics are not plotted for the zero value of the medium volume flow because a certain minimum is taken into account from which the device operates correctly.

In small installations, e.g. in single-family houses, differential pressure stabilizers may not find economic justification, but the fact is that in every installation type they ensure better operating conditions of radiator thermoregulators. In large systems, application of stabilizers should be a common practice because they enable energy consumption optimization, reduce operating costs and lengthen the installation life. Both pressure relief valves and differential pressure stabilizers, like radiator thermoregulators, are proportional self-actuated devices (unless they are equipped with

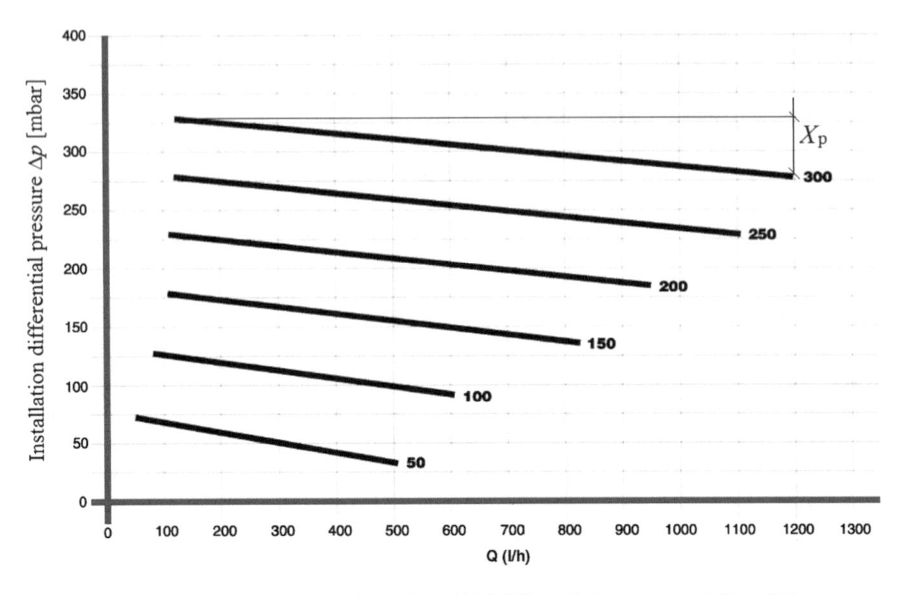

Fig. 2.60 Hydraulic characteristics of the *Herz 4007* differential pressure stabilizer [7]

additional drives). A change in the regulating plug shift is approximately proportional to the change in differential pressure on the plug/measuring membrane.

2.3.6 Automatic Flow Stabilizers

In terms of operation, flow stabilizers are basically differential pressure stabilizers. They operate based on the same principle as differential pressure stabilizers incorporated into a branch in series, taking pressure signals from an element with a set hydraulic resistance value, e.g. from a manual valve or orifice. The difference is that here the element is built into the valve and the differential pressure signal on it is supplied onto the valve membrane within the valve structure (internal differential pressure measurement). Differential pressure stabilization on the regulating element makes it possible to prevent oscillations in the medium mass/volume flow set by means of the valve if pressure oscillations arise in the pipework. It also enables operation at an almost unchanged operating regulation characteristic of the valve (cf. Sect. 3.3), as relatively constant and high values of outer and, consequently, total authority (a_z and a_c, respectively) are maintained. A diagram illustrating a typical plug flow stabilizer and a section thereof are shown in Fig. 2.61.

Figure 2.62 presents a ball flow stabilizer—a *PIQCV* valve (also available as C2..QP valve) made by *Belimo*. The device is intended for co-operation with an actuator setting the opening degree of the regulating element (the ball) and, thereby, the medium flow rate. The differential pressure measurement is in this case fully

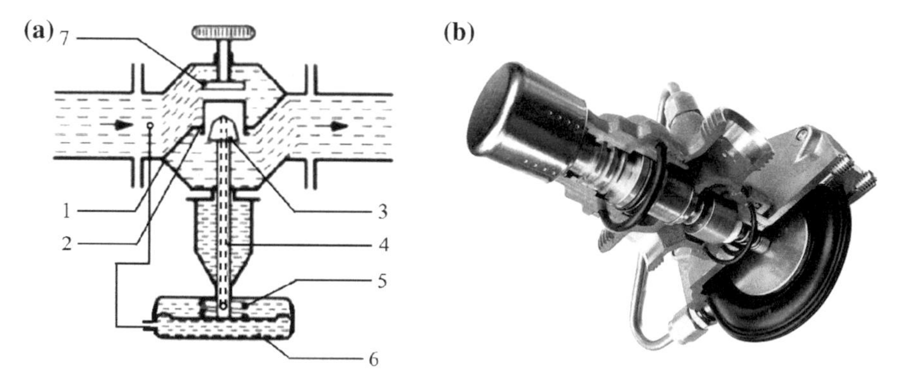

Fig. 2.61 **a** Flow stabilizer diagram [17]; **b** *Herz 4001* flow stabilizer (own materials); 1—valve body; 2—valve seat; 3—valve plug; 4—plug pin; 5—setting spring; 6—membrane mechanism; 7—built-in adjustable throttling element

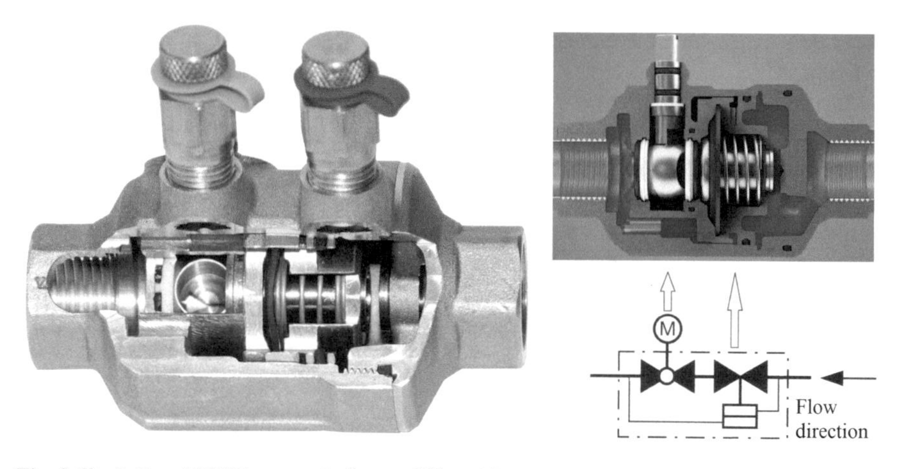

Fig. 2.62 *Belimo PIQCV* automatic flow stabilizer [4]

internal, i.e. both pressure signals, on both sides of the regulating element (the ball), are taken inside the valve body and no capillaries are used. The internal measuring-executive system (the measuring membrane with the setting spring) is designed to keep a set value of the medium flow rate at pressure oscillations on the pipework primary side in the range of 16–350 kPa (naturally—within the proportional range resulting from the device parameters).

This valve type—the automatic flow stabilizer—and the *Danfoss Dynamic Valve RA-DV* radiator thermoregulator presented in Sect. 2.2.3 are very similar in terms of structure. They draw on the same idea of integrating, in a single body, two devices often co-operating with each other in the installation.

Like automatic differential pressure stabilizers, automatic flow stabilizers are made as an option enabling current regulation of the stabilized parameter (differential pressure or flow rate, respectively), or they come in the variant with the parameter value set depending on the design. In the latter case, as they are not adjustable, if a change is needed in the stabilized parameter, the valve has to be replaced with another from a given series of types. This solution is less universal but cheaper, so it may be favourable if it is not necessary to change the value of the stabilized quantity in the installation given part. The operating parameters of an installation or a part thereof are determined during the thermal and hydraulic balancing process. Efforts are then made to keep the pressure and flow rate values constant. It is then not necessary to use adjustable devices. However, if the installation operating conditions are changed, e.g. due to the system extension, permanent shutdown of some of its parts, changes in the regulated building thermal load, a considerable increase in the pipework hydraulic resistance, fouling accumulating in pipes, etc., the possibility of introducing current on-line corrections to initially selected parameters is useful. For this reason, it is more advantageous to select devices where the set parameter value is adjustable.

In the case of large heating installations, elaborate automatic control systems are also frequently used that are responsible for current control of and corrections to operating parameters. Regulation and balancing valves equipped with actuators, e.g. automatic flow stabilizers with actuators as discussed above, are used for this purpose. But there are yet more advanced solutions of the valves. Co-operating with the control system, they enable further improvement in the installation operating conditions, better stabilization of the heated building temperature (preventing the temperature rise above set values in the first place) and a reduction in operating costs.

One example is the *Belimo EPIV* valve (which in fact is a set of combined devices). It is a regulation ball valve intended for co-operation with an actuator and it is combined with a flow meter. In such a set the valve plays the role of an automatic flow stabilizer. The flow meter is installed here in place of the membrane differential pressure stabilizer. Co-operating with an actuator, it is responsible for the flow rate stabilization, varying the regulating element opening degree appropriately. The flow measurement and the electronic control based on this parameter make it possible to couple the device with the global regulation system. Such a combination enables better co-operation with the installation other devices compared to individual action, which is the case if the flow rate is regulated by means of a built-in differential pressure stabilizer (e.g. the *PIQCV* valve). Moreover, the device makes it possible to set the controller so that the flow rate value is limited both from the bottom (of course within the limits resulting from the minimum available pressure in the system) and top.

In contrast to the *PIQCV* valve, the differential pressure required for the *EPIV* valve correct operation does not have a lower bound of 16 kPa. The 16 kPa condition is the effect of the structure and parameters of the membrane mechanism incorporated into the *PIQCV* valve. The *EPIV* valve does not have such a system and the minimum differential pressure value results directly from the valve flow capacity, i.e. the valve

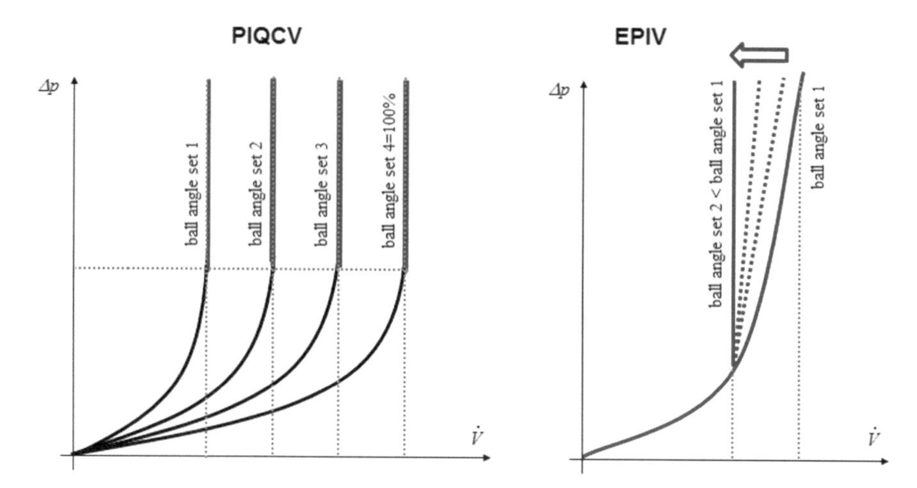

Fig. 2.63 Example regulation characteristics of the *Belimo PIQCV* and *EPIV* valves

size. For example, selecting a bigger valve, with a higher value of flow factor k_{vs}, at a set flow rate, a lower pressure loss and, consequently, a smaller differential pressure value in the system are obtained [cf. Formula (6.19)]. Values of the order of a few kPa can be achieved in practice.

As mentioned above, the *PIQCV* valve is a self-actuated proportional (P) controller. As such, it maintains the stabilized quantity value within deviations resulting from its proportional range. The *EPIV* valve is an improvement on the device—it is able to ensure an absolutely constant value of the stabilized quantity. The valve is fitted with a built-in flow meter measuring the medium flow rate on a continuous basis and sending a signal to the control system/actuator coupled with the ball. The measured value is compared to the set one and if deviations occur, the ball opening degree is corrected until the set flow rate is achieved. The basic difference then between the *PIQCV* and the *EPIV* valves is that in the former it is the ball that is responsible for the initial mass/volume flow value which is then stabilized by the membrane system. For example—if there is a rise in differential pressure in the installation, the membrane system plug closes and the pressure value downstream the valve is reduced, which prevents increments in the medium flow rate. In the latter, the ball both sets and stabilizes the initial flow rate (in co-operation with the measuring system, the controller and the actuator, of course). For example—if there is a rise in differential pressure in the installation, the ball opening degree is reduced and, thereby, pressure increments downstream the valve are eliminated, which prevents increments in the medium flow rate. In the *EPIV* valve, the ball opening degree is therefore dependent on the momentary differential pressure value in the installation, whereas in the *PIQCV* valve it is constant. This is illustrated graphically in Fig. 2.63.

A further improvement is fitting the unit with elements intended for differential temperature measurement. The *Energy Valve* made by *Belimo* is an example of such a solution. The temperature sensors (installed in pipe joints in between pipe sections) are mounted on the regulated installation supply and return pipe. Apart from stabilizing the flow rate, the valve protects the system against excessive increments therein and prevents a heat exchanger (e.g. a radiator) operation in its static thermal characteristic range in which increments in the medium flow rate have hardly any impact on the rise in the heat output. This makes it possible to reduce operating costs related to the work of circulation pumps. It is described in Sect. 2.1.2.4 and illustrated by computational example 2 that in some ranges of the radiator static thermal characteristic great increments in the medium flow rate result in small changes in the heat output—e.g. even if the medium volume flow value rises by a few hundred percent, the heat output increases merely by about 10. By contrast, the rise in the demand for power to drive the pump(s) is proportional to the cube of the increase in the medium volume flow value, as indicated by Eqs. (2.12)–(2.14). Consequently, quantitative regulation of the radiator output by means of changes in the medium flow rate in this range of the radiator static thermal characteristic generally does not have an essential impact on the temperature inside the building but it causes deterioration of the economic indicators of the heating system operation. The *Energy Valve* measures (at standard 30-s intervals) the difference in temperatures of the installation section it is installed on and compares the changes in the medium flow rate with changes in the medium cooling corresponding thereto, i.e. with changes in the radiator heat output. The smaller the difference in temperatures at a rise in the mass/volume flow value, the smaller the heat output increments (cf. Figs. 2.22 and 2.23 and computational example 2). Using collected data and the assumed initial criterial parameters, the data processing system is able to determine the level of the medium flow rate above which it is unprofitable to increase it further. Regardless of changes in the pipework secondary side hydraulic resistance, e.g. due to uncontrolled or unwanted opening of the radiator thermoregulators, it will not allow a rise in the medium flow rate. In practice, the point corresponding to this level of the medium volume flow is sometimes referred to as *the point of saturation with the heat exchanger thermal power*. It is worth noting that the flow rate should be determined not only based on measurements of the installation given section, i.e. the analysis of the thermal characteristics of the radiators used in it, but also with regard to prices of the energy carrier, i.e. electricity in this case, because the two criterial parameters practically decide about the optimal limit value of the medium flow rate. It can easily be noticed that if energy prices are low, the point will be located farther on the radiator static thermal characteristic. Considering that in most countries of the world energy prices are currently relatively high and taking account of the efforts aiming to minimize energy consumption for heating and cooling of buildings, such a system finds ample justification and is characterized by a short payback period.

The wider the thermal load range within which measurements are performed, the more accurate the limit working point determination because it is only at nominal loads, for which the installation is designed, that data needed to define the point parameters (the medium mass/volume flow values) are obtained. In practice, such parameters are achieved in the peak of the heating season, at the lowest outdoor temperatures. However, the control system database must also have access to non-nominal parameters so that relevant comparisons can be made and the limit value can be established. For this reason, the system should gather data for at least one entire heating season/one year.

The data gathered by the acquisition system can be analysed conveniently using a PC computer fitted with dedicated software and a convenient graphical user interface. This makes it possible to find out the real shape of the static thermal characteristic of the pipework given section and all operating parameters thereof. Selected criterial parameters can also be set at the user's discretion.

In 2017, in response to the customers' rising expectations and making more innovative improvements, the *Belimo* company launched a new version of the valve onto the market—the *Energy Valve V3*. The energy flow process, including too low a value of water cooling Δt_{w} in the heat receiver or "cool" receivers (e.g. fancoils), can be managed from the web server level using an Internet browser. A cloud-based platform is also made available where each *Energy Valve V3* can transmit data by means of a computer network. The data may be accessed by the user and (by the user's permission) by a group of the *Belimo* experts, who may prepare a report including a description of the level of the installation performance.

Another essential improvement on the valve previous versions, which in fact is a unique solution for control devices used in heating installation in general, is the function of checking and making real-time corrections to the resultant regulation characteristic (illustrating the dependence of the receiver thermal power on the valve relative opening degree, cf. Fig. 4.10b) of the installation section with the control device. As described in Chaps. 4 and 5, the valve regulation characteristic not only has an unfavourable shape in terms of co-operation with heat (or "cool") receivers, but it also varies depending on the pre-setting value, the pressure distribution among the installation elements, etc. Moreover, the heat receiver regulation characteristic demonstrates similar tendencies—its shape varies as a function of the thermal load degree (the cooling of the medium flowing through the receiver, cf. e.g. Fig. 4.9). It is also dependent on the receiver type [33, 34]. For this reason, the resultant characteristic usually has an unfavourable non-linear shape. *Energy Valve V3* makes it possible to eliminate this downside, introducing real-time corrections to selected parameters of the installation operation, and ensures that the final regulation characteristic shape is linear in a wide range of input parameters and thermal loads of the receiver. Considering the effect, the action is similar to the procedure of the regulation valve optimal sizing and the final regulation characteristic shaping presented in Sect. 4.3 (cf. 4.10). The advantage of *Energy Valve V3*, which is an elaborate automatic device of intermediate regulation, is that it is fully universal—it is not intended for just one specific heat receiver type or shape of its thermal characteristic. Instead, it can optimize the final regulation characteristic curve to match practically any type of the heat receiver.

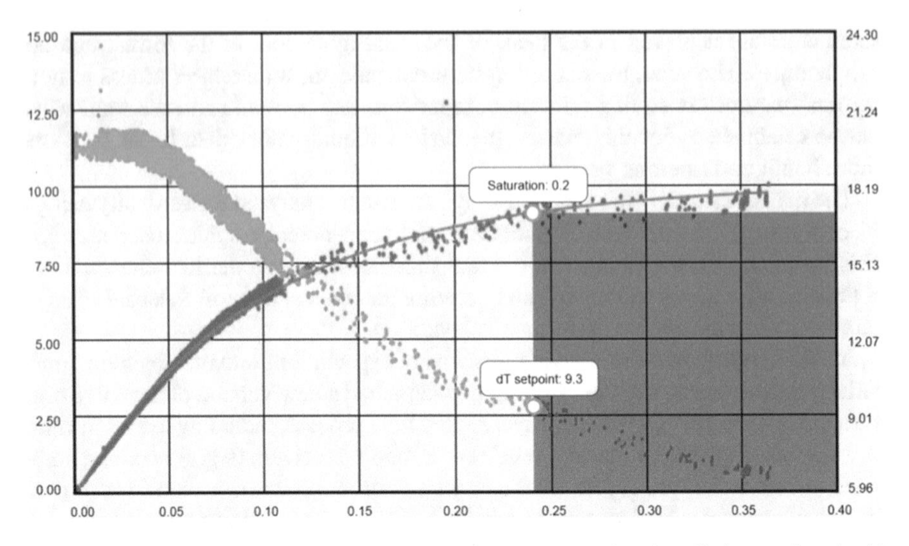

Fig. 2.64 Example report on monitored operating parameters of an installation equipped with *Energy Valve V3* [4]

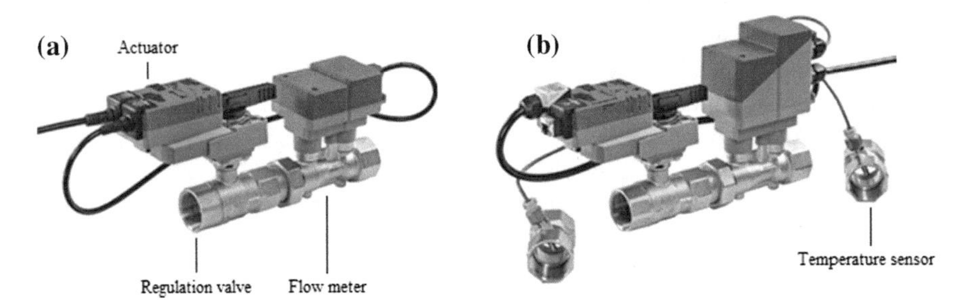

Fig. 2.65 *EPIV* (**a**) and *Energy Valve* (**b**) made by *Belimo* [4]

Like its previous versions, the valve is most often used in heating and cooling installations, fancoils, etc., as an element of current regulation. But it can also be used in heating installations to control a group of receivers and performing the function e.g. of the flow rate stabilizer or limiter which is additionally able to control the process of energy consumption management. Classic self-actuated flow rate stabilizers/limiters do not offer such opportunities.

Figure 2.64 presents an example report on the device operation to illustrate the wide range of functionality described above.

Figure 2.65 presents the two solutions discussed above and offered by the *Belimo* company, i.e. the *EPIV* and *Energy Valve*.

Flow controllers/stabilizers are used wherever it is necessary to maintain a relatively constant flow rate and prevent exceeding the set value thereof regardless of oscillations in pressure or hydraulic resistance occurring in the installation (within the range limited by the valve properties and the installation operating parameters).

It should be remembered that the use of a self-actuated flow stabilizer (with no actuator) in one section together with e.g. a radiator thermoregulator without any other appropriate devices is unfavourable because if the thermoregulator starts to open, trying to increase the flow rate, the flow stabilizer will tend to reduce the flow, which may result in the installation two-state, on-off action. This phenomenon can be avoided by using a two-way valve in between the two devices to connect the supply pipe to the return one. The valve will separate the devices creating separate circuits.

Like the devices discussed earlier, such valves should be mounted on the return pipe.

2.3.7 Automatic Flow Limiters

Automatic flow limiters are in fact flow stabilizers performing the function of the former. They may be self-actuated or driven by actuators, like e.g. the *Belimo* products: *EPIV* or *Energy Valve* discussed above. The medium flow rate stabilization means that the valve, in the case of a self-actuated element, keeps the set value within a specific proportional range. Thereby, it prevents excessive increments therein. A rise in the pipework primary side pressure involves an increase in the flow rate. This means a pressure drop on the additional throttling element. The pressure downstream the element is therefore lower compared to the upstream value. Due to the fact that lower pressure is carried over under the measuring membrane and higher pressure above it, if a rise occurs on the primary side, the valve starts to close and the flow rate is reduced. The same happens if the secondary side hydraulic resistance decreases (e.g. thermoregulators open). At a given primary side pressure value, this involves a rise in the flow rate and, consequently, a bigger pressure drop on the throttling element. The pressure drop is carried over onto the membrane, the valve starts to close and the flow rate is reduced. The range of the medium flow rate increment is thus determined by the adjustable throttling element hydraulic resistance.

Data sheets provided by manufacturers often include a slightly misleading description of the principle of the device operation. According to it, the valve closes completely once the set permissible mass/volume flow value is reached. However, such a description is incorrect because a closed valve eliminates any flow whatsoever, let alone its maximum value. The flow is stopped completely (assuming perfect tightness). The description and the principle of operation should be understood as follows: in a system with no flow limiter a rise in active pressure or a drop in the pipework hydraulic resistance involve a natural increase in the medium flow rate, which at a certain moment reaches the permissible value. If the factors causing the increase get stronger, the flow rate continues to rise. If a flow limiter is installed in such a system, it closes as the factors mentioned above get stronger, and the increment in the flow rate is reduced. The device closes completely once the factors reach

1. Casing
2. Flow regulator
3. Valve body
4. Valve insert
5. Hydraulically relieved valve plug
6. Valve pin
7. Built-in spring regulating the flow rate range
8. Pressure impulse channel
9. Actuator
10. Membrane
11. Union nut
12. Impulse tube

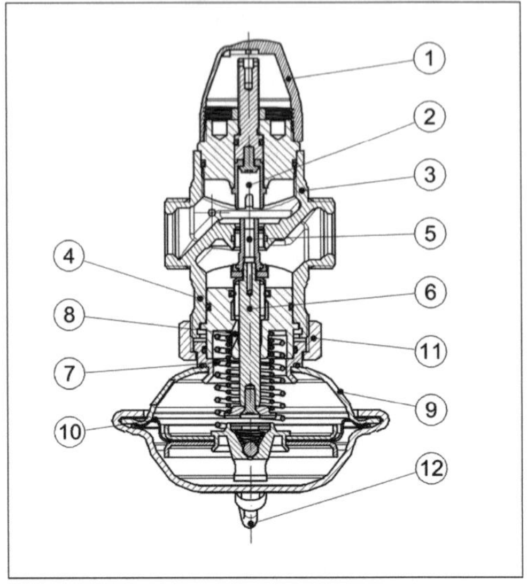

Fig. 2.66 Structure of the *Danfoss AVQ* flow limiter [5]

a level that might result in the permissible mass/volume flow value in the pipework without a flow limiter. This again is a simplification because when a valve is installed, even if not set or activated, additional hydraulic resistance is created, which already slightly reduces the flow rate to a value lower than the maximum one, before the element is actually installed. As a result, the values of the pipework pressure or of the pipework hydraulic resistance with an installed valve causing the valve to close completely must be corrected by the change created by the valve, compared to the pipework with no valve (pressure higher by the pressure drop on a fully open valve for the maximum permissible flow rate; or the pipework hydraulic resistance reduced by the added hydraulic resistance of the valve). The device cross-section is presented in Fig. 2.66.

2.3.8 Combined Differential Pressure and Flow Stabilizers

Combined differential pressure and flow stabilizers perform the function of the two individual devices. They are made basically as two mechanisms integrated with each other, with one regulating plug. They thus have two measuring membranes and a built-in setting throttling element from which the pressure loss signal is taken to stabilize the flow. Also in this case, there are valves with internal or external differential pressure measurement. A diagram of such a device is presented in Fig. 2.67.

The combined differential pressure and flow stabilizer should be mounted in the system in series.

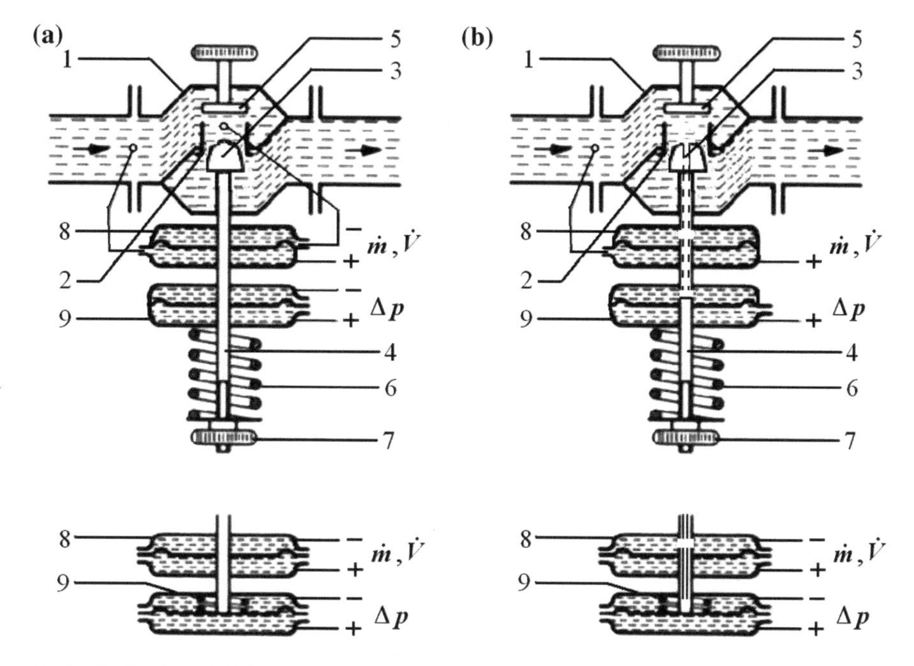

Fig. 2.67 Combined differential pressure and flow stabilizer with external and internal differential pressure measurement [(**a**) and (**b**), respectively]; 1—valve body; 2—valve seat; 3—valve plug; 4—plug pin; 5—setting throttling element; 6—setting spring 7—setting device 8—membrane mechanism of flow stabilization; 9—membrane mechanism of pressure stabilization [17]

For all the above devices intended for stabilization, a minimum pressure drop and/or the medium flow rate have to be ensured to enable stable and correct operation. The values thereof are usually specified by the manufacturer.

2.3.9 *Combined Differential Pressure Stabilizers and Flow Limiters*

These devices combine the differential pressure stabilizer and the flow limiter functions in the same way as the combined pressure and flow stabilizers described above. The view and the cross-section of such a device (in the industrial (high-pressure) version) are presented in Fig. 2.68.

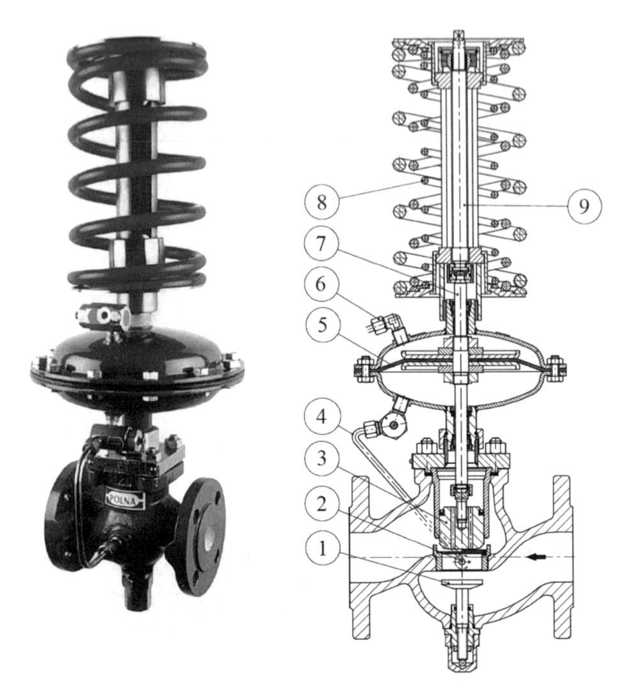

Fig. 2.68 *POLNA ZSN-6* combined differential pressure stabilizer and flow limiter [9]

References

1. Albers, J., Dommel, R., Montaldo-Ventsam, H., Nedo, H., Übelacker, E., Wagner, J.: Zentralheizungs- und Lüftungsbau für anlagenmechaniker SHK, 5. uberarbeitete auflage, Handwerk und Technik, Hamburg (2005)
2. Albers, J., Dommel, R., Montaldo-Ventsam H., Nedo, H., Übelacker, E., Wagner, J.: Systemy centralnego ogrzewania i wentylacji-poradnik dla projektantów i instalatorów (Central Heating and Ventilation Systems—A Guide for Designers and Installers). WNT, Warszawa (2007)
3. Ambrosius, R.: Untersuchungen an Regelvorrichtungen für Dampf und Wasserheizkörper, Beihefte zum Gesundh.-Ing., Reihe 1, Beiheft 14 Oldenbourg, München (1919)
4. Catalogue information of Belimo (on-line version, https://www.belimo.com)
5. Catalogue information of Danfoss (on-line version, https://www.danfoss.com/en)
6. Catalogue information of Heimeier (on-line version, https://www.imi-hydronic.com)
7. Catalogue information of Herz (on-line version, http://www.herz.com.pl)
8. Catalogue information of IMI Hydronic Engineering (on-line version, https://www.imi-hydronic.com)
9. Catalogue information of POLNA (on-line version, http://www.polna.com.pl)
10. Catalogue information of Socla (on-line version, https://armatura-socla.pl)
11. Catalogue information of Valvex (on-line version, http://valvex.com)
12. Chlipawski, T., Dura, M., Vinohradnik, S.: Armatura grzejnikowa (Radiator armature). Gas Water Sanitary Eng. **38**(6), 205–208 (1964)
13. European Standard EN 12831:2003: Heating Systems in Buildings. Method for Calculation of the Design Heat Load
14. European Standard EN 215:2004: Thermostatic Radiator Valves—Requirements and Test Methods

15. German Standard DIN 215: Thermostatische Heizkörperventile. Teil 1: Anforderungen und Prüfung
16. Kołodziejczyk, W.: Armatura regulacyjna w ogrzewaniach wodnych (Control armature in hydronic heating systems). Arkady, Warszawa (1985)
17. Kołodziejczyk, W.: Pomiary zużycia energii w budynkach (Measurements of Energy Consumption in Buildings). Centralny Ośrodek Informacji Budownictwa, Warszawa (1993)
18. Kołodziejczyk, W.: Termostatyczne zawory grzejnikowe w instalacjach centralnego ogrzewania (Thermostatic Radiator Valves in Central Heating Systems). Centralny Ośrodek Informacji budownictwa, Warszawa (1992)
19. Kołodziejczyk, W.: Zmodernizowana konstrukcja zaworu grzejnikowego z podwójną regulacją (Modernized Radiator Valve Structure with Double Regulation). District Heating Heating Vent. **3**(10), 298–303 (1971)
20. Kwapisz, J.: Automatyczny zawór grzejnikowy (Automatic radiator valve). Przegląd Informacyjny - Ciepłownictwo **3** (1970)
21. Kwapisz, J.: Badania laboratoryjne automatycznego zaworu grzejnikowego produkcji OBD-SPEC (Laboratory Tests of the Automatic Radiator Valve Produced by OBD-SPEC). Przegląd Informacyjny - Ciepłownictwo **4** (1970)
22. Kwapisz, J.: Piłatowicz, Z.: Opis patentowy 48274. Zawór lub kurek do centralnego ogrzewania zaopatrzony w wymienną kryzę lub dyszę (Patent description 48274. A central heating valve or tap provided with a replaceable orifice or nozzle). Urząd Patentowy Rzeczypospolitej Polskiej
23. Mielnicki, J.S.: Możliwości regulacji wstępnej i eksploatacyjnej za pomocą zaworów grzejnikowych (Possibility of preliminary and exploitation regulation by means of radiator valves). District Heating Heating Vent. **1**(3), 73–80 (1969)
24. Mizielińska, K., Rubik, M.: Ciepłownictwo: eksploatacja, projektowanie, inwestycje. Poradnik (District heating: Exploitation, Design, Investments. Guide), Fundacja Rozwoju Ciepłownictwa "Unia Ciepłownictwa". Filia Wydawnictwa Techniczne, Warszawa (1994)
25. Mizielińska, K.: Metody hydraulicznego oddzielenia obiegów kotłowych od obiegów grzewczych. District Heating Heating Vent. **29**(10), 8–14 (1997)
26. Mizielińska, K.: Nie bójmy się kotłów kondensacyjnych. Układy hydrauliczne w kondensacyjnych źródłach ciepła. Część 1. Polski Instalator **7–8**, 19–22 (1996)
27. Mizielińska, K.: Nie bójmy się kotłów kondensacyjnych. Układy hydrauliczne w kondensacyjnych źródłach ciepła. Część 2. Polski Instalator **9**, 11–15 (1996)
28. Mizielińska, K.: Roszenie kotła – nie! Część 1. Polski Instalator **3**, 60–63 (1996)
29. Mizielińska, K.: Roszenie kotła – nie! Część 2. Polski Instalator **4**, 27–32 (1996)
30. Mizielińska, K.: Układy hydrauliczne w źródłach ciepła z kilkoma kotłami. Polski Instalator **5**, 51–57 (1996)
31. Mizielińska, K.: Zastosowanie pionowego rozdzielacza hydraulicznego w modernizowanych, małych źródłach ciepła. Polski Instalator **6**, 12–19 (1996)
32. Mizielińska, K.: Zastosowanie specjalnych rozdzielaczy hydraulicznych. District Heating Heating Vent. **27**(1), 13–15 (1995)
33. Muniak, D.: Grzejniki w wodnych instalacjach grzewczych. Dobór, konstrukcja i charakterystyki cieplne (Radiators in Hydronic Heating Installations. Structure, Selection and Thermal Characteristics). WNT/PWN, Warszawa (2015)
34. Muniak, D.: Radiators in Hydronic Heating Installations Structure. Selection and Thermal Characteristics. Springer, Berlin (2017)
35. Muniak, D.: Wspomaganie komputerowe równoważenia hydraulicznego instalacji centralnego ogrzewania. Część I (Computer-aided hydraulic balancing in Central Heating Installations. Part II). District Heating Heating Vent. **43**(11), 480–486 (2012)
36. Polish Standard PN-M-75009:1991: Armatura instalacji centralnego ogrzewania. Zawory regulacyjne. Wymagania i badania (Central heating installation fittings. Regulation valves. Requirements and tests)
37. Polish Standard PN-M-75010:1990 Termostatyczne zawory grzejnikowe. Wymagania i badania (Thermostatic Radiator Valves. Requirements and Test)

38. Pyrkov, V.: Gidrawliczeskoje regulirowanije sistem otoplenija i ochłażdjenija. Teorija i praktika. Danfoss, Kijów (2005)
39. Pyrkov, V.: Regulacja hydrauliczna systemów ogrzewania i chłodzenia. Teoria i praktyka (Hydraulic Regulation of Heating and Cooling Systems. Theory and Practice). Systherm Serwis, Poznań (2007)
40. Roos, H.: Hydraulik der Wasserheizung, wydanie 3. Oldenbourg Verlag GmbH, Monachium (1995)
41. Roos, H.: Zagadnienia hydrauliczne w instalacjach ogrzewania wodnego (Hydraulic Issues in Water Heating Installations). PNT CIBET, Warszawa (1997)
42. Rozporządzenie Ministra Infrastruktury z dnia 12 kwietnia 2002 r. w sprawie warunków technicznych, jakim powinny odpowiadać budynki i ich usytuowanie (Dziennik Ustaw Nr 75, Poz. 690), z późniejszymi zmianami (Regulation of the Minister of Infrastructure of 12 April 2002 on the technical conditions to be met by buildings and their location, Dz.U. (Journal of Laws) 02.75.690 as amended
43. Schlott, S.: Ubertragungsverhalten der hydraulishen weiche in einer zweikesselanlage, HLH 3/2993
44. Sikorski, M.: Kryzy dławiące do regulacji instalacji centralnego ogrzewania (Throttling orifices for central heating system regulation). Gas Water Sanitary Eng. **39**(4), 133–137 (1965)
45. Szkarowski, A., Łatowski, L.: Ciepłownictwo (District Heating). WNT, Warszawa (2012)
46. Wąsowski, J.: Badania zaworów ze stałą kryzą dławiącą (Testing of valve with fixed throttling orfice). Gas Water Sanitary Eng. **40**(1), 28–30 (1966)
47. Word Wide Web site: http://www.centrum.pemp.pl/
48. Xu, B., Fu, L., Di, H.: Dynamic simulation of space heating systems with radiator controlled by TRVs in buildings. Energy Build. **40**(9), 1755–1764 (2008)
49. Xu, B., Huang, A., Fu, L., Di, H.: Simulation and Analysis on Control Effectiveness of TRVs in District Heating System. Energy Build. **43**(5), 1169–1174 (2011)

Chapter 3
Pressure Losses in the Heating Installation Pipework and Hydraulic Resistance

3.1 Pressure Losses in the Heating Installation Pipework

The flow of the medium through the pipework is affected by pressure losses. The loss concerns static pressure and consequently—total pressure. Dynamic pressure in a given section of the pipework with a constant cross-section, being the effect of the medium velocity and density, is unchanged because the two parameters remain constant if the water flowing in the system is in the liquid state. The considerations presented in this chapter concern the water flow through circular ducts with diameters smaller than those of what is referred to as mini-channels, for which some equations of classical fluid mechanics are not valid.

The working medium pressure loss on a straight section in the pipework *i*th section is defined as the sum of pressure losses on the straight section and on local obstacles using the following formula:

$$\Delta p_{\mathrm{c,i}} = \Delta p_{\mathrm{l,i}} + \Delta p_{\mathrm{m,i}}, \tag{3.1}$$

$$\Delta p_{\mathrm{l,i}} = \frac{1}{2}\rho_{\mathrm{i}} \cdot w_{\mathrm{i}}^2 \cdot \lambda_{\mathrm{i}} \cdot \frac{l_{\mathrm{i}}}{d_{\mathrm{i}}}, \tag{3.2}$$

$$\Delta p_{\mathrm{m,i}} = \frac{1}{2} \cdot \rho_{\mathrm{i}} \cdot w_{\mathrm{i}}^2 \cdot \sum \zeta_{\mathrm{i}}. \tag{3.3}$$

Term $\frac{1}{2} \cdot \rho_{\mathrm{i}} \cdot w_{\mathrm{i}}^2$ represents dynamic pressure. The pressure losses in a given circuit, which in the steady state are equal to active pressure, are defined as the sum of pressure losses of individual sections connected in series and making up the circuit, according to the following formula:

$$\Delta p_{\mathrm{ob}} = \sum_{j} \Delta p_{\mathrm{c,i}} = \sum_{j} \left(\frac{1}{2} \cdot \rho_{\mathrm{i}} \cdot w_{\mathrm{i}}^2 \cdot \lambda_{\mathrm{i}} \frac{l_{\mathrm{i}}}{d_{\mathrm{i}}} + \frac{1}{2}\rho_{\mathrm{i}} \cdot w_{\mathrm{i}}^2 \cdot \sum \zeta_{\mathrm{i}} \right). \tag{3.4}$$

D. P. Muniak, *Regulation Fixtures in Hydronic Heating Installations*, Studies in Systems, Decision and Control 187, https://doi.org/10.1007/978-3-030-03128-2_3

Local resistances arise due to the installations fixtures, regulation, cut-off and/or mixing valves, knees, bends, offset pipes, distributors, branches, etc., which cause a change in the medium flow direction or create the jet local swirling.

In the case of the installation pipework sizing, relation (3.4) is written as:

$$\Delta p_{\mathrm{ob}} = \sum_{i} (R_{\mathrm{i}} \cdot l_{\mathrm{i}} + Z_{\mathrm{i}}), \tag{3.5}$$

It follows from the comparison between relations (3.4) and (3.5) that pressure losses on straight sections can be expressed by means of the following formula:

$$\Delta p_{\mathrm{l,i}} = R_{\mathrm{i}} \cdot l_{\mathrm{i}} = \frac{1}{2} \cdot \rho_{\mathrm{i}} \cdot w_{\mathrm{i}}^2 \cdot \lambda_{\mathrm{i}} \frac{l_{\mathrm{i}}}{d_{\mathrm{i}}}. \tag{3.6}$$

Considering that:

$$w_{\mathrm{i}} = \frac{\dot{V}_{\mathrm{i}}}{A_{\mathrm{i}}} = \frac{4 \cdot \dot{V}_{\mathrm{i}}}{\pi \cdot d_{\mathrm{i}}^2}, \tag{3.7}$$

the following is obtained:

$$R_{\mathrm{i}} = \frac{1}{2} \cdot \rho_{\mathrm{i}} \cdot \left(\frac{4 \cdot \dot{V}_{\mathrm{i}}}{\pi \cdot d_{\mathrm{i}}^2}\right)^2 \frac{\lambda_{\mathrm{i}}}{d_{\mathrm{i}}} = \dot{V}_{\mathrm{i}}^2 \cdot \lambda_{\mathrm{i}} \frac{8 \cdot \rho_{\mathrm{i}}}{\pi^2 \cdot d_{\mathrm{i}}^5}. \tag{3.8}$$

A comparison between the expressions in relations (3.4) and (3.5) for the pressure loss due to local resistances gives the following notation:

$$\Delta p_{\mathrm{m,i}} = Z_{\mathrm{i}} = \frac{1}{2} \cdot \rho_{\mathrm{i}} \cdot w_{\mathrm{i}}^2 \cdot \sum \zeta_{\mathrm{i}} = \dot{V}_{\mathrm{i}}^2 \cdot \sum \zeta_{\mathrm{i}} \frac{8 \cdot \rho_{\mathrm{i}}}{\pi^2 \cdot d_{\mathrm{i}}^4}. \tag{3.9}$$

For the circuit given section with the medium volume flow $\dot{V}_{\mathrm{i}}$ (the flow of an incompressible fluid—a liquid), the following can be written:

$$\Delta p_{\mathrm{c,i}} = \dot{V}_{\mathrm{i}}^2 \cdot \left(\lambda_{\mathrm{i}} \frac{8 \cdot l_{\mathrm{i}} \cdot \rho_{\mathrm{i}}}{\pi^2 \cdot d_{\mathrm{i}}^5} + \sum \zeta_{\mathrm{i}} \frac{8 \cdot \rho_{\mathrm{i}}}{\pi^2 \cdot d_{\mathrm{i}}^4}\right). \tag{3.10}$$

Analysing formula (3.10), it can be seen that the only unknown quantities are λ_{i} and ζ_{i}. The pressure loss is proportional to the product of the square of the medium volume flow and the term included in the parentheses. If the term is a constant value, a square function is obtained which on the chart of pressure losses plotted in linear coordinates as a function of the flow is illustrated by a parabola with its zero point in the zero point of the coordinate system. As presented in Figs. 3.1, 3.2 and 3.3, in an isothermal flow (which is the case considered herein) coefficient λ depends on the flow nature characterized by the *Reynolds* number and the pipe relative roughness e. The roughness is calculated using the following formula:

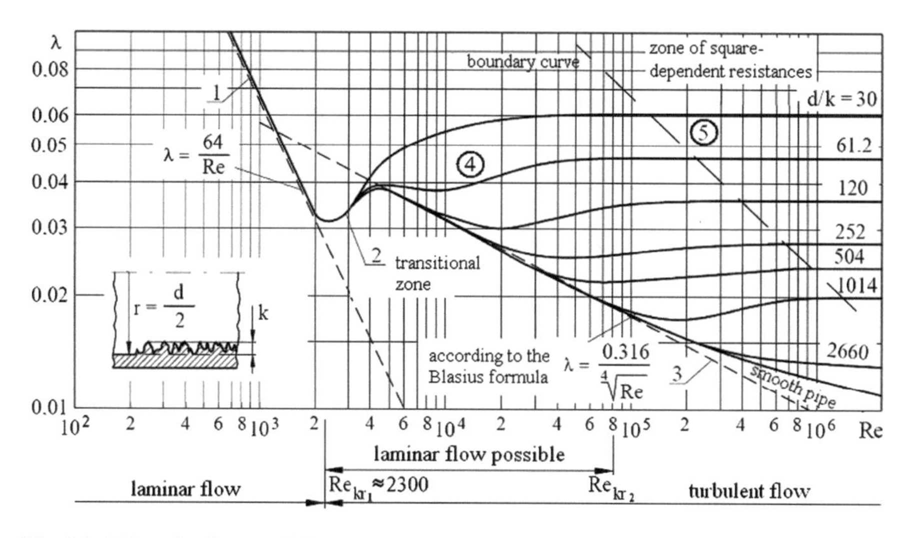

Fig. 3.1 *Nikuradse* diagram [10]

$$e = k/d \tag{3.11}$$

The *Reynolds* number is the ratio between inertial forces (momentum) and the forces of viscosity, and it is expressed as:

$$Re = \frac{w \cdot d}{\nu} = \frac{w \cdot d \cdot \rho}{\mu}. \tag{3.12}$$

If the fluid flow is dominated by viscosity forces, it is laminar. If inertial forces dominate, the flow is turbulent. The value of the *Reynolds* number that separates the two is referred to as the critical *Reynolds* number. In the laminar (layered) flow, the fluid individual layers move in parallel to each other, without mixing. In the turbulent flow, the layers get mixed so the velocity vectors of the fluid individual particles are not parallel to each other. Coefficient λ is found by means of different formulae, depending on the flow nature (the *Reynolds* number), the pipe relative roughness e and the resulting thickness of the fluid boundary layer. These are usually empirical or semi-empirical relations, and in the case of the laminar (layered) flow they can be derived fully by means of analytical methods. An extensive theoretical basis for this issue can be found e.g. in [10, 24].

Determination of specific *Reynolds* numbers characterizing the flow and implying relations appropriate for the calculation of the coefficient of linear pressure losses λ is complicated by the fact that the laminar-to-turbulent flow transition takes place at a higher value of the *Reynolds* number than the transition in the opposite direction. For this reason, two characteristic *Reynolds* numbers are distinguished that separate the laminar flow from the turbulent one: the critical lower *Reynolds* number $Re_{\mathrm{kr},1}$ and the critical upper *Reynolds* number $Re_{\mathrm{kr},2}$ ("critical" understood as the one separating

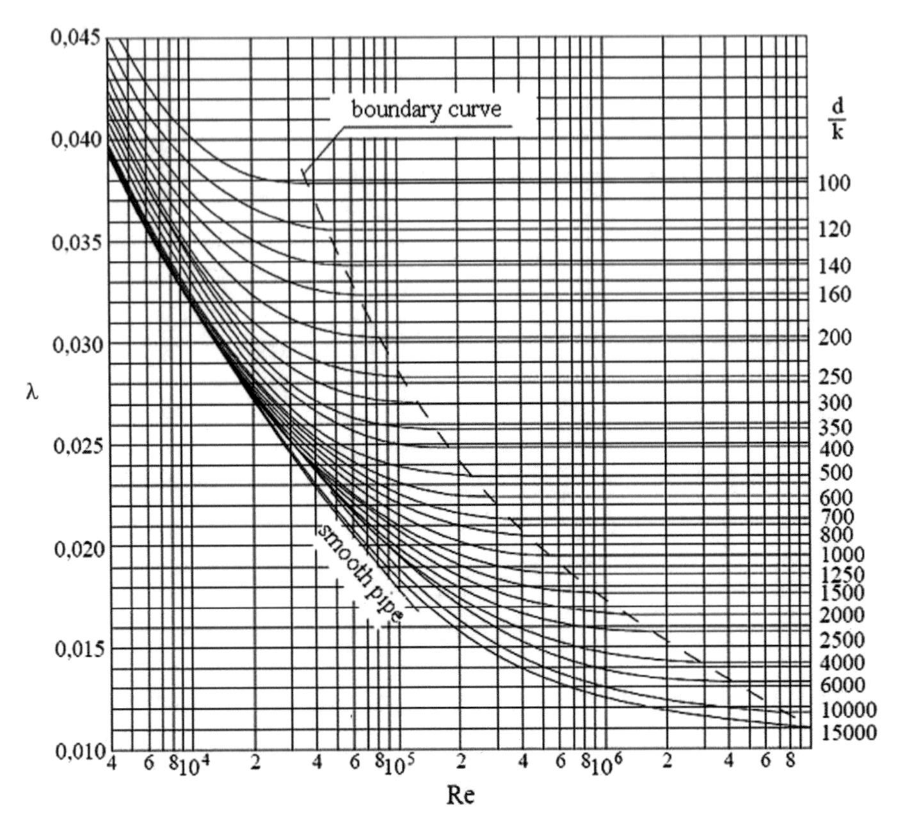

Fig. 3.2 Linear pressure loss coefficient depending on the *Reynolds* number according to *Colebrook–White* [10]

the two regions). In the transitional range between the two values, the flow can be laminar. But if the jet is disturbed slightly, the flow becomes turbulent and does not return to the previous state. Such laminar flow is therefore unstable. In laboratory conditions, laminar flows were observed at the *Reynolds* number of about $Re_{gr,2} \approx 80{,}000$ [10]. In technical practice, however, the upper boundary is usually adopted at $Re_{gr,2} \approx 4000$. Below the lower boundary value of the *Reynolds* number of $Re_{gr,1} \approx 2300$, the flow through a circular pipe is always laminar.

The works on determination of the friction factor (the coefficient of linear pressure losses) were conducted, among others, by *Nikuradse* in the 1930s. The results of his analyses gave rise to the *Nikuradse* diagram, aka the *Nikuradse* harp, as presented in Fig. 3.1. *Nikuradse* used brass pipes with diameters from 25 to 100 mm. In order to set specific relative roughness values, he used sand with the grain size of 0.2–3.2 mm glued onto the pipe inner surface. For this reason, the term “sand (uniform) roughness” is used in contrast to technical, natural roughness encountered in reality. The relative roughness values obtained from the applied grain size totalled $0.001 < e < 0.035$ [25]. The friction factor (the linear pressure loss coefficient) for

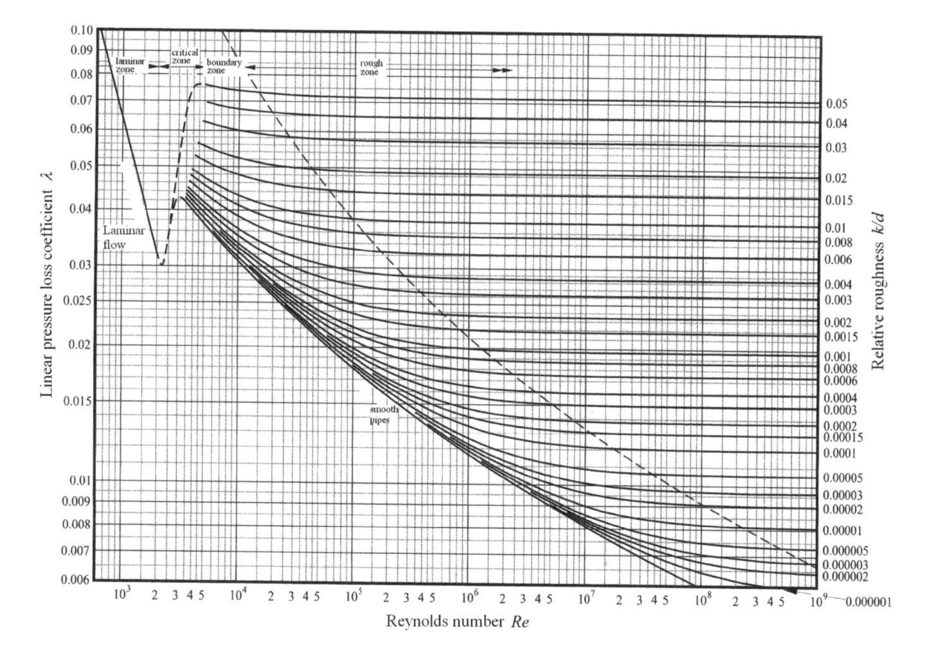

Fig. 3.3 *Moody* chart [21]

pipes with sand roughness in a non-laminar flow is calculated using the *Nikuradse* semi-empirical formula [25]:

$$\lambda = \frac{1}{\left(1.14 + 2\log\frac{1}{e}\right)^2}. \tag{3.13}$$

Friction factor λ for steel and cast-iron circular pipes with diameters of 20–600 mm, at different values of natural roughness (unlike the *Nikuradse* tests using sand uniform roughness) in non-laminar flows was also studied and measured by *Colebrook* and *White* [10]. Integrating the equation describing a pressurized medium turbulent flow through a circular pipe, they obtained the following formula:

$$\frac{1}{\sqrt{\lambda}} = -\frac{\ln 10}{C_3\sqrt{8}} \cdot \log\left(\frac{2.535 \cdot C_1}{Re\sqrt{\lambda}} + \frac{C_2 \cdot k/d}{0.111}\right). \tag{3.14}$$

The coefficient constants were determined by means of laboratory testing as: $C_1 = 0.099$, $C_2 = 0.03$, $C_3 = 0.407$. Substitution of these values leads to the semi-empirical equation published in 1937 and most often referred to in literature for the range between the lines marked as 3 and the boundary curve in Fig. 3.1 (rough hydraulic pipes, non-fully developed turbulent flow), which is known as the *Colebrook-White* formula:

$$\frac{1}{\sqrt{\lambda}} = -2\log\left(\frac{2.51}{Re\sqrt{\lambda}} + \frac{e}{3.71}\right). \tag{3.15}$$

This is the formula recommended in the PN-76/M-34034 standard: *Pipelines. Principles of calculating pressure losses* [19]. Its graphical illustration is referred to as the *Colebrook-White* diagram, as presented in Fig. 3.2.

The formula has a confounded character and in practice, instead of being found analytically, the coefficient of linear pressure losses λ is determined using read-outs from an extended *Colebrook-White* diagram, referred to as the *Moody* chart (e.g. Appendix 3 for [19]), which is presented in Fig. 3.3. The chart is a combination of the diagram of pressure losses for the laminar flow with the *Colebrook-White* diagram. It can be seen that in zone 4 it differs from the *Nikuradse* diagram. This is due to the fact that in the case of pipes with natural roughness, the linear pressure loss coefficient decreases in the entire range of the turbulent flow together with a rise in the *Reynolds* number until it reaches a certain constant value, irrespective of relative roughness.

The boundary curve marked in the two figures, which separates the zones of variable and constant values of coefficient λ depending on the *Reynolds* number, can be determined using the empirical relation developed by *Rouse* [11]:

$$Re_{\mathrm{gr}} = \frac{200}{\sqrt{\lambda}} \cdot \frac{1}{e}. \tag{3.16}$$

Using the *Nikuradse* and the *Moody* charts, it is possible to separate a few zones [10, 11, 15]:

- the laminar flow zone ($Re < 2300$), marked as 1. The boundary layer thickness is in this case bigger than the height of the pipe wall non-uniformity (absolute roughness k). In this range of flows, the linear pressure loss coefficient depends only on the *Reynolds* number, according to the *Hagen-Poiseuille* formula:

$$\lambda = \frac{64}{Re}. \tag{3.17}$$

- the transitional zone (between the laminar and the turbulent flow) ($2300 < Re < 4000$) marked as 2, with a strictly non-defined character of the flow, where the linear pressure loss coefficient is determined depending on the actual character of the flow through the pipeline according to the rules valid for adjacent zones. In design calculations, the existence of the less favourable condition is assumed.
- the turbulent flow zone for hydraulically smooth pipes ($4000 < Re < 1/e$ according to [15], or $Re > 4000$ and $e < e_{gr}$ according to [19]), i.e. for the boundary layer thickness bigger than the pipe wall absolute roughness. The zone is marked as 3. The turbulent flow occurs in the fluid volume beyond the boundary layer. The coefficient of linear pressure losses depends here on the *Reynolds* number only.

The value of λ can be calculated using the *Blasius* and the *Nikuradse* formulae, or those developed by *Prandtl-von Karman* or *Konakov*.

Blasius tested brass pipes in the range of the *Reynolds* number of $2300 < Re < 10^5$ [26]. The line in the figure is drawn according to the *Blasius* formula:

$$\lambda = \frac{0.3164}{\sqrt[4]{Re}}. \tag{3.18}$$

If $10^5 \leq Re \leq 10^8$, the *Nikuradse* formula is applied:

$$\lambda = 0.0032 + \frac{0.221}{Re^{0.257}}. \tag{3.19}$$

The *Prandtl-von Karman* formula (or the *Colebrook-White* formula if $k/d = 0$, i.e. for hydraulically smooth pipes, corrected slightly by *Nikuradse*) [10, 19] has the following form:

$$\frac{1}{\sqrt{\lambda}} = -2\log\left(\frac{2.51}{Re\sqrt{\lambda}}\right) = 2\log\left(Re\sqrt{\lambda} - 0.8\right). \tag{3.20}$$

The *Konakov* formula [15] is expressed as:

$$\lambda = \frac{1}{\left(2\log\frac{5.62}{Re^{0..}}\right)^2} = \frac{1}{(1.8\log Re - 1.5)^2} = (1.8\log Re - 1.5)^{-2} \tag{3.21}$$

- the transitional zone of the turbulent flow in hydraulically rough pipes ($1/e < Re < 2000/e$ according to [15], or $Re < Re_{gr}$ and $e > e_{gr}$ according to [19]), marked as 4, where the coefficient of linear pressure losses depends both on the *Reynolds* number and relative roughness e, generally falling down at the beginning and then levelling off at a certain value. The boundary layer thickness starts to decrease as much as to reveal partly but increasingly the pipe surface irregularities causing more resistances to arise. The value of λ can be calculated from the *Colebrook* formula with a non-confounded form [15]:

$$\lambda = \frac{1}{\left[2\log\left(\frac{e}{3.71} + \frac{5.76}{Re^{0.9}}\right)\right]^2}. \tag{3.22}$$

- the zone of a fully developed turbulent flow through hydraulically rough pipes ($Re > 2000/e$ according to [15], or $Re > Re_{gr}$ and $e > e_{gr}$ according to [19]), marked as 5, where the boundary layer thickness is smaller than the relative roughness value, which causes turbulence of the flow in the entire cross-section. The coefficient of linear pressure losses depends here on relative roughness e only. This results from the fact that in this range the pressure losses are proportional to the square of the fluid flow mean velocity in the pipe cross-section.

The value of λ can be determined using the *Prandtl-Nikuradse* formula, which is also referred to as a simplification of the *Colebrook-White* formula, which technically is the *Colebrook-White* formula without the term taking account of the impact of the *Reynolds* number because in this range of the medium flows $Re \rightarrow \infty$ [15]. The formula is expressed as:

$$\lambda = \frac{1}{\left(2\log\frac{e}{3.71}\right)^2} = \frac{1}{(1.14 - 2\log e)^2} = (1.14 = 2\log e)^{-2}. \tag{3.23}$$

The formulae presented above can be used as relations simplified in comparison to the *Colebrook-White* confounded formula. They give results which deviate by about 5% from those obtained by means of the latter and can also be used to determine the initial value of λ in the procedure for calculating the coefficient real value iteratively. Apart from that, in the case of pipes with technical (natural) roughness in the range of turbulent flows, the following relations are used occasionally:

- the *Altšul* formula [10]:

$$\lambda = 0.11 \cdot \left(\frac{68}{Re} + e\right)^{0.25}, \tag{3.24}$$

- the *Moody* formula [25]:

$$\lambda = 0.0055 \cdot \left[1 + \left(20000 \cdot e + \frac{10^6}{Re}\right)^{1/3}\right], \tag{3.25}$$

- the *Walden* formula [25]:

$$\frac{1}{\sqrt{\lambda}} = -2\log\left(\frac{6.1}{Re^{0.91}} + 0.268 \cdot e\right). \tag{3.26}$$

The boundary layer thickness, and, consequently, whether a given pipe is classed in the case of turbulent flows as hydraulically smooth or rough, can be established by comparing boundary roughness e_{gr} to the analysed pipe relative roughness e [19]:

- in pipes with uniform natural roughness the *Filonienko-Altšul* formula can be applied:

$$e_{gr} = \frac{18\log Re - 16.4}{Re}, \tag{3.27}$$

or, if $Re \leq 10^5$, the *Blasius* formula can be used:

$$e_{gr} = 17.85 \cdot Re^{-0.875}. \tag{3.28}$$

– in pipes with non-uniform natural roughness the *Altšul-Ljacer* formula can be applied:

$$e_{gr} = \frac{23}{Re}. \tag{3.29}$$

If $e < e_{gr}$, the pipe is hydraulically smooth. If $e > e_{gr}$, the pipe is hydraulically rough.

As previously mentioned, the relations presented above that can be used to calculate the coefficient of linear pressure losses and, consequently, total pressure losses, relate to a fully developed and isothermal flow. In reality, however, water flowing in the piping cools down irrespective of the thermal insulation effectiveness. As a result, the water viscosity changes in the first place because viscosity is substantially temperature-dependent. Another effect is a change in the *Reynolds* number and in the value of λ. Due to that, the changes in pressure losses on the pipework straight sections are non-linear and their total amount differs to a certain degree from the value calculated under the adopted assumptions. Relevant relations that make it possible to determine pressure losses in such a situation more accurately can be found for example in [10].

3.2 Hydraulic Resistance

Considering recommended velocity values of the medium flow through pipes, diameters of the ducts and the share of pressure losses on local resistances in total pressure losses [cf. formula (3.40)], the central heating pump installations are most often operated in the range of the fluid turbulent flow. However, for most of the heating season the installation is operated under a partial load. Therefore, if apart from qualitative regulation (changeable temperature of the installation supply water) quantitative regulation (changeable value of the medium volume flow) is additionally used, the medium flows can also occur in the laminar and transitional range. For this reason, it cannot be assumed that in straight sections coefficient λ_i and, thereby, the $(\lambda_i \cdot 8 \cdot \rho_i)/(\pi^2 \cdot d_i^5)$ term in formula (3.10) are constant, which would be the case if flows occurred only in the range of a fully developed turbulent flow. In the system of linear coordinates, the hydraulic characteristic of such an element of the pipework is therefore not a parabola but a curve composed of three parts corresponding to the three basic areas in the *Moody* and the *Nikuradse* chart. The first part is a straight line because in this region the coefficient of linear pressure losses is linearly dependent on the *Reynolds* number, and in this way—also on the flow velocity. According to formulae (3.2) and (3.17), it can therefore be written as follows:

$$\Delta p_{l,i_i} = \frac{1}{2} \cdot \rho_i \cdot w_i^2 \cdot \lambda_i \frac{l_i}{d_i} = \frac{1}{2} \cdot \rho_i \cdot w_i^2 \cdot \frac{64}{Re} \cdot \frac{l_i}{d_i} = \frac{1}{2} \cdot \rho_i \cdot w_i^2 \cdot \frac{64 \cdot \nu_i}{w_i \cdot d_i} \cdot \frac{l_i}{d_i}$$
$$= \rho_i \cdot w_i \cdot \frac{32 \cdot \nu_i \cdot l_i}{d_i^2}. \tag{3.30}$$

Next, it should be established whether the pipe is hydraulically smooth or rough, according to formulae (3.27)–(3.29). In the case of a smooth pipe, a function of pressure losses is obtained with the exponent of 1.75 resulting from the application of the *Blasius* formula (3.18) to find the value of λ:

$$\Delta p_{l,i} = \frac{1}{2} \cdot \rho_i \cdot w_i^2 \cdot \frac{0.3164}{\sqrt[4]{Re}} \cdot \frac{l_i}{d_i} = 0.1582 \cdot \rho_i \cdot l_i \frac{w_i^{1.75} \cdot \nu_i^{0.25}}{d_i^{1.25}}. \tag{3.31}$$

If the pipe is hydraulically rough (technical, natural roughness), the *Colebrook-White* formula (3.15) for example can be used at the beginning. It can then be replaced by formula (3.23), which is deprived of the part that takes account of the impact of the *Reynolds* number at the inlet of the area where the flow is fully turbulent (this is identical with the situation where in the *Colebrook-White* formula $Re \to \infty$). In such a case, the exponent of a given pipe hydraulic characteristic varies in the range of 1.75–2.00. Therefore, the resultant characteristic is a broken curve connecting the straight segment and a section of the parabola with an increasing slope. This means that if the function describing flow-related pressure losses is created according to formulae (3.2) and (3.10), account should be taken of both the variability of the exponent at w_i as a function of w_i, and, thereby, at $\dot{V}_i$, and the variability of the resultant coefficient that the term is multiplied by. However, such notation would be rather complex. To simplify it, the hydraulic resistance concept is introduced to make it possible to construct a possibly simple mathematical model that could be interpreted practically (without the use of computer programs). It is then a characteristic parameter, constant for a given section of the pipework with straight pipes. Considering that the slope of the curve illustrating pressure losses changes depending on the flow of the medium, a proposal is made to write the part of the equation related to the medium pressure losses on a straight section in the following form [14, 22, 23]:

$$\Delta p_{l,i} = \dot{V}_i^2 \cdot \lambda_i \frac{8 \cdot l_i \cdot \rho_i}{\pi^2 \cdot d_i^5} = r_{l,i} \cdot \dot{V}_i^m, \tag{3.32}$$

Using such notation, the changes in pressure losses depend only on changes in the volume flow of the medium. The characteristic correction exponent m is a function of the medium flow velocity and the pipe inner diameter. Another factor is the type of the pipe material and the resulting roughness of the pipe surface. The correction exponent is usually included in the range of $1.75 < m < 2$. Owing to such an approach, hydraulic resistance can be treated as a constant value, which is characteristic of a given section of the pipework with straight pipes. This is useful, for example, in the analysis of phenomena that occur when the medium is throttled by regulating valves with known characteristics. The value of parameter m can be determined having two points of measurement of pressure losses depending on the medium flow through the pipe [14, 22, 23]. The following can be written for the two points:

$$\Delta p_{l,1,i} = r_{l,i} \cdot \dot{V}_{1,i}^m,$$
$$\Delta p_{l,2,i} = r_{l,i} \cdot \dot{V}_{2,i}^m.$$

Dividing the equations by sides, the following is obtained:

$$\frac{\Delta p_{l,1,i}}{\Delta p_{l,2,i}} = \left(\frac{\dot{V}_{1,i}}{\dot{V}_{2,i}}\right)^m \Rightarrow m = \frac{\log \frac{\Delta p_{l,1,i}}{\Delta p_{l,2,i}}}{\log \frac{\dot{V}_{1,i}}{\dot{V}_{2,i}}} = \frac{\log \frac{R_{1,i}}{R_{2,i}}}{\log \frac{\dot{V}_{1,i}}{\dot{V}_{2,i}}} = \frac{\log \frac{R_{1,i}}{R_{2,i}}}{\log \frac{\dot{m}_{1,i}}{\dot{m}_{2,i}}}. \tag{3.33}$$

The formula indicates that if the hydraulic characteristic is plotted in a chart scaled logarithmically on both axes (the characteristic is then a straight line), the exponent is the straight line slope with respect to the horizontal axis.

The value of the local loss coefficient ζ is most often found experimentally by measuring the total loss of pressure on an obstacle and dividing it by the measured value of dynamic pressure according to the relation resulting from formula (3.3):

$$\Delta p_{m,i} = \zeta_i \cdot \frac{1}{2} \cdot \rho_i \cdot w_i^2 \Rightarrow \zeta = \frac{\Delta p_{m,i}}{\frac{1}{2} \cdot \rho_i \cdot w_i^2}. \tag{3.34}$$

It should be mentioned that the total pressure drop determined in this way in fact includes also the pressure drop on the pipes connected to the system and stabilizing the flow for measuring purposes, and the pressure drop on straight sections of the connector of the obstacle, such as the valve port for example. In the former case, the measuring error impact can be taken into account and the result can be corrected, which is rather difficult in the latter case. Anyway, the error resulting therefrom is small and may be ignored. Among others, due to these factors, the coefficient of local resistances is characterized by a certain variability depending on the *Reynolds* number. In the heating installation design practice, however, it is considered as a constant quantity [14, 22–24] because the variability is much smaller compared to sections of straight pipes. The transitional range between the laminar and the turbulent flow for a local resistance is much smaller than for the pipe due to the changes in the fluid flow direction, sudden changes in the cross-section, etc., which leads to creation of local vortices and the jet separation from the wall. Another fact that needs considering is that the coefficients of local resistances are determined and given for the medium fully developed flow (with a defined velocity profile).

In reality, it is possible that local resistances are installed at too short a distance from each other and the medium flow in between them cannot develop fully. In this situation, summing up individual values of coefficients of the local resistances will give wrong results. A local resistance affects the jet along a length that can be determined from the following formula [10]:

$$l_i = 0.5 \cdot \frac{\zeta_i}{\lambda_i} \cdot d_i = 0.5 \cdot l_{z,i}, \tag{3.35}$$

If the distance between two local resistances installed in the pipework exceeds $0.5l_z$, it is assumed that individual coefficients of local resistances may be summed up without the risk of errors. Otherwise, it is recommended that the quantity equivalent value should be determined.

If it is assumed that the local pressure loss coefficient ζ is constant, like in the case of the coefficient of linear pressure losses (λ), the pressure drop caused by local resistances can be expressed as:

$$\Delta p_{\mathrm{m,i}} = r_{\mathrm{m,i}} \cdot \dot{V}_{\mathrm{i}}^{2}, \tag{3.36}$$

The total pressure loss in a given section of the pipework can then be defined by the following relation:

$$\Delta p_{\mathrm{c,i}} = r_{\mathrm{l,i}} \cdot \dot{V}_{\mathrm{i}}^{m} + r_{\mathrm{m,i}} \cdot \dot{V}_{\mathrm{i}}^{2}. \tag{3.37}$$

If the medium flow is turbulent, all the quantities in formulae (3.4), (3.10) are constant and known, so the $(\lambda_{\mathrm{i}} \cdot l_{\mathrm{i}} \cdot 8 \cdot \rho_{\mathrm{i}})/(\pi^2 \cdot d_{\mathrm{i}}^5) + (\sum \zeta_{\mathrm{i}} \cdot 8 \cdot \rho_{\mathrm{i}})/(\pi^2 \cdot d_{\mathrm{i}}^4)$ term can be treated as a certain constant value of the total hydraulic resistance, marked as $r_{c,i}$ and defined as follows:

$$r_{\mathrm{c,i}} = \lambda_{\mathrm{i}} \frac{8 \cdot l_{\mathrm{i}} \cdot \rho_{\mathrm{i}}}{\pi^2 \cdot d_{\mathrm{i}}^5} + \sum \zeta_{\mathrm{i}} \frac{8 \cdot \rho_{\mathrm{i}}}{\pi^2 \cdot d_{\mathrm{i}}^4}. \tag{3.38}$$

The result is then the following equation:

$$\Delta p_{\mathrm{c,i}} = r_{\mathrm{l}} \cdot \dot{V}_{\mathrm{i}}^{m} + r_{\mathrm{m,i}} \cdot \dot{V}_{\mathrm{i}}^{2} = r_{\mathrm{c,i}} \cdot \dot{V}_{\mathrm{i}}^{n}, \tag{3.39}$$

The resultant correction exponent of the characteristic (n ($m<n < 2$)) is related to the whole of a given part of the pipework, i.e. to the straight sections and local resistances installed therein. Its value is found depending on the share of the local resistances in total hydraulic resistances of the pipework given section, according to the following formula [22, 23]:

$$n = \frac{\ln\left[a_{\mathrm{m}} \cdot \dot{V}_{\mathrm{i}}^{2} + (1 - a_{\mathrm{m}}) \cdot \dot{V}_{\mathrm{i}}^{m}\right]}{\ln \dot{V}_{\mathrm{i}}}, \tag{3.40}$$

If the share increases, the value of exponent n rises and approaches $n=2$. The higher the share of the pipework local resistances, the shorter the straight segment on the linearly scaled hydraulic characteristic. In practice, due to the usual high share of local resistances ($a_{\mathrm{m}} > 0.6$) and like in the case of the exponent of the hydraulic characteristic of the valve, it is often assumed that $n=2$, which can be noticed analysing the nomograms provided by the product manufacturer. In a typical installation, the straight segment in the chart is usually negligibly short, and in a linear system of coordinates the pressure loss line depending on the flow is a square parabola. formula (3.39) then comes down to the following:

$$\Delta p_{\mathrm{c,i}} = r_{\mathrm{c,i}} \cdot \dot{V}_{\mathrm{i}}^{2}. \tag{3.41}$$

Consequently, relations (3.32), (3.36) and (3.41) can be brought to a general form that enables determination of a given element hydraulic resistances based on the known value of the medium volume flow and on the known pressure drop caused by it, according to the following formula:

$$r_{\mathrm{i}} = \frac{\Delta p_{\mathrm{i}}}{\dot{V}_{\mathrm{i}}^{2}}. \tag{3.42}$$

The equation relates three parameters, so two additional equations can be written:

$$\dot{V}_{\mathrm{i}} = \sqrt{\frac{\Delta p_{\mathrm{i}}}{r_{\mathrm{i}}}}, \tag{3.43}$$

$$\Delta p_{\mathrm{i}} = r_{\mathrm{i}} \cdot \dot{V}_{\mathrm{i}}^{2}. \tag{3.44}$$

If the flow of an incompressible medium is considered, and this is the case if water is in the liquid state (the change in density as a function of pressure can be omitted), the volume flow in the relations presented above may be replaced with the mass flow. The following formulae are then obtained:

$$r_{\mathrm{i}} = \frac{\Delta p_{\mathrm{i}}}{\dot{m}_{\mathrm{i}}^{2}}. \tag{3.45}$$

$$\dot{m}_{\mathrm{i}} = \sqrt{\frac{\Delta p_{\mathrm{i}}}{r_{\mathrm{i}}}}, \tag{3.46}$$

$$\Delta p_{\mathrm{i}} = r_{\mathrm{i}} \cdot \dot{m}_{\mathrm{i}}^{2}. \tag{3.47}$$

Naturally, in this situation hydraulic resistance has to be related to the mass flow and not to the volume flow, which is the case in formulae (3.42)–(3.44).

In practice, pressure losses in pipes are calculated using nomograms provided by manufacturers. The nomograms usually illustrate unit linear pressure losses *R* as a function of the flow velocity or the medium mass or volume flow rate, depending on the pipe nominal diameter. The nomograms are usually double-logarithmic scale graphs which, due to linear interpolation, present straight lines in the entire range, and the described phenomenon of a change in the slope of the line representing pressure losses at the laminar-to-turbulent flow transition cannot be observed. But there are nomograms in which the phenomenon is visible. Figures 3.4, 3.5, 3.6 and 3.7 present nomograms for both cases.

It should also be noted that such nomograms concern the flow of water with a specific temperature given by the manufacturer and because of that they may not be used directly in different conditions of the pipe operating temperature. Although temperature does not affect water density much, it does have a considerable impact on the medium viscosity, which demonstrates significant changeability and decreases with a rise in temperature. As follows from formula (3.12), this quantity, by conditioning the values of the *Reynolds* number and, thereby, coefficient λ, has a direct impact on

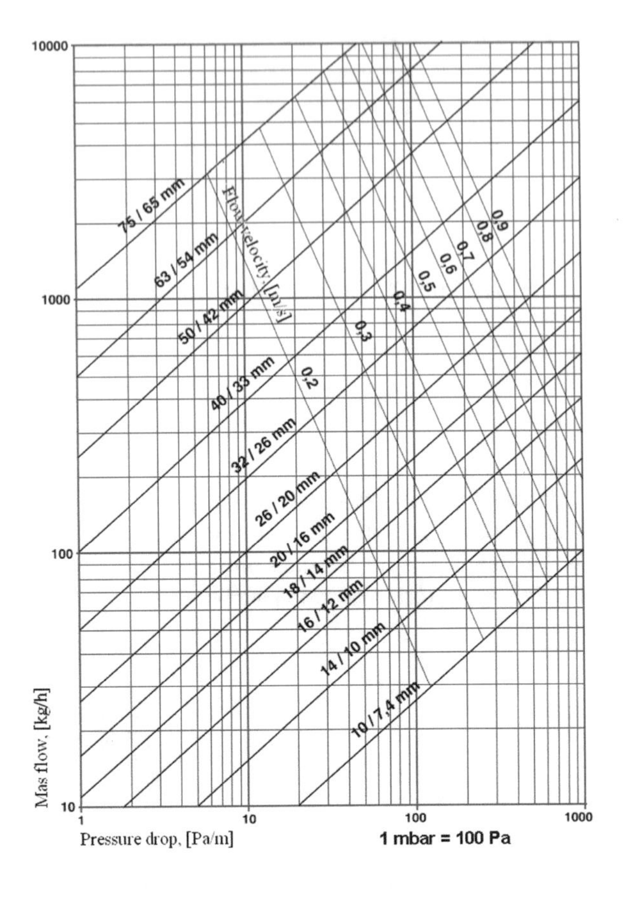

Fig. 3.4 Unit linear pressure loss R in *Herz* multilayer pipes with absolute roughness $k = 0.007$ mm (water mean temperature: 70 °C) [4]

the value of unit linear pressure losses R in formula (3.8) and hydraulic resistance r in formula (3.32). Therefore, the higher the water temperature, the smaller the value of the pipe hydraulic resistance. If the necessary calculations of the medium viscosity and, consequently, of the hydraulic resistance value with respect to temperature are not performed, errors may arise in the calculation results.

3.2.1 Combinations of Hydraulic Resistances

Considering the medium spread in the system and the hydraulic resistance seen from the pipework beginning, the pipework creates a system of individual hydraulic resistances with a certain equivalent total hydraulic resistance value. The value of this quantity is necessary to analyse, for example, co-operation between the circulation pump and the pipework at variable values of the pump pressure, changes in the throttling of regulation valves, etc. It is calculated, if possible, by "curling up" the diagram to a single element with a resultant, equivalent hydraulic resistance.

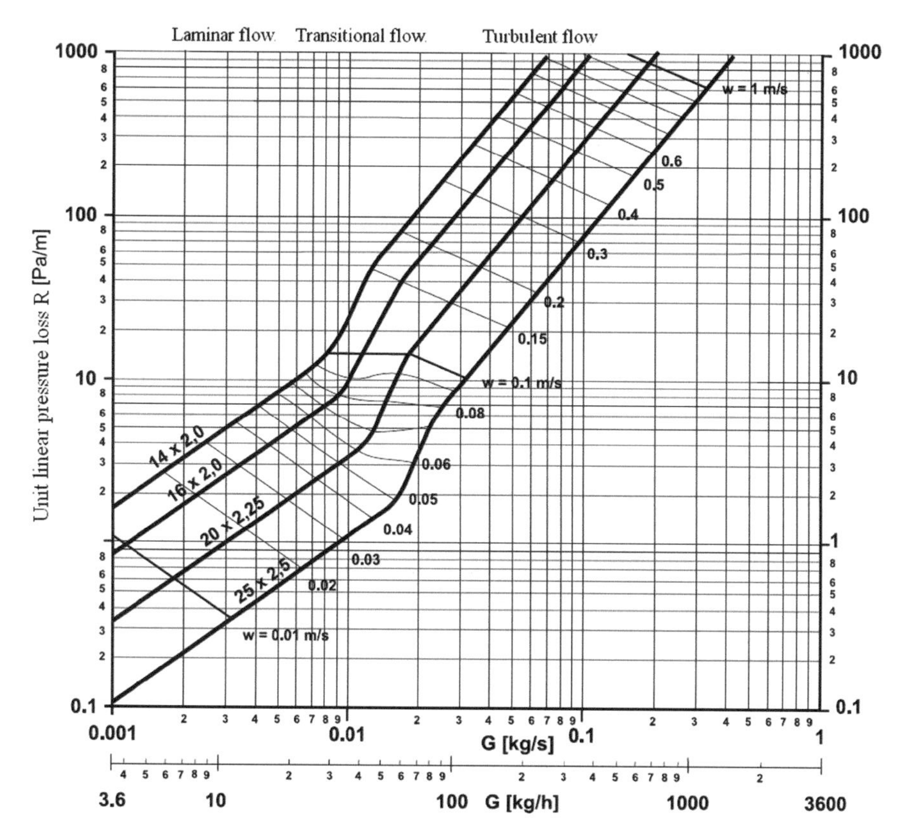

Fig. 3.5 Unit linear pressure loss R in *KISAN* multilayer pipes with absolute roughness $k=0.003$–0.005 mm (water mean temperature: 70 °C) [5]

Depending on the way in which the pipework elements are connected, the following can be distinguished: a serial combination, a parallel combination and a complex combination, where the mechanism of just "curling up" into a single equivalent hydraulic resistance cannot be applied.

3.2.1.1 Serial Combination of Hydraulic Resistances

A serial combination of two hydraulic resistances is shown in Fig. 3.8a. In such a system, the medium mass/volume flow through both resistances is the same and equal to the total flow. The pressure drops on individual resistors can be different because they depend on the mutual ratio of resistances (they are equal if the resistances are identical). Nonetheless, together they give the total pressure loss. This can be expressed as follows:

(I) $\dot{m}_c = \dot{m}_1 = \dot{m}_2$
(II) $\Delta p_c = \Delta p_1 + \Delta p_2$

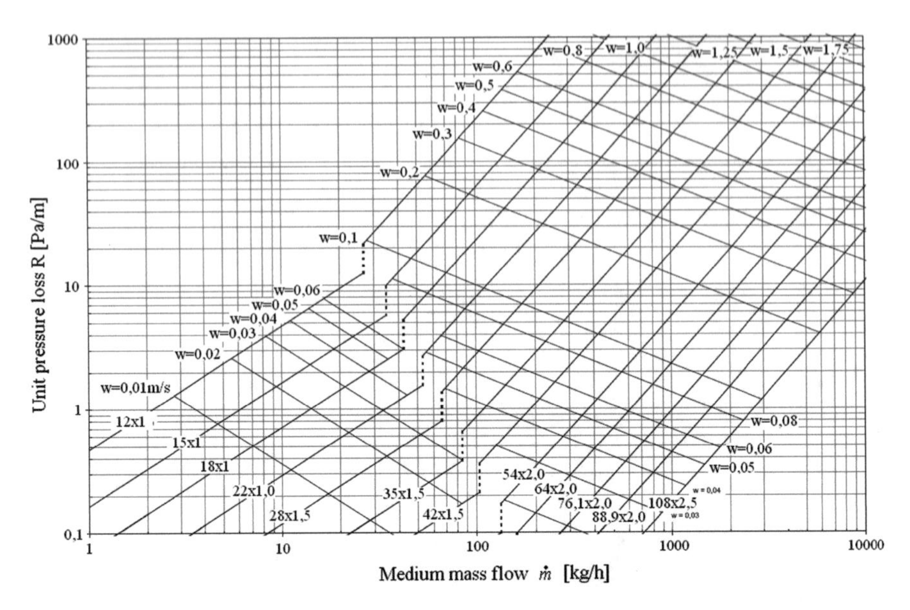

Fig. 3.6 Unit linear pressure loss R in copper pipes with absolute roughness $k = 0.01$ mm (water mean temperature: 70 °C) [8]

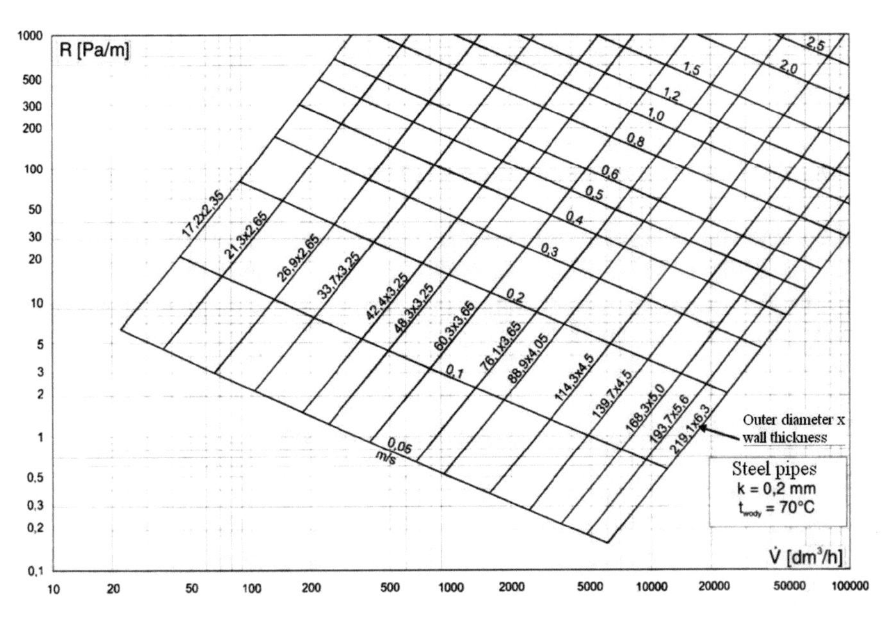

Fig. 3.7 Unit linear pressure loss R in steel pipes (water mean temperature: 70 °C) [27]

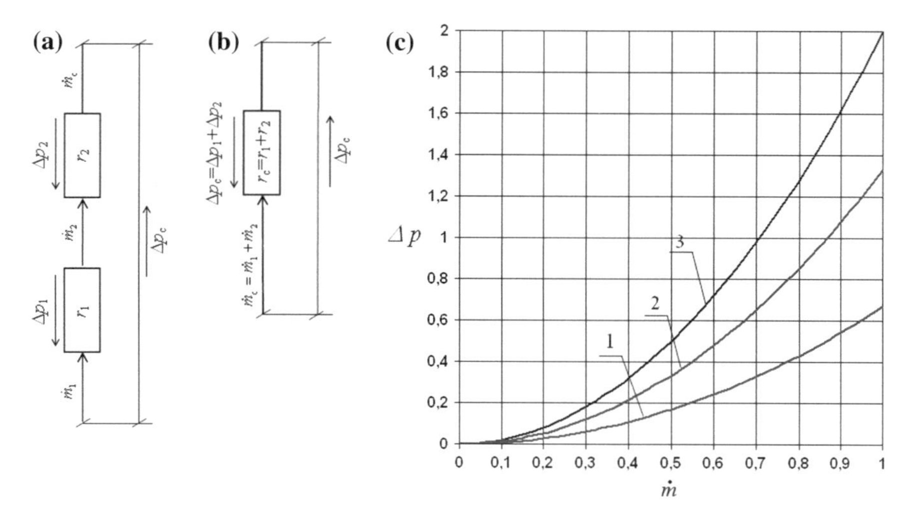

Fig. 3.8 **a** Diagram of two hydraulic resistances connected in series; **b** equivalent hydraulic resistance; **c** hydraulic characteristics of individual hydraulic resistances (lines 1, 2) and the resultant characteristic (line 3)

The pressure loss can be related to the mass flow using the sought parameter, i.e. hydraulic resistance (assuming a square relation of flow-related pressure losses for simplification), using the following formulae:

(III) $\Delta p_c = r_c \cdot \dot{m}_c^2$
(IV) $\Delta p_1 = r_1 \cdot \dot{m}_1^2$
(V) $\Delta p_2 = r_2 \cdot \dot{m}_2^2$

Substituting equations III–V in formula II, the result is as follows:

(VI) $r_c \cdot \dot{m}_c^2 = r_1 \cdot \dot{m}_1^2 + r_2 \cdot \dot{m}_2^2$

Considering equation I, the following is obtained:

(VII) $r_c = r_1 + r_2$

In the case of the *i*th number of resistances, the following equation is obtained:

(VIII) $r_c = r_1 + r_2 + \cdots + r_i$

In a serial combination, the resultant hydraulic resistance is then the sum of individual resistances. There is a simple analogy here with electric circuits. The equivalent of the potential difference (voltage) is the differential pressure in the system, the equivalent of the current is the medium mass/volume flow, and electrical resistance corresponds to hydraulic resistance. The only essential difference is that the pressure drop is a square function of the medium mass/volume flow, whereas the voltage drop in an electric circuit is a linear function of the current flowing through a resistance.

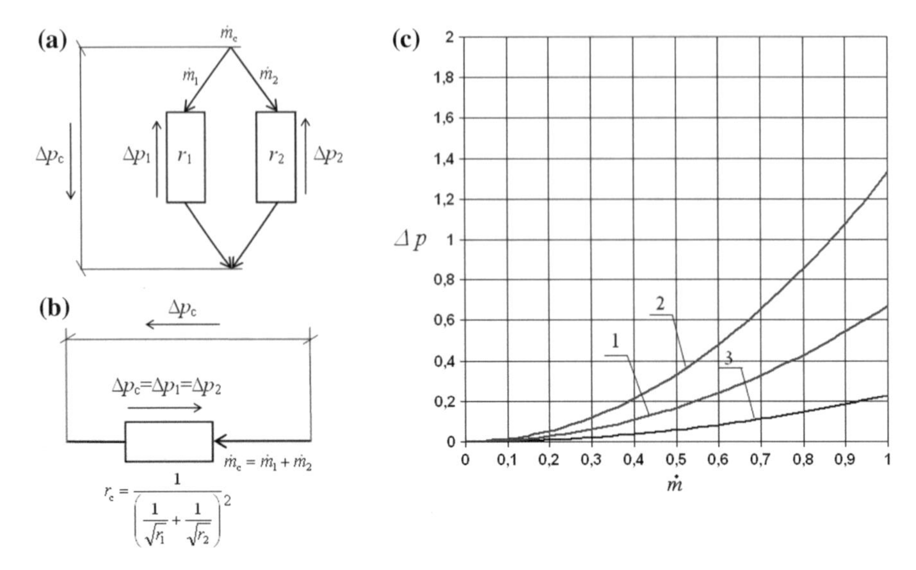

Fig. 3.9 **a** Diagram of two hydraulic resistances connected in parallel; **b** equivalent hydraulic resistance; **c** hydraulic characteristics of individual hydraulic resistances (lines 1, 2) and the resultant characteristic (line 3)

Figure 3.8b illustrates the diagram "curling-up" to a single equivalent hydraulic resistance. Figure 3.8c, scaled in conventional (dimensionless) units, presents the creation of the resultant hydraulic characteristic if two hydraulic resistances are connected in series. Resistance r_2 is set as double resistance $r_{1.}$ At a given mass flow of the medium, the pressure drops are summed up for individual curves which correspond to individual hydraulic resistances. The slope of the resultant curve is therefore bigger compared to the curves illustrating losses on individual resistances. Naturally, this does not mean that the value of the characteristic exponent is changed. The only value that changes is the base of the power of the function describing the characteristic, which is represented by total equivalent resistance r_c.

3.2.1.2 Parallel Combination of Hydraulic Resistances

A parallel combination of two hydraulic resistances is shown in Fig. 3.9a. In such a system, pressure losses on both resistances are the same and equal to total pressure. The medium mass flows through the two resistances can be different and they depend on the mutual ratio of the values of the resistances (they are equal if the resistances are identical). Nonetheless, together they give the total mass flow value. This can be expressed as follows:

$$\text{(I}'\text{)} \quad \dot{m}_c = \dot{m}_1 + \dot{m}_2$$
$$\text{(II}'\text{)} \quad \Delta p_c = \Delta p_1 = \Delta p_2$$

The pressure loss can be related to the mass flow using the sought parameter, i.e. hydraulic resistance (assuming a square relation of flow-related pressure losses for simplification), using the following formulae:

(III′) $\Delta p_c = r_c \cdot \dot{m}_c^2$
(IV′) $\Delta p_1 = r_1 \cdot \dot{m}_1^2$
(V′) $\Delta p_2 = r_2 \cdot \dot{m}_2^2$

Substituting equations III′–V′, transposed for mass flow $\dot{m}$, in formula I', the result is as follows:

(VI′) $\sqrt{\frac{\Delta p_c}{r_c}} = \sqrt{\frac{\Delta p_1}{r_1}} + \sqrt{\frac{\Delta p_2}{r_2}}$

Considering condition (II′), the following is obtained:

(VII′) $\sqrt{\frac{\Delta p_c}{r_c}} = \sqrt{\frac{\Delta p_c}{r_1}} + \sqrt{\frac{\Delta p_c}{r_2}}$

The equation can be simplified by dividing the term under the roots by Δp_c. The following relation is then obtained:

(VIII′) $\sqrt{\frac{1}{r_c}} = \sqrt{\frac{1}{r_1}} + \sqrt{\frac{1}{r_2}}$

This gives:

IX′) $r_c = \frac{1}{\left(\sqrt{\frac{1}{r_1}} + \sqrt{\frac{1}{r_2}}\right)^2} = \frac{1}{\left(\frac{1}{\sqrt{r_1}} + \frac{1}{\sqrt{r_2}}\right)^2}$

In the case of the ith number of resistances, the following equation is obtained:

(X′) $r_c = \frac{1}{\left(\frac{1}{\sqrt{r_1}} + \frac{1}{\sqrt{r_2}} + \ldots + \frac{1}{\sqrt{r_i}}\right)^2}$

This means that if hydraulic resistances are connected in parallel, the form of the formula for equivalent resistance is qualitatively different compared to the electric circuit. Analysing formula VIII′, it can be seen that the difference lies only in the root, under which are the inverses of individual resistances, whereas in electric circuits no such root occurs. The cause is simple—in a hydraulic system the pressure drop is square-dependent on the medium mass flow and therefore a change in hydraulic resistance does not have a linear effect on changes in the mass flow (which is the case in an electric circuit, where a change in electrical resistance involves a change in the flowing current), but affects it with the square root.

Figure 3.9c presents the creation of the resultant hydraulic characteristic if two hydraulic resistances are connected in parallel. At a given total pressure/pressure loss value, the mass flows are summed up for individual curves corresponding to individual hydraulic resistances. The slope of the resultant curve is therefore smaller. Like previously, this does not mean that the value of the characteristic exponent is changed. The only value that changes is the base of the power of the function describing the characteristic, which is represented by total equivalent resistance r_c.

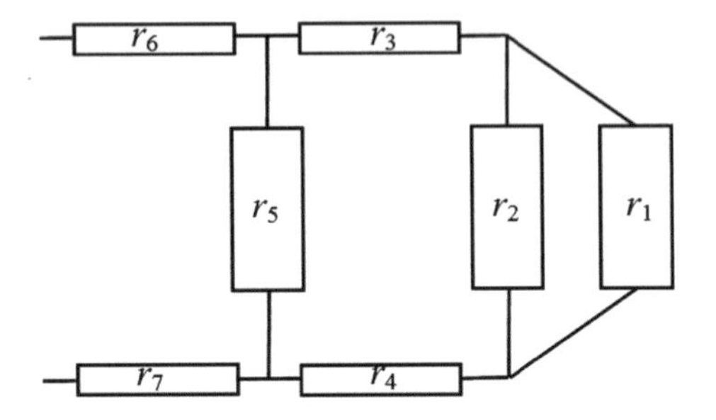

Fig. 3.10 Example diagram of the hydraulic system

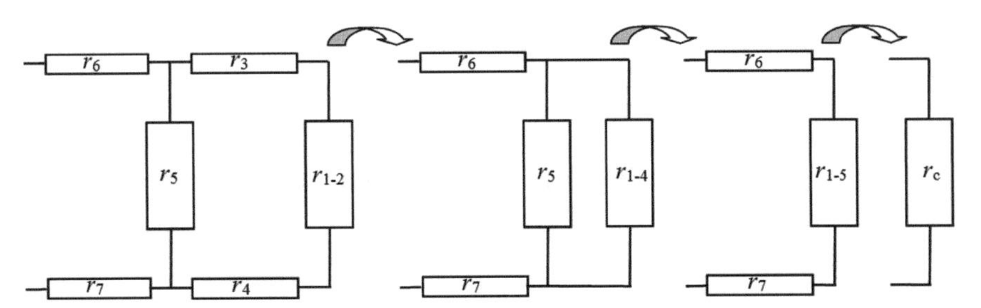

Fig. 3.11 "Curling up" of the hydraulic system diagram

3.2.1.3 Complex Combinations of Hydraulic Resistances

In every heating installation individual hydraulic resistances representing individual components of the network (pipes, valves, radiators, etc.) are connected to each other in series, in parallel or they make up a series-parallel combination. Equivalent hydraulic resistance should then be determined according to the method of finding resistance in the parallel and in the series combination, "curling up" the diagram accordingly. An example diagram of such a system is presented in Fig. 3.10. Although insignificant from the point of view of the equivalent system hydraulic analysis, a distinction is introduced: the small rectangles represent pipes and the big ones—heat receivers.

In this system resistances r_1 r_2 are connected in parallel. The r_{1-2} resultant resistance is then connected in series to resistances r_3 and r_4. Next, the branch r_{1-4} equivalent resistance is connected to resistance r_5 in parallel. The r_{1-5} resultant resistance is then connected in series to resistances r_6 and r_7, creating total resistance r_c. The diagram can of course be "curled up" in this way, i.e. step by step, replacing subsequent groups of elements with a single equivalent element which is then connected to another, creating another equivalent element, etc. The "curling" is in this case performed starting from the farthest resistance (in relation to the system inlet side—the pressure source), as presented in Fig. 3.11.

However, in less extensive systems, and having appropriate skills, the relation describing the total equivalent resistance can be written without intermediate steps.

It is then favourable to apply the curling procedure starting from the system inlet. In this case, the final relation obtained for the diagram above has the following form:

$$r_c = r_6 + r_7 + \frac{1}{\left(\frac{1}{\sqrt{r_5}} + \frac{1}{\sqrt{r_3 + r_4 + \frac{1}{\left(\frac{1}{\sqrt{r_1}} + \frac{1}{\sqrt{r_2}}\right)^2}}}\right)^2}$$

The presented method of "curling up" diagrams can only be used if individual hydraulic resistances are connected in parallel or in series. In order to establish that the former is the case, the pressures on two analysed receivers that are to be replaced with a single equivalent pressure have to be identical. In a serial combination, it is the medium mass flow that has to be the same. In some situations, however, such a distinction is impossible to make and therefore the presented formulae cannot be used for systems connected in series and in parallel. It is typical then to connect elements in what is referred to as the *Tichelmann* system, which is a variant of a parallel two-string structure where the distribution lengths are the same for all branches, irrespective of the distance from the location of the pressure source. The system is sometimes used as an alternative to the "ordinary" parallel structure, where subsequent elements, being farther and farther away from the supply source, operate in longer and longer branches and the pressure on them is falling systematically. Making these lengths equal, which is the aim of the *Tichelmann* system application, makes it possible to bring pressure losses to a level similar for each branch and to ensure a similar pressure value on every receiver. Theoretically, there is then no need to use additional balancing valves lowering pressure on earlier receivers compared to the last one, for which the pump pressure is selected to make up for pressure losses on the longest pipe sections. In practice, however, equal pressure values on receivers are not achieved because pressure losses arise not only in pipes but also on local obstacles (here: on three-way pipes). And the coefficients of local pressure losses of these elements are different depending on whether the situation is the medium mixing (inflow) or distribution (outflow) [1–3]. Example diagrams of the typical installation and of the *Tichelmann* system are presented in Fig. 3.12a, b, respectively.

It can be seen that in the *Tichelmann* system the medium flowing through each receiver flows through the same number of hydraulic resistances representing given sections of the pipework. But in this case the diagram cannot be "curled up". It is true that resistances r_1 and r_3, for example, are connected in series to the rest of the system, but it is impossible to translate this part into equivalent resistance using the relations valid for in-series or parallel combinations. At first sight, resistances r_7 r_8 and r_6 are connected in series and they are then connected to resistance r_5 in parallel. But this is not so. In order to sum up the resistances of resistors $r_{7,}$ r_8 and r_6 according to the principle valid in the in-series combination, the medium mass flow in the receivers has to be identical. Although the values of the mass flow through receivers r_7 and r_8 are identical and the receivers resistances can be added to each other, this is not the mass flow through receiver r_6, because this quantity is

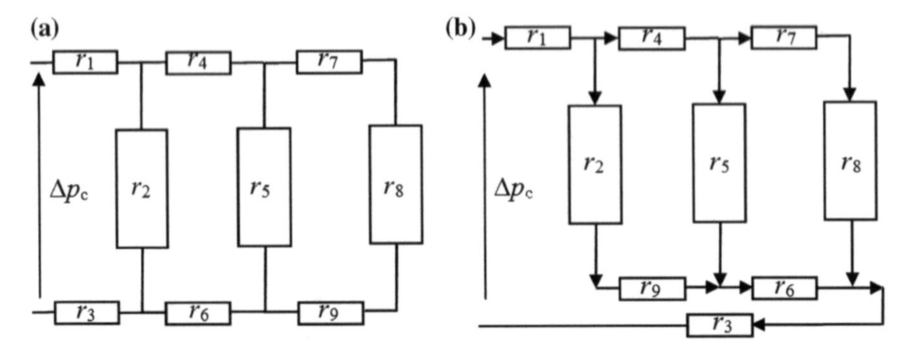

Fig. 3.12 Hydraulic system diagram: **a** classical configuration, **b** *Tichelmann* system

made up by the mass flows through receivers r_9 (and consequently r_2) and r_5 (and consequently r_4). The situation is the same for the system subsequent parts.

Apart from the *Tichelmann* system of carrying pipes, the impossibility of determining equivalent hydraulic resistance by means of "curling up" the diagram is typical of ring networks or systems with more than one pressure source, i.e. with more than one point of the medium supply. Such situations are quite common in district heating networks but rather scarce in typical heating installations. Different approaches can be used to solve such systems. As previously mentioned, the system of hydraulic resistances corresponds to the circuit with electrical resistances. Such problems can therefore be solved using the matrix method of mesh currents (here: the medium mass flows), based on Kirchoff's second law, or the method of nodal potentials (here: pressures), based on Kirchoff's first law. Both methods enable determination of the medium mass flow in individual branches, and, thereby, pressure values in individual points. Iterative methods are also used, where certain initial values of the medium mass flow and directions of the medium flow in individual branches are assumed, and the system is balanced step by step. One of them is the *Hardy Cross* method, described e.g. in [6, 9, 12, 16].

The presented approach, defining and using a constant parameter that characterizes the pipework hydraulically, enables a simple and analytical perspective and analysis of phenomena taking place in a given circuit or a part or element thereof. It makes it possible to calculate the medium spreads, changes in the medium flow values at changes in the medium pressure values and distribution, etc. There are direct analogies here with the electric circuit with its resistances, where differential pressure is the equivalent of the electric potential (voltage) difference, the medium mass flow is the equivalent of the flowing current and hydraulic resistance corresponds to electrical resistance. The analysis of hydraulic systems then proceeds like in the case of electric circuits with resistors and a source of voltage. In fact, the only difference is that in an electric circuit the voltage drop is linearly dependent on the current, whereas in a hydraulic system, under the simplifications described earlier, the pressure drop depends on the square of the medium mass flow. From the practical point of view, this approach is usually sufficient. Taking full account of the phenomena described previously and related to the variability of the coefficients of linear pressure losses λ and local resistances ζ as a function of the medium mass flow and velocity, it turns out that the effects of the circuit quantitative regulation are described by confounded equations, and such problems cannot be solved using an analytical approach. This is illustrated in [10], for example.

The approach using hydraulic resistance in the analysis of the operation of hydraulic systems is less common than the approach using electrical resistance in electric circuits. This is due to a few reasons. The basic one is that hydraulic resistance can be related to different quantities, including different multiples and submultiples thereof. It can be related to the mass flow, expressed in kg/s, kg/h, etc., the volume mass flow, expressed in m^3/s, m^3/h, dm^3/s, dm^3/h, etc., or the medium flow velocity. Considering also that the mass flow, at a given volume flow value, depends on the medium density, the element hydraulic resistance does not have a single value. It differs depending on whether it is related to the mass flow of water (in the liquid or in the gaseous state), oil, etc., despite the fact that it concerns the same element and the element performance for the same value of the volume flow. In order to unify the analysis, for each case the correct hydraulic resistance value should be determined that corresponds to a given medium density, or hydraulic resistance should be used in relation to the volume flow only. The units in which the parameter is to be expressed should also be specified and, if needed during the analysis, they should each time be converted appropriately. Moreover, as previously described, hydraulic resistance is not an absolutely constant parameter. The effect of all these factors is that the unit unification by means of the ohm, representing electrical resistance, is troublesome.

3.3 Regulation Valve Flow Factor and Control Characteristic

The valve control characteristic, whether the closing or the throttling one, can be plotted using a few couples of parameters. One of them is the opening degree, which can be measured for example by the pre-setting value, the plug lift over the valve seat, the turn of the valve knob, the control shutter opening degree, etc., and the other—the values of the medium mass flow, hydraulic resistance, the coefficient of local pressure losses or the factor flow. In the case of valves, and regulation valves in particular, the basic flow capacity parameter commonly applied in practice is the flow factor (k_v). For this reason, this quantity is usually used to present control characteristics. Moreover, the characteristic may operate on absolute or relative values. Absolute values relate directly to measured values, taking account of relevant units; relative values relate the value measured for the valve given opening degree to the maximum measured value, e.g. for the maximum opening degree. Whichever the case, the valve flow capacity has to be determined for a given opening degree of the valve.

In fluid mechanics, pressure losses on valves are usually found by means of computational relations like for a local obstacle, i.e.:

$$\Delta p_z = \zeta \cdot \frac{1}{2} \cdot \rho \cdot w^2. \tag{3.48}$$

But the process of the regulation valve sizing and selection for heating installations is based on a different approach. The quantity that de facto has a decisive impact on the valve flow capacity is the medium flow surface area A. It can be calculated by transforming relation (3.48) to:

$$\Delta p_z = \zeta \cdot \frac{1}{2} \cdot \rho \cdot \frac{\dot{V}^2}{A^2}. \tag{3.49}$$

This gives the following notation:

$$A = \dot{V} \cdot \sqrt{\frac{\zeta \cdot \rho}{2 \cdot \Delta p_z}}. \tag{3.50}$$

Because coefficient ζ depends on the valve design and the medium flow surface area A, the relation has a confounded form. It can therefore be solved by means of iterative methods. But iteration is not used in practical valve sizing. The starting point is a relation resulting from the transformation of the *Bernoulli* theorem describing the flow of an incompressible fluid, which gives the following equation:

$$\dot{V} = \alpha \cdot A \cdot \sqrt{\frac{2\Delta p_z}{\rho}}. \tag{3.51}$$

If the product in the formula of the constant quantities and those directly related to the valve is expressed as:

$$k_v = \alpha \cdot A \cdot \sqrt{2}, \tag{3.52}$$

Equation (3.51) can be written as follows:

$$\dot{V} = k_v \cdot \sqrt{\frac{\Delta p_z}{\rho}}, \tag{3.53}$$

where k_v is the flow factor. Analysing formula (3.53), it can be seen that the parameter written in this way has the surface area unit (m^2). This is the equivalent surface area of the cross-section of the medium jet. If $\Delta p_z/\rho = 1$, the flow factor value is equal in numbers to the medium mass flow.

Transforming Eq. (3.51) for the sought value A, the following is obtained:

$$A = \frac{\dot{V}}{\alpha} \cdot \sqrt{\frac{\rho}{2 \cdot \Delta p_z}}. \tag{3.54}$$

Like Eq. (3.50), the formula has a confounded form. No proportional relation exists between quantities A and $\dot{V}$ (because the valve discharge coefficient α depends on surface area A). Therefore, there is a different value of coefficient α corresponding to each value of parameter A. This is one of the reasons for the introduction of quantity k_v, referred to as the *flow factor* and including in itself variable α, as indicated by formula (3.52).

The flow factor is the measure of the valve flow capacity. It is defined as the value of the reference medium (water with the temperature of 5–40 °C [20]) volume flow in m^3/h through a valve, assuming a pressure drop on the valve of 1 bar for any position of the throttling/closing element. For a fully open valve (100% opening degree), the flow factor is marked as k_{v100}. The parameter mean value determined for the tested series of valves is marked as k_{vs}.

The *flow factor* concept was introduced in the USA in 1944. Originally, it was denoted as c_v, a symbol still used today. But expressed in the US and not the SI units (gallon/min), it could not be adopted directly in countries with a different system of weights and measures. The flow factor was introduced into the SI system in 1957 by a German—K.H. Früh [7], who marked it as k_v (*Koeffizient—coefficient* and *Ventil—valve*). The definition differed from the present one in the range of units. The flow was defined in l/min, and the pressure drop on the valve—in *at* (technical atmosphere).

Marking density for the reference medium as ρ_0, the pressure loss on the valve as Δp_0 and taking account of the definition above, relation (3.51) can be written in the following form:

$$k_{v,z} = \alpha \cdot A \cdot \sqrt{\frac{2\Delta p_0}{\rho_0}}. \tag{3.55}$$

The measure of the flow factor written in this way is the volume flow rate expressed in m^3/s. Dividing (3.51) by (3.55), the following is obtained:

$$\frac{\dot{V}}{k_{v,z}} = \sqrt{\frac{\rho_0 \cdot \Delta p_z}{\rho \cdot \Delta p_0}}. \tag{3.56}$$

If the medium for which the pressure loss on the valve is measured is water, i.e. the reference medium for which the flow factor is defined, then $\rho_0 = \rho$ and $\Delta p_0 = 1$. In this case, after relevant transformations and assuming that water density is constant, the result is:

$$k_{v,z} = \frac{\dot{V}}{\sqrt{\Delta p_z}}. \tag{3.57}$$

And this is the relation used in practice. Although the measure of the quantity written in this way is m^3/(h · bar$^{1/2}$), the generally accepted unit is m^3/h.

It follows from the analysis of formula (3.49) that all the quantities on the formula right side, except $\dot{V}$, are constant and/or set for a given opening degree of the valve. In their entirety then, they may be treated as a certain characteristic parameter, i.e. the element hydraulic resistance. The change in the pressure loss on the valve is then proportional to the square of the change in the medium volume flow. This assumption is made in practice, where the valve nomograms provided by manufacturers show lines of a square dependence of the pressure loss on the volume flow (in the logarithmic scale the lines are straight). It may therefore be written [cf. formula (3.49)] that:

$$\Delta p_{\mathrm{Z}} = r_{\mathrm{Z}} \cdot \dot{V}^2, \tag{3.58}$$

where r_{Z} is the valve hydraulic resistance related to the medium volume flow. Similar relations can be written relating hydraulic resistance to the medium mass flow or to the flow velocity because for water in the liquid state (negligible compressibility) the quantities are proportional to each other.

$$\Delta p_{\mathrm{Z}} = r_{\mathrm{Z}} \cdot \dot{m}^2, \tag{3.59}$$

$$\Delta p_{\mathrm{Z}} = r_{\mathrm{Z}} \cdot w^2, \tag{3.60}$$

Naturally then, the hydraulic resistance value and the unit thereof have to be related to a given flow parameter.

Using this notation, it is possible to find the relation between the valve flow factor k_{v} at the valve given opening degree and the valve hydraulic resistance r_{Z}. In such a case, however, one parameter should be expressed by means of quantities used in the other. Taking the flow factor dimension as the starting point and expressing the medium volume flow in m^3/h, the pressure loss in bar and, as a result, hydraulic resistance in (bar $\cdot$ h^2)/m^6, the following equation is correct:

$$\frac{\dot{V}^2}{k_{\mathrm{v,Z}}^2} = r_{\mathrm{Z,x}} \cdot \dot{V}^2 \Rightarrow r_{Z,x} = \frac{1}{k_{\mathrm{v,Z}}^2}. \tag{3.61}$$

Hydraulic resistance is thus inversely proportional to the square of the flow factor. The relation describing the flow factor as a function of hydraulic resistance can be written in the same manner:

$$k_{v,x} = \frac{1}{\sqrt{r_{\mathrm{Z,Z}}}}. \tag{3.62}$$

For computational purposes, it is possible to determine the relation between the valve flow factor and the coefficient of local pressure losses. Using the equation describing the medium mass flow through the valve port, the following formula is obtained:

$$\dot{m} = \rho \cdot A \cdot w, \tag{3.63}$$

from which it follows that:

$$w = \frac{4 \cdot \dot{m}}{\rho \cdot \pi \cdot d^2}. \tag{3.64}$$

Considering that pressure losses on a local obstacle (the valve in this case) are defined by relation (3.48), after relevant transformations and substitution in the equation above, the result is:

$$\sqrt{\frac{2\Delta p_Z}{\zeta \cdot \rho}} = \frac{4 \cdot \dot{m}}{\rho \cdot \pi \cdot d^2}, \tag{3.65}$$

which means that:

$$\zeta = \frac{\Delta p_Z \cdot \rho \cdot \pi^2}{8} \cdot \left(\frac{d^2}{\dot{m}}\right)^2 = 1.234 \cdot \Delta p_Z \cdot \rho \cdot \left(\frac{d^2}{\dot{m}}\right)^2, \tag{3.66}$$

where all quantities should be substituted in the SI system units. However, it is more convenient to use the inner diameter of the valve connector pipe and the water mass flow expressed in mm and kg/h, respectively. It can therefore be written that:

$$\zeta = 1.234 \cdot \left(0.001^2 \cdot 3600\right)^2 \cdot \Delta p_Z \cdot \rho \cdot \left(\frac{d^2}{\dot{m}}\right)^2 = 16 \cdot 10^{-6} \cdot \Delta p_Z \cdot \rho \cdot \left(\frac{d^2}{\dot{m}}\right)^2. \tag{3.67}$$

Assuming that the density of water with the temperature of 20 °C is at the level of $\rho = 998.2$ kg/m^3, the equation presented above takes the following form:

$$\zeta = 15.9 \cdot 10^{-3} \cdot \Delta p_Z \cdot \left(\frac{d^2}{\dot{m}}\right)^2. \tag{3.68}$$

The formula relating the flow factor to the coefficient of local resistances can be written using the transformed version of formula (3.57), noting that the pressure drop on the valve Δp_z should then be expressed in Pa:

$$k_v = \frac{\dot{m}}{\rho} \cdot \frac{1}{\sqrt{10^5 \Delta p_Z}} \Rightarrow \dot{m} = k_V \cdot \rho \cdot \sqrt{10^5 \Delta p_Z}. \tag{3.69}$$

Substituting the above in formula (3.68), at the water density value assumed earlier, the following equation is obtained:

$$\zeta = 15.9 \cdot 10^{-3} \cdot \Delta p_Z \cdot \left(\frac{d^2}{k_V \cdot \rho \cdot \sqrt{10^5 \Delta p_Z}}\right)^2 = 1.6 \times 10^{-3} \left(\frac{d^2}{k_V}\right)^2. \tag{3.70}$$

The sizing of radiators and of the heating installation is performed for design conditions, i.e. for the highest thermal and hydraulic loads resulting therefrom. Throttling is therefore quantitative regulation consisting in a reduction in the medium mass flow compared to the design value. This involves a drop in the radiator heat output. The dependence, however, is not linear [17, 18], which in fact is the desired state considering the accuracy and stability of the regulation process in its entire range. The requirement is related mainly to static characteristics of the radiator thermostatic heads (as presented in Fig. 2.14), which are elements of the current local regulation of the heat output. Due to that, the response in the form of a change in the position of the thermostatic head follower (h), and as a result—of the regulation valve plug, will be almost proportional to the input function in the form of a change in the sensor temperature t_c, which corresponds to the room temperature t_i, as the measured value. The radiator (static) thermal characteristics are not linear—the curves illustrating them are close to logarithmic and, for a given radiator, depend additionally on the water cooling degree [17, 18]. As a result, if such a regulated object is connected to a regulating element with a linear characteristic, the system response will be non-linear, i.e. the regulation quality and accuracy will deteriorate. Due to that, it is necessary to use such executive elements (plugs and flow channels in regulation valves) that will correct the radiator thermal characteristic non-linearity to obtain a linear final regulation curve. In the radiator regulation technique theory, the most common characteristics of the valve are linear and equal-percentage ones (also referred to as exponential or logarithmic). In the case of a linear (proportional) characteristic, the following can be written:

$$\frac{\mathrm{d}\dot{V}_{\mathrm{x}}}{\mathrm{d}h_{\mathrm{x}}} = \mathrm{const}, \tag{3.71}$$

which means that for each and every pre-set and elementary relative increment in the opening degree h_x there is a corresponding relative increment in the medium volume flow $\dot{V}_x$. In other words, an increment in the volume flow is linearly dependent on the rise in the valve opening degree. A double increase in the valve opening degree involves a twice higher volume flow of the medium. In practice, for flows in the turbulent range, the increment in the volume flow rate is proportional to the increment in the medium flow cross-section surface area A_x, which can be observed in Fig. 3.14. Transformation and integration of the equation presented above, with respect to the flow factor definition, give the following relation:

$$k_{\mathrm{vx}} = \mathrm{const} \cdot h_{\mathrm{x}}. \tag{3.72}$$

The following relation is then valid for a fully open valve:

$$k_{v100} = \mathrm{const} \cdot h_{100}. \tag{3.73}$$

Dividing the two equations by each other, the following is obtained:

$$\frac{k_{\mathrm{vx}}}{k_{\mathrm{v100}}} = \frac{h_{\mathrm{x}}}{h_{100}}. \tag{3.74}$$

In the case of an equal-percentage characteristic, the following can be written:

$$\frac{\mathrm{d}(\dot{V}_{\mathrm{x}}/\dot{V}_{100})}{\mathrm{d}(h_{\mathrm{x}}/h_{100}) \cdot \dot{V}_{\mathrm{x}}/\dot{V}_{100}} = c. \tag{3.75}$$

This means that for each and every pre-set and elementary absolute increment in the opening degree h_x there is, in the entire range of the characteristic, a corresponding constant relative increment in the medium volume flow $\dot{V}_{\mathrm{x}}$ totalling $c\%$. Hence the name of the characteristic. Rearranging Eq. (3.75), the following can be written:

$$\frac{\mathrm{d}(\dot{V}_{\mathrm{x}}/\dot{V}_{100})}{\dot{V}_{\mathrm{x}}/\dot{V}_{100}} = c \cdot \mathrm{d}(h_{\mathrm{x}}/h_{100}). \tag{3.76}$$

Integration of the above gives:

$$\ln(\dot{V}_{\mathrm{x}}/\dot{V}_{100}) = c \cdot h_{\mathrm{x}}/h_{100} + \mathrm{Const}. \tag{3.77}$$

Integration constant Const can be determined from the following boundary condition:

$$\text{for}\, h_{\mathrm{x}}/h_{100} = 1,\ \dot{V}_{\mathrm{x}}/\dot{V}_{100} = 1. \tag{3.78}$$

Substituting formula (3.78) in (3.77), the following is obtained:

$$\ln 1 = c + \mathrm{Const} \Rightarrow c = -\mathrm{Const}.$$

Taking the condition into account, it can be written that:

$$\ln(\dot{V}_{\mathrm{x}}/\dot{V}_{100}) = c \cdot (h_{\mathrm{x}}/h_{100} - 1). \tag{3.79}$$

Taking account of the logarithm and the flow factor definitions, the following expression is obtained:

$$k_{\mathrm{vx}}/k_{\mathrm{v100}} = \exp[c \cdot (h_{\mathrm{x}}/h_{100} - 1)]. \tag{3.80}$$

The characteristics plotted based on the relations derived above are presented in Fig. 3.13. As it can be noticed, the equal-percentage characteristic, irrespective of parameter c, is unable to ensure that the flow is cut off completely because at the zero-opening degree it is still characterized by a certain non-zero value of the medium volume flow referred to as leakage. For this reason, the valve flow channel cannot be calculated based on this curve type for the plug lift entire range (opening degree) of

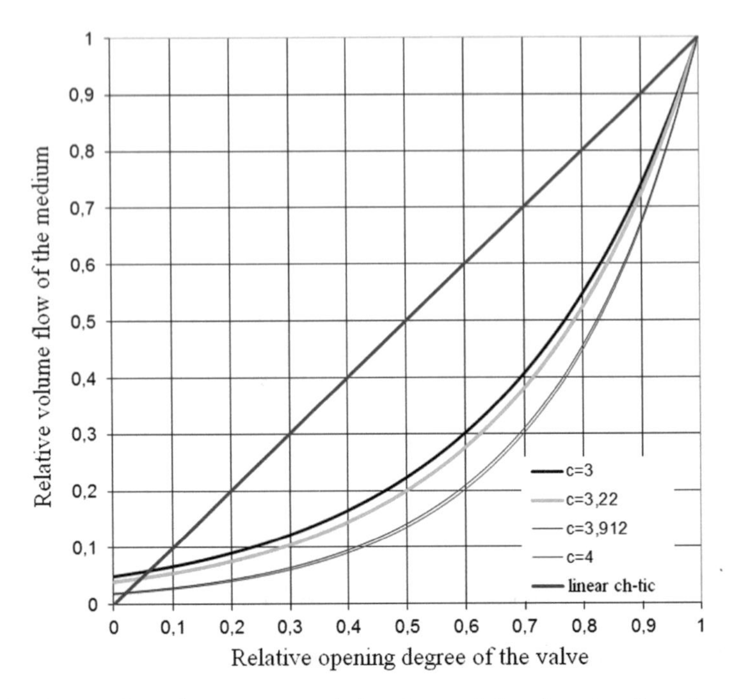

Fig. 3.13 Linear and equal-percentage regulation characteristics of a valve (for a few different values of coefficient c)

h_{100}. Instead, the channel should be shaped according to a different function ensuring that the flow can be cut off completely.

For $h_x/h_{100} = 0\%$, the flow factor is defined as $k_{vx} = k_{v0}$. In this case, relation (3.80) gives:

$$\ln(k_{v100}/k_{v0}) = c. \tag{3.81}$$

The maximum-to-minimum flow factor ratio is referred to as the theoretical regulation ratio [22, 23] or the amplification factor (k) [13], according to the following relation:

$$k_{v100}/k_{v0} = k_z. \tag{3.82}$$

The higher the value of k, the bigger the curvature of the characteristic and the smaller the leakage. For example, for $k = 50$ and $k = 25$, the respective values are $k_{v0}/k_{v100} = 2\%$, $c = 3.91$ and $k_{v0}/k_{v100} = 4\%$, $c = 3.22$.

Using the formulae presented above, Table 3.1 illustrates, for parameter $c = 50$, the meaning of the principle of equal-percentage increments. The first few initial steps are presented only to keep the size of the table within reasonable limits. For an elementary, infinitely small increment $d(h_x/h_{100})$, the value of c is obtained for

Table 3.1 Relative increments in the flow factor values for different assumed increments in the valve opening degree

$k=50$, $c=3.912$ (for $d(h_x/h_{100}) \rightarrow 0$)

h_x/h_{100}	k_{vx}/k_{v100}	$c\%$	h_x/h_{100}	k_{vx}/k_{v100}	$c\%$	h_x/h_{100}	k_{vx}/k_{v100}	$c\%$
0	0.02	–	0	0.02	–	0	0.02	–
0.1	0.0295751	47.87	0.01	0.020798	3.989	0.001	0.0200784	3.919
0.2	0.0437345	47.87	0.02	0.0216276	3.989	0.002	0.0201571	3.919
0.3	0.0646727	47.87	0.03	0.0224905	3.989	0.003	0.0202361	3.919
0.4	0.0956352	47.87	0.04	0.0233877	3.989	0.004	0.0203154	3.919
0.5	0.1414213	47.87	0.05	0.0243208	3.989	0.005	0.0203950	3.919
0.6	0.2091279	47.87	0.06	0.0252911	3.989	0.006	0.020475	3.919
0.7	0.3092495	47.87	0.07	0.0263001	3.989	0.007	0.0205552	3.919
0.8	0.4573050	47.87	0.08	0.0273494	3.989	0.008	0.0206358	3.919
0.9	0.6762433	47.87	0.09	0.0284405	3.989	0.009	0.0207167	3.919
1	1	47.87	0.1	0.0295751	3.989	0.01	0.0207979	3.919
			0.11	0.0307551	3.989	0.011	0.0208794	3.919
			0.12	0.031982	3.989	0.012	0.0209613	3.919
			0.13	0.033258	3.989	0.013	0.0210434	3.919
			0.14	0.0345848	3.989	0.014	0.0211259	3.919
			0.15	0.0359646	3.989	0.015	0.0212087	3.919

subsequently set opening degrees of the valve. The higher the ratios, the bigger the deviation. Nonetheless, though, the same relative increments in the volume flow, expressed by the ratio between the flow factors in subsequent rows and converted to $c\%$, correspond to constant absolute increments in the opening degree.

The derived expressions describe the relative volume flow and the relative flow factor depending on the relative opening degree of the valve for the case where the medium total pressure loss falls on the valve regulating element which constitutes the only hydraulic resistance in the system. In this case volume flows are equal in numbers to flow factors, which makes it possible to use these quantities interchangeably, as presented in the derived equations. However, the situation where the medium total pressure loss falls on the valve regulating element as the only hydraulic resistance never occurs in the system. The regulating element is installed in the valve body with a specific hydraulic resistance. Moreover, in the case of double-regulation valves, there is usually an additional regulating-throttling element also characterized by a certain hydraulic resistance changing as a function of the pre-setting. Furthermore, the valve operates within the pipework with another specific value of hydraulic resistance which may vary. In both cases, the additional hydraulic resistance, which is not subject to current regulation, is added to the regulating element hydraulic resistance value, and this involves deformation of the original regulation characteristic. The deformation always proceeds "upwards", i.e. for the valve given relative opening degree, a bigger relative value of the medium volume flow is obtained. This impact is described by the concepts of *inner authority* and *outer authority*, and as a resultant factor—*total authority*. The appropriate curve is selected depending on the regulated object characteristic, the impact of the elements that deform the two characteristics and the shape of the expected final regulation curve.

Considering the above, three situations can be distinguished that correspond to the valve given regulation characteristic:

- the valve regulating element original regulation characteristic, as an illustration of changes in the element flow factor $k_{v,reg,x}/k_{v,reg,100}$ depending on the opening degree h_x/h_{100}. This is the original regulation characteristic. It is affected only by the shape of the interface between the plug (or another regulating element) and the valve seat.
- the valve internal regulation characteristic, as an illustration of changes in the entire valve flow factor $k_{v,x}/k_{v,100}$ depending on the regulating element opening degree h_x/h_{100}. This is the original regulation characteristic deformed due to hydraulic resistance of the valve body and of the throttling element (if any) used to select the pre-setting value.
- the valve operation regulation characteristic, as an illustration of changes in the medium volume flow $\dot{V}_x/\dot{V}_{100}$ through the valve depending on the regulating element opening degree h_x/h_{100}. This is the original regulation characteristic deformed due to hydraulic resistance of the valve body, of the throttling element (if any) and of the network of connected pipes.

It is impossible to obtain a specific constant shape of the regulation characteristic of a valve operating in the pipework. Even if the plug-valve seat interface is properly shaped to obtain the original characteristic after its deformation due to the impact of the valve body, the curve is still subject to variable deformation due to the effect of the throttling element depending on the hydraulic resistance set on it. If the device is a single-regulation valve without the possibility of setting initial throttling, deformation arises due to the impact of the network of connected pipes. These are some of the reasons why manufacturers so rarely design valves for specific shapes of regulation characteristics. The valve regulating elements are usually designed without any specific assumptions concerning their characteristics. The factors which are taken into consideration concern assembly- or structure-related issues. An extensive discussion of this problem and of the valve *authority* concept is presented in Chap. 4.

Considering the regulation quality, the factors of great significance are the value of and the variability in the ratio between the increment in the fluid flow cross-section surface area and the increment in the opening degree: $\Delta A_x/\Delta h_x$. The smaller the ratio and the less it varies, the more precisely the valve can be regulated. This is affected not only by the type of the regulation characteristic but also by the regulating element range of travel h_{max}. Its higher value naturally translates into higher accuracy of regulation. Analysing the equation describing the linear characteristic, it can be seen that in the lower range of the plug lift a change in the plug position causes bigger relative changes in the medium flow cross-section surface area and, thereby, in the medium volume flow, compared to the plug movement close to the upper limit position. This is due to the fact that the downside of the valve linear characteristic is too high and too low sensitivity in the lower and in the upper range, respectively,

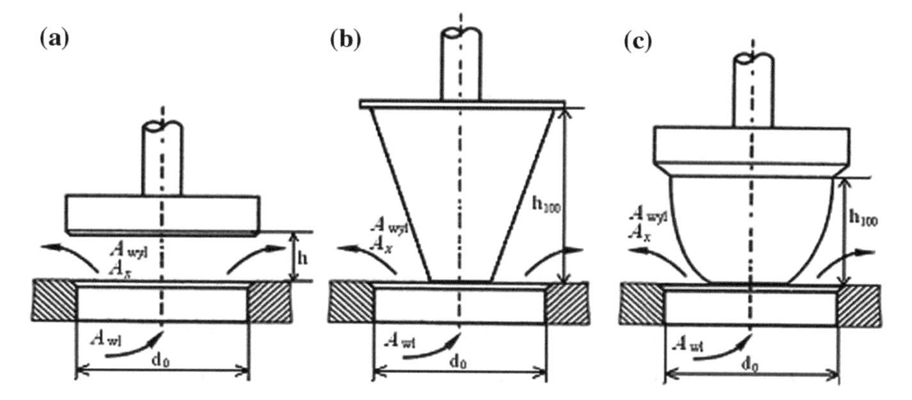

Fig. 3.14 Solutions of the valve regulating element with proportional and equal-percentage changeability in the medium free flow surface area as a function of the increment in the plug lift: **a** plug moving above the valve seat, **b** valve moving inside the valve seat, **c** plug shaped to obtain the original equal-percentage regulation characteristic

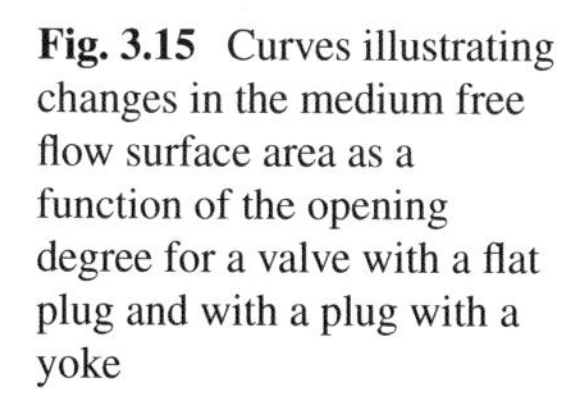

Fig. 3.15 Curves illustrating changes in the medium free flow surface area as a function of the opening degree for a valve with a flat plug and with a plug with a yoke

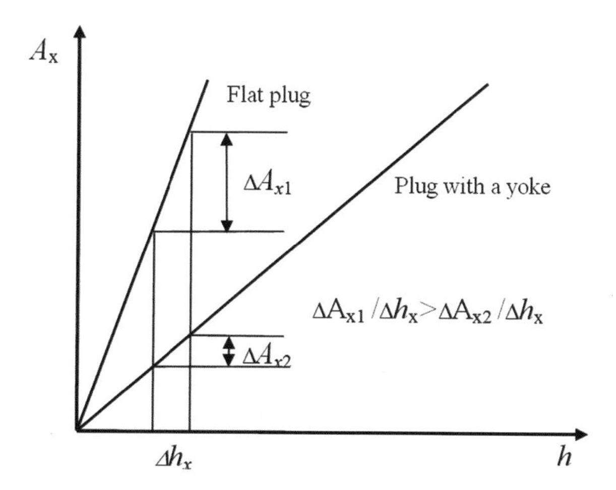

which in some conditions may result in unstable operation of the circuit regulated by such a valve. This occurs if the valve is controlled e.g. by thermostatic or electronic heads, or actuators, due to play in the mechanisms. A counter-measure in this case may be to design the valve—at a given maximum surface area of the medium flow cross-section A_{100}—for an appropriately bigger plug lift h_{max} and as a valve with a plug with a yoke or with a full plug moving in the valve seat, and not with a flat plug, as shown in Fig. 3.14a. Additionally, Fig. 3.14c presents a profile of the valve closing plug that guarantees the original equal-percentage regulation characteristic.

A comparison between example characteristics illustrating the two solutions mentioned as counter-measures is presented in Fig. 3.15.

The equal-percentage characteristic is free from the downside described above. It ensures constant changeability, irrespective of the range of the plug travel.

The presented issues related to regulation valve characteristics, as illustrations of changes in the medium volume flow depending on the valve opening degree, concern both one-way straight-through and two-way valves. In the former case, the images of changes in the medium volume flow upstream and downstream the valve are naturally identical because this is the same branch and the same regulation circuit. It is represented by the final operation regulation characteristic of a given regulating element of the valve. In the latter case, due to the division of the medium flow into two branches, i.e. two connector pipes serving two separate circuits, the image of the changes in the medium volume flow in each of them is different from the image at the valve inlet. This is so because (if the valve is not fully open or fully closed) the total volume flow is divided into two component volume flows and the value of each of them is smaller than the total volume flow. Moreover, the mutual relation between the volume flow values in these two branches may be different, i.e. a double increase in the volume flow in one of them for example does not have to mean a double drop in the other, and therefore, if summed up, the changes do not have to give the initial volume flow value if one of the branches is completely cut off and the medium flows only through the other. This will be the case if the final regulation characteristics in the valve both branches are different, i.e. if their original forms are different, and/or the deformation of the characteristics in the two branches is different under the influence of the network of connected pipes. The causes of such deformation are discussed in Chap. 4. Figure 3.16 presents a graphical illustration of the situation. If, for example, two regulation characteristics for two connector

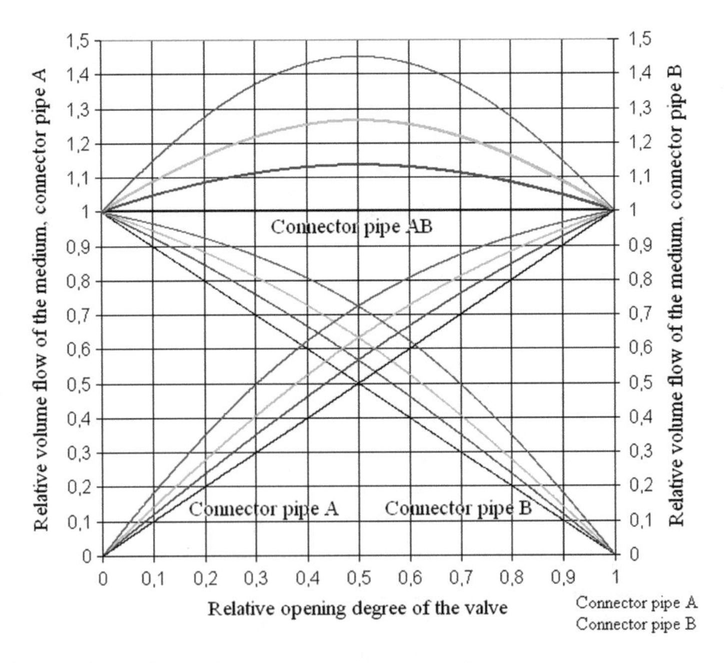

Fig. 3.16 Example regulation characteristics of two-way valves

pipes (A and B) are linear (black lines in the figure), the sum of the volume flows in them, in the collective connector pipe AB, is constant, for each opening degree of the valve, in every point of the characteristic. It is also equal to the volume flow in one of the connector pipes if the flow is completely cut off in the other. If the regulation characteristics are not linear, the sum of the volume flows depending on the valve opening degree cannot be linear or constant, as illustrated in the figure by the coloured lines. In this case, these are also the curves illustrating the original linear characteristic (black lines), but deformed in the same manner in both branches due to what is referred to as *total authority* a_c discussed further below.

References

1. Amanowicz, Ł., Wojtkowiak J.: Badania eksperymentalne wpływu zmiany sposobu zasilania powietrznego gruntowego wymiennika ciepła typu rurowego na jego charakterystykę przepływową-część 1: Równomierność rozpływu (Experimental investigations concerning influence of supply system of ground earth-to-air pipe-type heat exchangers on its flow characteristics. Part 1. Flow uniformity). District Heating Heating Vent. **41**(6), 208–212 (2010)
2. Amanowicz, Ł., Wojtkowiak, J.: Badania eksperymentalne wpływu zmiany sposobu zasilania powietrznego gruntowego wymiennika ciepła typu rurowego na jego charakterystykę przepływową-część 2: Straty ciśnienia (Experimental investigations concerning influence of supply system of ground earth- to-air pipe-type heat exchanger on its flow characteristics. Part 2. Pressure losses). District Heating Heating Vent. **41**(7–8), 263–266 (2010)
3. Amanowicz, Ł., Wojtkowiak, J.: Straty ciśnienia w gruntowych powietrznych wielorurowych wymiennikach ciepła o kącie odejścia 45 stopni (Pressure losses in ground earth-to-air multipipe heat exchangers with connection angle of 45°). District Heating Heating Vent. **41**(12), 451–454 (2010)
4. Catalogue information of Herz (on-line version, http://www.herz.com.pl)
5. Catalogue information of Kisan (on-line version, http://www.kisan.pl)
6. Cross, H.: Analysis of Flow in Networks of Conduits or Conductors. Engineering Experiment Station, Bulletin No. 286 (1936)
7. Früh, K.H: Berechnung des durchflusses In regelventilen mit hilfe des "kv-Koeffizienten". Regelungstechnik **5**(9) (1957)
8. Górecki, A., Fedorczyk, Z., Płachta, J., Płuciennik, M., Rutkiewicz, A., Stefański, W., Zimmer, J.: Instalacje wodociągowe, ogrzewcze i gazowe, na paliwo gazowe, wykonane z rur miedzianych. Wytyczne stosowania i projektowania (Water, Heating and Gas Installations, for Gaseous Fuel, Made of Copper Pipes. Application and Design Guidelines). Biblioteka Polskiego Centrum Promocji Miedzi (wydanie elektroniczne/electronic issue) (2009)
9. Hodge, B.K., Taylor Robert, P.: Analysis and Design of Energy Systems, 3rd edn. Prentice Hall (1999)
10. Jeżowiecka-Kabsch, K., Szewczyk, H.: Mechanika płynów (Fluid Mechanics). Oficyna Wydawnicza Politechniki Wrocławskiej, Wrocław (2001)
11. Koczyk, H. (ed.): Ogrzewnictwo praktyczne. Projektowanie, montaż, eksploatacja (Practical Heating. Design, Installation, Operation). SYSTHERM SERWIS, Poznań (2015)
12. Kogut, K., Bytnar, K.: Obliczanie sieci gazowych (Calculation of Gas Pipeworks). Uczelniane Wydawnictwa Naukowo-Dydaktyczne AGH, Kraków (2007)
13. Kołodziejczyk, W.: Armatura regulacyjna w ogrzewaniach wodnych (Control Armature in Hydronic Heating Systems). Arkady, Warszawa (1985)
14. Kopp, L.: Die Wasserheizung. Springer, Berlin (1958)

15. Książyński, K.W.: Hydraulika. Zestawienie pojęć i wzorów stosowanych w budownictwie (available on line: https://suw.biblos.pk.edu.pl/downloadResource&mId=524581) Hydraulics. List of Concepts and Formulae Used in Construction, Kraków (2008)
16. Lopes, A.M.G.: Implementation of the hardy-cross method for the solution of piping networks. Comput. Appl. Eng. Educ. **12**(2), 117–125 (2004)
17. Muniak, D.: Grzejniki w wodnych instalacjach grzewczych. Dobór, konstrukcja i charakterystyki cieplne. Radiators in Hydronic Heating Installations. (Structure, Selection and Thermal Characteristics). WNT/PWN, Warszawa (2015)
18. Muniak, D.: Radiators in Hydronic Heating Installations. (Structure, Selection and Thermal Characteristics). Springer, Berlin (2017)
19. Polish Standard PN-76/M-34034: Rurociągi. Zasady obliczeń strat ciśnienia (Pipelines. Principles of Calculation of Pressure Losses)
20. Polish Standard PN-79/M-42063: Zawory regulacyjne. Normy i określenia (Regulation Valves. Standards and Definitions)
21. Rennels, D.C., Hudson, H.M.: Pipe Flow—A Practical and Comprehensive Guide. Wiley, New Jersey (2012)
22. Roos, H.: Hydraulik der Wasserheizung, wydanie 3. Oldenbourg Verlag GmbH, Monachium (1995)
23. Roos, H.: Zagadnienia hydrauliczne w instalacjach ogrzewania wodnego (Hydraulic Issues in Water Heating Installations). PNT CIBET, Warszawa (1997)
24. Turkiewicz, K., Warchoł, E.: Ogrzewnictwo: pomoce do ćwiczeń (Heating: Exercise Aids). Wydawnictwa Politechniki Śląskiej, Gliwice (1982)
25. Walden, H.: Mechanika Płynów (Fluid Mechanics). Wydawnictwa Politechniki Warszawskiej, Warszawa (1991)
26. Weinerowska, K. (ed.): Laboratorium z mechaniki płynów i hydrauliki (Laboratory of Fluid Mechanics and Hydraulics). Politechnika Gdańska, Gdańsk (2004)
27. Word Wide Web site: http://instalacje.gep.com.pl/

Chapter 4
Regulation Valve Co-operation with the Pipework

4.1 Regulation Valve *Authority*

This section presents a discussion of issues related to the valve co-operation with the pipework in which the valve is installed. The analysis of the effects of regulation by means of a valve, i.e. by changing the medium mass/volume flow due to a change in the valve setting (opening degree) is related not only to the valve regulation characteristic but also to an essential additional factor referred to as the valve *authority*.

4.1.1 Qualitative Description and Physical Sense of the Regulation Valve Authority

The term *authority* comes from the Latin and then French words *auctoritas*, *l'autorité*. It was adopted in literature and engineering practice to determine one of the basic parameters associated with the valve operation. Qualitatively speaking, the quantity informs about the share of the valve, which—by definition—is the fundamental regulating element, in the regulation process and about the effect of changes in the valve setting on the medium volume flow. The higher the impact compared to a valve operating without taking account of the unfavourable weakening effect of the pipework, the greater the valve authority.

Each regulation valve has a certain regulation characteristic. After the valve is installed in the pipework, the characteristic gets deformed due to hydraulic resistances. The deformation of the original (initial) regulation characteristic is defined by the valve authority. The greater the valve authority, the smaller the deformation of the characteristic and the smaller the change in the valve initial control properties. The valve impact on the regulation process rises with an increase in authority. The

D. P. Muniak, *Regulation Fixtures in Hydronic Heating Installations*, Studies in Systems, Decision and Control 187, https://doi.org/10.1007/978-3-030-03128-2_4

term *authority* comes directly from this particular feature—the capacity for exerting influence on the hydraulic system during the regulation process.

Nowadays, the term *valve authority* has several meanings and is rather imprecise. But it still relates to the valve regulation capacity. Such a definition is justified, even though it may raise some terminological concerns. How well a valve is able to perform regulation in a given circuit colloquially means how well it can cope with and govern the regulated (sub)system. In this light, *regulating authority*, or simply *authority* seems an appropriate term. The parameter has a substantial impact on the heating installation performance. The use of correct algorithms for the regulation valve sizing and selection and for the installation hydraulic balancing translates into the installation energy efficiency, minimizing operating costs and lengthening the element life.

The *valve authority* relates formally to valves regulating the medium mass/volume flow. The parameter can thus be ascribed to any valve because every valve offers the opportunity for the medium flow regulation. A change in the valve setting involves a change in the medium flow. However, some valve types, e.g. ball cut-off valves (different solutions thereof exist—cf. Sect. 2.3.1) are by principle not intended for the medium regulation process but for on/off operation—either cutting off the flow completely or enabling possibly lowest pressure losses if fully opened. As there is no need in this case for current regulation of the medium mass flow, they are not designed to comply with the regulation characteristic suitable for co-operation with radiators. They do not enable a precise setting, or read-out thereof, either. For this reason, *authority* is not defined for such valves and the concept is reserved exclusively for valves intended for current regulation, especially—for the radiator regulation valve.

A few basic tasks performed by the radiator regulation valves are specified in Sect. 2.1. They include as follows:

- balancing pressure drops with active pressure in the installation controlled circuits for a given value of the medium mass/volume flow,
- ensuring the design value of the medium mass/volume flow, according to the results of calculations of the process of the installation hydraulic and thermal balancing,
- ensuring the possibility of automatic current regulation, i.e. of the room temperature regulation by changing the radiator heat output.

It is generally accepted that in order to perform the last of the tasks mentioned above properly and to ensure the installation stable and smooth operation under a wide range of loads, the valve should be characterized by an appropriately high authority value included, according to generally accepted requirements [35, 25, 26, 12, 14, 10, 11], in the following interval:

$$\mathbf{0.3-0.7}$$

This recommended range of values can be found in almost every book or study on issues related to the heating installation regulation valves. It results from the assumed maximum permissible deformation of the valve initial regulation characteristic. Too low values, especially if the valve initial regulation characteristic is linear, will result in unstable operation of the thermostatic assembly, which in fact will operate as a non-continuous on-off device and not a continuous regulator. The effects of that can be as follows:

- non-decaying oscillations in the installation,
- hydraulic impacts in the installation,
- lower efficiency of the installation in the entire season,
- lower parameters of the heated rooms thermal comfort,
- faster wear of elements or equipment installed in the pipework,
- higher operating costs of the installation.

Exceeding the recommended upper limit will involve too high financial outlays in relation to the achieved results because in such a situation the pipes will have too big diameters, i.e. they will be oversized (cf. Formulae 4.1 and 4.2), or the required value of the circulating pump pressure will be higher. Consequently, investment costs will rise and the installation operating costs may get higher due to:

- high unit heat flux losses in pipes with big diameters, or
- increased demand for power needed to drive the pump.

Moreover, in the case of thermostatic valves, this may result in a considerable increase in the value of the medium mass flow through the radiator to a value exceeding the design limit determined for the nominal opening degree. Considering a number of factors and taking account of the economic calculation and optimization, the above-mentioned range of variability in the values of authority of the radiator regulation valves is generally accepted. But it turns out that this condition is often not satisfied in reality, which means that the essential arguments based on which it is formulated are ignored. This is due to two reasons. Firstly, the commonly adopted method of calculation of the authority of regulation valves—whether manual, thermostatic or balancing—is faulty. Secondly, the condition is practically impossible to meet if the parameter value is determined correctly.

The heating installation hydraulic issues were studied in Poland by e.g. *Wojciech Kołodziejczyk*, *Jan Stefan Mielnicki* and *Jerzy Kwapisz*. *Kołodziejczyk's* works focus on regulation devices and fixtures, the radiator valves in particular. They constitute a significant contribution to the development of the knowledge of the regulating capacity of valves, their hydraulic characteristics and effective co-operation with radiators. *Kołodziejczyk* discusses theoretical basics of the radiator regulation valves co-operation with heat receivers (radiators), putting forward new methods of the valve sizing and selection. His work devoted to regulation fixtures in heating installations [13] also presents the derivation process of formulae and relations needed to calculate the profile of the radiator valve plug.

Studies in this field have also been carried out by other European scientists. The most significant authors in recent years are Roos [28, 29] and *Pyrkov* and *Szaflik* [25, 26]. None of the studies by Mielnicki [18–20], Kołodziejczyk [13] or Roos [28, 29] defines *inner authority*. Still, the Polish researchers offer a qualitative description of the parameter. Both *Mielnicki* and *Kołodziejczyk* clearly indicate the causes of the deformation of regulation characteristics of radiator valves depending on their design, associating it rightly with the impact of what recently has been referred to as *inner authority*. Moreover, *Mielnicki* explained the phenomenon on the example of Polish-design valves, the manufacturers of which, keeping the phenomena in mind, shaped the structure of the valve regulating elements (plugs) accordingly (cf. description of the ideas by *Kwapisz* and *Pilatowicz* in Sect. 2.1.1). However, like in the case of *Roos*, the two researchers failed to offer a quantitative description in the form of relevant computational relations. Nor did they explain how this translates into the valve sizing process. The quantity is either ignored (*Roos*) or only partly taken into account in design considerations and relations (*Kołodziejczyk*), which makes the description incomplete. This results in considerable differences between the processes occurring in real heating installations and those described by means of the algorithms proposed by the scientists mentioned above. *Pyrkov* and *Szaflik* [28, 29] are in the group of those researchers who were the first to focus in their works on the discussion of the valve inner authority, both qualitatively and quantitatively, putting forward appropriate mathematical models. As proved by experimental verification, though, their concepts display some weaknesses, which means that, if applied, they will cause errors in calculation results.

As stated in Sect. 3.3, in reality the characteristics presented in Figs. 3.13 and 3.16 get deformed, which is due to two basic factors related to a change in the pressure distribution caused by additional hydraulic resistances produced in addition to the resistance of the valve regulating element. These are hydraulic resistances inside the valve, which do not take part in the flow current regulation, and hydraulic resistances of the circuit pipework. Figure 4.1 presents the charts illustrating the two types of regulation characteristics described in Sect. 3.3, after taking account of hydraulic resistances, which—as described—are then referred to as the valve *operating characteristics*. The curves are plotted based on the formulae derived and discussed extensively in the further part devoted to the computational method proposed herein.

It can be seen that the lower the authority value, i.e. the higher the value of the additional hydraulic resistance not subject to current regulation, the bigger the original characteristic deformation. The important thing is that the deformation always proceeds "upwards", i.e. for the valve given opening degree, the relative value of the medium volume flow is higher compared to the initial value. In the case of an equal-percentage characteristic, this additionally results in a higher value of leakage.

Such and similar characteristics, plotted in relative coordinates, require a comment to those unfamiliar with the method of making them. In Fig. 4.1a and b, line 1 concerns a valve which is not installed in the pipework; the other lines concern a valve installed in it. As it can be observed, at a given opening degree of the valve, they give higher values of the medium volume flow. This may raise some doubts because the volume flow through a valve installed in the pipework, at the pipework given pressure, cannot be higher than the volume flow through the valve itself, as the flow is additionally limited by the pipework hydraulic resistances. Due to that, its value will always be lower than that for the valve alone. If for example, at the maximum opening degree,

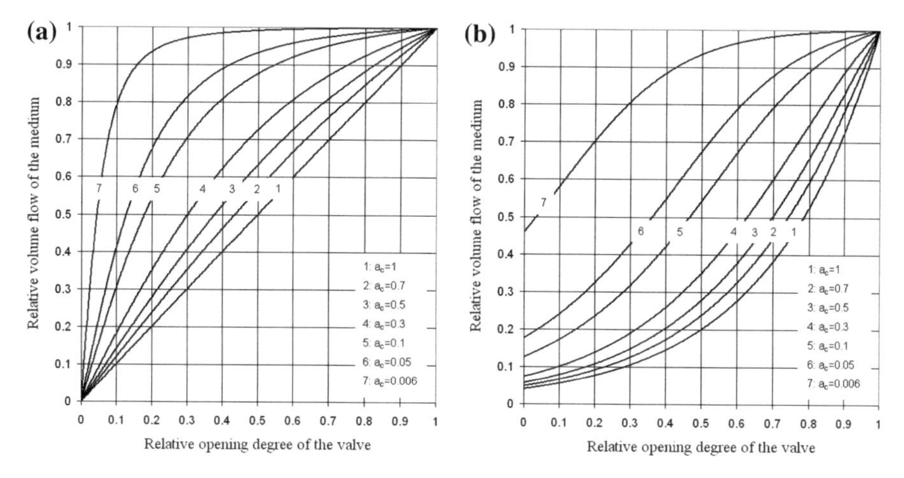

Fig. 4.1 Impact of the valve (total) authority on the valve regulation characteristic shape **a** for the original (initial) linear curve, **b** for the original (initial) equal-percentage curve, for c = 3.22

the volume flow is 3 (conventional units), after the valve is installed in the pipework, the value will total—for example—2 (conventional units). But characteristics plotted in relative coordinates assume that in both cases these are reference values totalling thereby 100% (i.e. 1) of the range of their variability. The volume flow values for other, smaller opening degrees are divided by the values for the maximum opening degree, which gives relative values from the range of 0–100% (0–1). For the data presented above, in the case of a valve not installed in the pipework, 50% of the volume flow equal 1.5 (conventional units), and for a valve installed in the pipework—50% of the medium volume flow total 1 (conventional unit). It can be seen that in fact only the relative values are equal (here: 50% = 0.5) whereas the absolute ones are different. For a valve not installed in the pipework, the absolute value of the medium volume flow is bigger. The situation can be analysed on the example of the charts presented in Figs. 5.12 and 5.13. Calculations are performed only to make it possible to compare the shape of the considered characteristics directly in a single chart. For this purpose, the curves need to start and end in the same points of the chart.

Such a perspective is usually adopted to present also the values placed on the chart other axis, e.g. the valve opening degree h_x/h_{100}. But some issues need to be made more specific here. The same relative (percentage) changes in and values of the medium volume flow may correspond to different absolute changes and values, as described above. This may also be the case with the valve opening degree h_x/h_{100}. Regulation characteristics, e.g. those presented in Sect. 3.3 or in this chapter, created based on this approach do not provide information about absolute values of the parameters under consideration. This is not a big problem in the case of the medium volume flow. But for the valve opening degree, it may result in significant inaccuracies. In the analyses presented above (e.g. Figs. 3.13 and Fig. 4.1), the valve relative opening degrees correspond to its identical absolute values. For each characteristic in the figures, the valve operates within the same range of the plug travel defined as 100% of the entire available range. It may happen, however, that for a given valve subsequent characteristics will be plotted for different available ranges of travel of the valve plug, e.g. 100, 80, 60%, due to a limitation to the travel range imposed by a pre-setting. But in relative coordinates, according to the purpose of their application, these values will always be reference values equal to 100% (i.e. 1). However, only the first of them will be the characteristic related to 100% of the entire available range of the plug travel, which will be the maximum relative value h_{100} and the maximum absolute value h_{max} at the same time. For the two other characteristics, the 100% opening degree h_x/h_{100} will only be a relative opening degree because the absolute values related to the maximum available opening degree will total $h_{100}/h_{max} = 80\%$ and $h_{100}/h_{max} = 60\%$.

Therefore in the analysis of mathematical descriptions used to plot such characteristics (e.g. formulae 3.74 and 3.80), it should be considered whether e.g. the valve full opening degree h_{100} relates to the full maximum absolute range of the valve plug travel h_{max} or the full but relative range, being a part of h_{max}. For a clear-cut distinction between the two cases, all the quantities related to the maximum absolute range of travel h_{max} should be marked in the same way, e.g. using the subscript "max". This would affect for example the flow factor k_v, which would then be marked as $k_{v,max}$, and not $k_{v,\,100}$ or $k_{v,\,s}$, the medium volume flow $\dot{V}$, then marked as $\dot{V}_{max}$, instead of $\dot{V}_{100}$, etc. Such unification of all quantities would significantly improve the clarity of the analysis presented also in the further part of this book. However, this would require a change in the fundamental marking related to the valve—the flow factor k_v, and the value of the valve maximum flow factor is always given as $k_{v,\,100}$ or $k_{v,\,s}$. For this reason, a decision was made herein to stick to this marking, and in situations where a distinction is required between the maximum absolute range of the plug travel h_{max} and the full relative range h_{100}, a relevant comment is added.

The charts are plotted assuming a constant value of active pressure in the circuit. In practice, however, as previously stated, the value varies, rising with a decrease in the medium volume flow, even if the system is equipped with pressure-stabilizing devices. Because the intensity of this phenomenon depends on the shape of the throttling characteristic of the pump and, if any, the pressure stabilizers, it cannot be captured universally in the charts mentioned above. However, the phenomena cause further deformation of the original shape of the characteristics.

Due to the fact that a change in hydraulic resistance and the medium pressure drop in the radiator double-regulation valve may occur both on the closing and on the throttling element, each of these two parts can have its own authority. This is because according to the computational relations describing the valve authority, the pressure drop on the throttling or the closing element can be related to the active pressure value in the circuit. But the radiator heat output current regulation depends only on the closing element, controlled either manually or automatically. The closing element and its characteristic are then essential from the point of view of the radiator thermal characteristic and the final regulation curves. For this reason, the value and the changeability of this particular element authority are important, not of the authority of the throttling part responsible for lasting and assembly-related balancing of hydraulic resistances in the circuit. The reasoning presented in the further part will therefore be related to the closing element unless stated otherwise.

Some of the assumptions that became the pillars of the current state of knowledge of regulation valves, including elements used in heating installations, were formulated at the beginning of the 20th century. As early as in 1909, Gramberg [8] stated that a valve in a central heating circuit displayed fair regulating capacity if its hydraulic resistance in the circuit total hydraulic resistance, corresponding to the pressure drop, totalled 50%, and if that value was generated in the valve regulating part. Especially the latter remark is of great importance. Later on, the condition was presented in two notations as [18]:

- the valve superior role (marked a', to avoid a denotation conflict):

$$a' = \frac{\Delta p_{z,x}}{\Delta p_{ob}} = \frac{\Delta p_{z,x}}{\Delta p_{z,x} + \Delta p_{str,x}}. \tag{4.1}$$

- the throttling criterion, the equivalent of today's *outer authority* a_{z100}:

$$a_{dł} = \frac{\Delta p_{z,100}}{\Delta p_{z,0}} = \frac{\Delta p_{z,100}}{\Delta p_{z,100} + \Delta p_{str,100}}. \tag{4.2}$$

Apart from very few cases, neither of the formulae presented above is able to capture and represent the physical sense of the notion it describes, and the description offered by them is most often incorrect. Even though it includes the pressure loss on the entire valve, Formula (4.1) cannot be treated as a relation defining total authority, but only as one describing the pressure distribution in the pipework for a given position of the closing and the throttling part. This is described in detail in further subsections devoted to the proposed method of the parameter determination. Formula (4.2), on the other hand, is a direct expression defining *outer authority* but only for a fully open valve, without any pre-setting, which originates from the work of historically first manual valves. It is assumed here that the circuit active pressure is constant and that the valve cuts off the flow completely. This also explains the equality between the denominators in the two forms of the formula.

Moreover, the requirement to meet the number condition of 0.3–0.7, still formulated at best for the outer authority, should be imposed on the total authority instead because the latter informs about the valve regulating part impact on and share in the shape and size of changes in the medium pressure and volume flow in a given circuit as regulation proceeds. The total authority value is related to the deformation of the original regulation characteristic of the closing element and not of the entire valve, as indicated by formula (4.1). For this reason, the results obtained from it, like in the case of formula (4.2), cannot be used as the basis for comparison with the above-mentioned formal requirement expressed in numbers and imposed on regulation valves. Further below, alternative computational relations are proposed that take account of the factors affecting the deformation of the valve original regulation characteristic and referred to as the valve *inner authority* and *outer authority*.

4.1.2 Regulation Valve Inner Authority

The valve plug is designed to obtain a certain regulation characteristic and the hydraulic resistance variability corresponding thereto depending on the opening degree h_x/h_{100}. The plug and the valve flow channel, making up the closing element responsible for current regulation, are mounted in the valve body, which may also include other throttling elements, such as those responsible for the valve presetting and initial throttling. The body and the parts mentioned above create an additional hydraulic resistance, which means a higher value of the total resistance and a reduced impact of the regulated resistance of the valve closing element on changes in the medium volume/mass flow. Arising inside the valve, the impact is referred to as the valve *inner authority*. In qualitative terms, the issue concerns the degree of the closing element characteristic deformation as early as inside the valve. Quantitatively, it is expressed as the ratio between the pressure loss occurring on the valve element responsible for current regulation and the total pressure loss on the valve, at constant available pressure, according to the following formula:

$$a_w = \frac{\Delta p_{reg}}{\Delta p_z} = \frac{r_{reg}}{r_z}. \tag{4.3}$$

The relation would suggest that the value of inner authority a_w of a valve operating at a certain range of the regulating element position h_{100} varies, because as closing proceeds, the value of the numerator in the formula gets bigger. A similar conclusion can be drawn analysing the expressions describing outer authority, which are presented further below. But as previously explained, inner authority is the image of the closing element original characteristic deformation due to additional resistances created on the valve. Because the resistances are not subject to current regulation and have a constant value, the inner authority value for a certain range of the regulating element position h_{100} is also constant. The characteristic is therefore affected by constant deformation (determined by the value of a_w) compared to the original shape of the curve for every relative opening degree h_x/h_{100}. In order to become a proper representation of the considered parameter as the degree of the original characteristic deformation, the relation presented above and illustrating the ratios of pressure drops and hydraulic resistances should be interpreted in an appropriate way. The numerator denotes the value not for any position of the closing element, but for the position from which the necessary regulation will begin. The value can thus be calculated for the full in the absolute scale, 100% opening of the currently regulated cross-section, or for any other opening degree (x%) provided that the range of the element position is treated as a reference value, being therefore the full 100% opening degree, but a relative one. The values of inner authority a_w and the degrees of the original regulation characteristic deformation obtained in the two cases will naturally be different. But in both situations the obtained results will represent relative deformation of characteristics, which (in the sense resulting from the value of authority a_w) is constant for the entire available range of the position of the element of the second stage of regulation, which complies with the physical sense of the parameter.

A similar situation occurs in the case of outer authority a_z. For a typical situation, i.e. "downward" regulation, if closing begins from a given range of the closing element position h_{100}, changes in the medium volume flow will occur according to the curve defined by the authority value determined for this range of travel of the closing part. Similarly then, if for the circuit certain working point, determined by a given position of the closing element, the element opening degree is raised above the set design value, denoting 100% relative opening, increments in the medium volume flow will occur according to the value of inner authority a_w determined for this range of the element position. The changes will result from a reduction in the closing element hydraulic resistance, at the other hydraulic resistances unchanged, and this proportion determines the value of inner authority a_w. It should be noted, however, that this does not mean that the new regulation curve, determined for the new maximum range of the valve regulating element position h_{100} will be described by the same value of inner authority a_w. For natural reasons, i.e. due to the different range of changeability of the function independent valuables and values, the characteristic, scaled in relative values, will have a different curvature and will be described by a different value of inner authority a_w corresponding to the regulating element new maximum relative range of position $h_{100.}$ Nonetheless, the results obtained from the calculations of the medium volume flows for a given absolute opening degree, and thereby—for different relative values thereof, will coincide. The first characteristic, sketched in the range of values corresponding to the second, will be a part of the second characteristic and will coincide with it. Such calculations performed for an example valve are presented in Sect. 4.2.1. They are confirmed experimentally in Sect. 5.3.

The described phenomenon of raising the regulating element opening degree above the design value is typical of the radiator thermoregulators, whose working point is defined for the partial opening resulting from the valve plug position. This is even the case in double-regulation thermostatic valves with one adjustable section of the medium flow. The reason for that is the fact that the reduced range of the closing part travel caused by the throttling element can still be wider than the closing part range of travel established for $X_p = 2$ K (cf. description of the *Herz TS-90-V* valve). In single-regulation thermostatic valves (no element of the regulation first stage) the values of Δp_{reg} and r_{reg} should be related to the maximum (100%) and not the set—partial—opening degree of the element of the second stage of regulation, resulting from the principle of the device operation. In double-regulation thermostatic valves with one adjustable section of the medium flow (e.g. a double plug)—to the maximum opening degree of the element of the regulation second stage at a set degree of opening of the element of the first stage of regulation (e.g. set value of the pre-setting). For the two types of valves, the possibility of additional opening results from the fact that the closing plug design position, if a need arises to operate with e.g. thermostatic heads, is usually set to a partial opening degree (<100%). For this reason, measurement data obtained for such conditions cannot be used to define inner authority. The data for the closing element maximum lift are then necessary. In manual valves, the values of Δp_{reg} and r_{reg} correspond directly to the maximum (no reduction) opening of the element of the second stage of regulation, at a set opening degree of the throttling element (at a given pre-setting value).

The natural thing is that in the process of reducing the available range of the closing element travel, the hydraulic resistance of this element section of the medium flow rises. This involves a decrease in the volume flow and—thereby—the medium smaller pressure losses on the body and on the throttling element, the resistance of which is set at a certain level. The initial volume flow value can be restored by appropriately raising the pressure supplied to the valve. The restored initial volume flow will then make pressure losses arise again on the body and on the throttling element, like before. The pressure loss on the closing element will then be higher because the element will throttle the excess of the higher total pressure value. It follows directly that inner authority a_W will get bigger. It also follows that if the valve for the full-closing position (complete limitation to the range of travel) tends to cut off the flow completely, the entire available pressure drop in the circuit tends to occur on the closing element. Consequently, inner authority a_W approaches one. This is illustrated graphically in Fig. 4.2. Using these comments and the computational relation, it can be seen that all hydraulic resistances occurring inside the valve but generated beyond the element with a section subject to current regulation will decrease the valve authority and thus deform its original regulation characteristic. This will naturally deteriorate the valve regulating capacity. In relation to the presented designs of the radiator regulation valve, it can be seen that co-operation between the two stages of regulation may have a twofold effect. In situations where the two elements make up one regulated section of the medium flow, one regulated hydraulic resistance will occur. A rise in throttling on one naturally involves a rise in throttling in the same section where the other regulation stage is located, which gives the same relation for inner authority a_W. In this way, a rise in the valve hydraulic resistance is related to a rise in inner authority a_W, just like in the case of outer authority a_Z. This occurs for example in the *Herz TS-90-V*, or the *Danfoss MSV-I* valve. On the other hand, in valves where the double-regulation principle is realized by the action of elements making up two separate sections (only one of which is subject to current regulation), the opposite relation will be the case. Even though it involves a rise in outer authority a_Z, an increase in the valve hydraulic resistance arising due to a rise in the resistance of the throttling part is related to a decrease in inner authority a_W. This results in a drop in the value of total authority a_c, which is the case for most currently manufactured radiator thermostatic valves.

Most theoretical studies do not take account of the existence of the described phenomenon of the distribution of pressure inside the valve, which is the same as assuming that $a_W = 1$. It is unfavourable in that in reality the parameter, for full opening of both regulating parts of the valve, usually takes values not exceeding $a_W \approx 0.4$, reaching levels even by a few orders lower ($a_W < 0.001$) in the case of double-regulation valves with two adjustable sections of the medium flow, for the smallest values of the pre-setting (cf. Sects. 5.3 and 5.4). The researcher who was one of the first to present the described phenomenon in numbers was *Pyrkov* (in collaboration with *Szaflik*) [25, 26]. Although he assumes, and rightly so, that a_w is always smaller than 1 ($a_w < 1$), translating this into characteristics of valves and

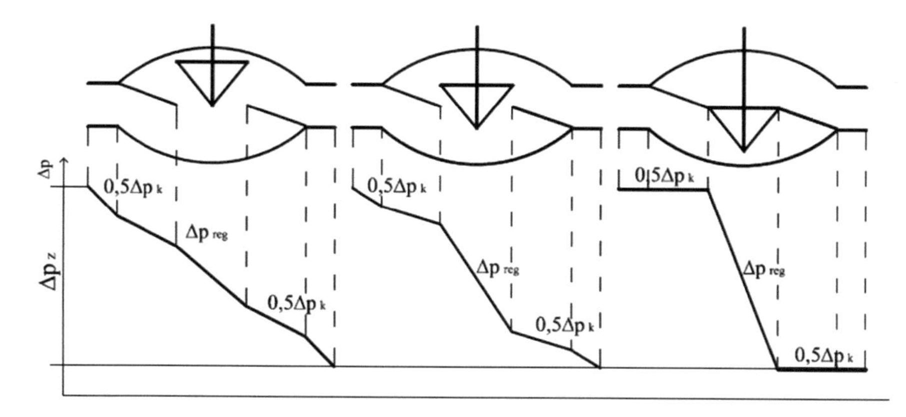

Fig. 4.2 Pressure distribution in a single-regulation valve and a double-regulation valve with one adjustable section of the medium flow (symbolic division of the regulation first- and second-stage elements capable of changing their mutual position marked on the plug)

final regulation curves, he takes no account of the fact that the parameter value is not constant as a function of the available relative range of the closing element travel (not to be confused with the opening degree for this travel range). As indicated by Fig. 4.2 and formula 4.3, a_w rises as the available range of travel is reduced, and the minimum value occurs if there is no reduction therein. Therefore the assumption of the inner authority constancy, made for example in [13, 25, 26], is wrong and leads to incorrect results of the regulation characteristic calculations. Moreover, it does not enable correct analytical determination of the valve plug required geometry. Furthermore, *Pyrkov* and *Szaflik* describe the phenomenon and calculation results for valves made by the *Danfoss* company, achieving results that will be right for the type of the double-regulation principle solution used in them. In the case of thermostatic valves, they will be qualitatively right for devices with two adjustable sections of the medium flow, but if the authors' methodology is applied to some radiator valves manufactured for example by the *Hertz* company, incorrect results will be obtained that contradict the physics of the phenomenon.

Figure 4.3 presents an example pressure loss distribution inside a double-regulation valve with two adjustable sections of the medium flow.

Considering hydraulic resistances and pressure distributions occurring inside a double-regulation valve, formula 4.3 can be expressed as:

- if there is only one adjustable section of the medium flow:

$$a_w = \frac{\Delta p_{reg}}{\Delta p_z} = \frac{\Delta p_{reg}}{\Delta p_{reg} + \Delta p_k} = \frac{\Delta p_{II} + \Delta p_I}{\Delta p_{II} + \Delta p_I + \Delta p_k}, \tag{4.4}$$

or, using hydraulic resistances:

$$a_w = \frac{r_{reg}}{r_z} = \frac{r_{reg}}{r_{reg} + r_k} = \frac{r_{II} + r_I}{r_{II} + r_I + r_k}. \tag{4.5}$$

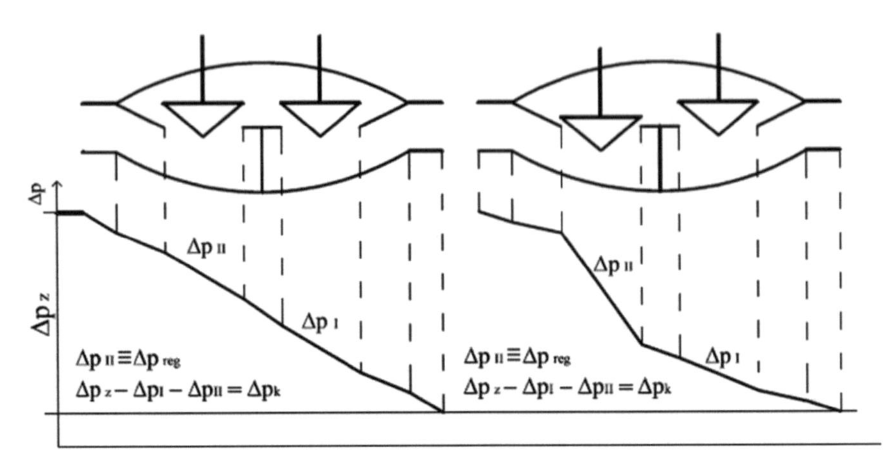

Fig. 4.3 Pressure distribution in a double-regulation valve with two adjustable sections of the medium flow (subscripts I and II, respectively, denote the first and the second stage of regulation)

- if there are two adjustable sections of the medium flow:

$$a_w = \frac{\Delta p_{reg}}{\Delta p_z} = \frac{\Delta p_{reg}}{\Delta p_{reg} + \Delta p_I + \Delta p_k} = \frac{\Delta p_{II}}{\Delta p_{II} + \Delta p_I + \Delta p_k}, \tag{4.6}$$

or, using hydraulic resistances:

$$a_w = \frac{r_{reg}}{r_z} = \frac{r_{reg}}{r_{reg} + r_I + r_k} = \frac{r_{II}}{r_{II} + r_I + r_k}. \tag{4.7}$$

The hydraulic resistance of the part subject to current regulation decreases with a rise in the opening degree, but never reaches zero. This means that even at the maximum available opening degree and the regulation second-stage element travel range $h_{100} = h_{max}$, the resistance causes a certain pressure drop in the medium flow through the valve. This can be understood easily looking at the design of the radiator valves presented in Figs. 2.27–2.29. For example, the *Herz TS-E* single-regulation valve is shown there in a fully open position, and the available range of the plug travel in this case is about 2.4 mm, even though it could be a few times bigger. For the other radiator valves presented therein, the situation is similar. For the plug full lift and the maximum travel range h_{max}, the medium flow surface area A_{100} is much smaller compared to the case with no plug in the valve body. Example structures that make it possible to obtain an almost maximum available surface area of the medium flow at the full opening degree are as follows: gates, ball cut-off valves and butterfly valves. The inner authority value in those devices is rather low at full opening, which causes a significant deformation of the original closing characteristic and results in the phenomena, in which the regulation in such cases is generally realized at a position close to full closing. This can be observed easily in the case of commonly used ball valves for example.

For the full available travel range (100%, in the absolute scale) of the regulating element, where the inner authority value a_w reaches the minimum, and for any other range, $x\%$ in the absolute and 100% in the relative scale, corresponding to a given pre-setting n_i, it can be written that:

- if there is only one adjustable section of the medium flow:

$$a_{w,100} \equiv a_{w,\min} = \frac{\Delta p_{reg,100}}{\Delta p_{z,100}} = \frac{\Delta p_{reg,100}}{\Delta p_{reg,100} + \Delta p_k} = \frac{\Delta p_{II,100} + \Delta p_{I,100}}{\Delta p_{II,100} + \Delta p_{I,100} + \Delta p_k}, \tag{4.8}$$

$$a_{w,x} = \frac{\Delta p_{reg,x}}{\Delta p_{z,x}} = \frac{\Delta p_{reg,x}}{\Delta p_{reg,x} + \Delta p_k} = \frac{\Delta p_{II,100} + \Delta p_{I,i}}{\Delta p_{II,100} + \Delta p_{I,i} + \Delta p_k}, \tag{4.9}$$

or, using hydraulic resistances:

$$a_{w,100} \equiv a_{w,\min} = \frac{r_{reg,100}}{r_{z,100}} = \frac{r_{reg,100}}{r_{reg,100} + r_k} = \frac{r_{II,100} + r_{I,100}}{r_{II,100} + r_{I,100} + r_k}, \tag{4.10}$$

$$a_{w,x} = \frac{r_{reg,x}}{r_{z,x}} = \frac{r_{reg,x}}{r_{reg,x} + r_k} = \frac{r_{II,100} + r_{I,i}}{r_{II,100} + r_{I,i} + r_k}. \tag{4.11}$$

- if there are two adjustable sections of the medium flow, pre-throttling is set on the throttling element, whose operation is independent of the closing part. The element available travel range does not change because there is no need for that, and in the absolute scale its value for a given valve is constant in every case. For that reason, the drop in flow factor k_v is related to the rise in hydraulic resistance r_I and, thereby, to the drop in inner authority a_w. It follows that the inner authority value changes from $a_{w,\max}$, reached for the throttling element maximum available travel range h_{max} at $r_{I,100}$, through $a_{w,x}$ at the ith position of the element of the first stage of regulation, to $a_{w,\min}$, which is obtained at $r_{I,0}$ for the throttling element minimum opening degree. Consequently, if the flow is completely cut off, $a_w = 0$. It may therefore be written that:

$$a_{w,100} \equiv a_{w,\max} = \frac{\Delta p_{reg,100}}{\Delta p_{z,100}} = \frac{\Delta p_{reg,100}}{\Delta p_{reg,100} + \Delta p_{I,100} + \Delta p_k} = \frac{\Delta p_{II,100}}{\Delta p_{II,100} + \Delta p_{I,100} + \Delta p_k}, \tag{4.12}$$

$$a_{w,x} = \frac{\Delta p_{reg,100}}{\Delta p_{z,x}} = \frac{\Delta p_{reg,100}}{\Delta p_{reg,100} + \Delta p_{I,i} + \Delta p_k} = \frac{\Delta p_{II,100}}{\Delta p_{II,100} + \Delta p_{I,i} + \Delta p_k}, \tag{4.13}$$

$$a_{w,\min} = \frac{\Delta p_{reg,100}}{\Delta p_{z,100}} = \frac{\Delta p_{reg,100}}{\Delta p_{reg,100} + \Delta p_{I,0} + \Delta p_k} = \frac{\Delta p_{II,100}}{\Delta p_{II,100} + \Delta p_{I,0} + \Delta p_k}, \tag{4.14}$$

or, using hydraulic resistances:

$$a_{w,100} \equiv a_{w,\max} = \frac{r_{reg,100}}{r_{z,100}} = \frac{r_{reg,100}}{r_{reg,100} + r_{I,100} + r_k} = \frac{r_{II,100}}{r_{II,100} + r_{I,100} + r_k}, \tag{4.15}$$

$$a_{w,x} = \frac{r_{reg,100}}{r_{z,x}} = \frac{r_{reg,100}}{r_{reg,100} + r_{I,i} + r_k} = \frac{r_{II,100}}{r_{II,100} + r_{I,i} + r_k}, \tag{4.16}$$

$$a_{w,\min} = \frac{r_{reg,100}}{r_{z,100}} = \frac{r_{reg,100}}{r_{reg,100} + r_{I,0} + r_k} = \frac{r_{II,100}}{r_{II,100} + r_{I,0} + r_k}. \tag{4.17}$$

If there is no element of the regulation first stage, 0 should be substituted for the quantities corresponding to it.

Determination of the distribution of hydraulic resistances and pressure losses inside a radiator valve (or any other valve with a plug), as one of the first assumptions in the valve sizing process—unlike calculations for a finished structure—is not an easy task. The resistance of the closing part created between the valve seat and the plug for full opening, like the changes in the element resistance, can be found using known mathematical relations. But the valve body resistance cannot be determined precisely in an analytical manner, especially if the element can take any shape. The situation becomes more complicated in the case of double-regulation valves with a built-in additional regulating part—the throttling element. Hydraulic resistances can be established using a few, both direct and indirect, methods and by means of auxiliary CFD analyses.

The next subsections propose and describe two methods of inner authority determination—a direct and an indirect one.

4.1.2.1 Direct Method of the Regulation Valve Inner Authority Determination

The method consists in measuring, on the target structure, the values of the quantities needed for determination of flow factor k_v or the valve hydraulic resistance r and used later on to calculate inner authority a_w. The valve body flow factor $k_{v,k}$, or the body hydraulic resistance r_k, should be determined without the closing element by measuring the medium volume flow at a set stabilized differential pressure on the valve. Analysing the two methods of the double-regulation principle realization, the following is then obtained:

- for a single-regulation valve and a double-regulation valve with one adjustable section of the medium volume flow:

$$k_{v,k} = \frac{\dot{V}_k}{\sqrt{\Delta p_k}}, \tag{4.18}$$

  or writing the valve hydraulic resistance, according to formula (3.42), as:

$$r_k = \frac{\Delta p_k}{\dot{V}_k^2}. \tag{4.19}$$

  Hydraulic resistance can also be calculated using formula (3.61), which relates the quantity to flow factor k_v.
- for a double-regulation valve with two adjustable sections of the medium flow, according to the physical sense of inner authority a_w, the quantities mentioned for

the valve body should be measured with an installed throttling element and with defined *i*th degrees of its opening, i.e. the pre-setting values. Naturally, the medium volume flows through the body and the throttling element are identical because the two elements are connected in series. For this reason, there is no need to write two separate expressions defining this quantity. The result is then the following equation:

$$k_{v,k+I,i} = \frac{\dot{V}_{k+I,i}}{\sqrt{\Delta p_k + \Delta p_{I,i}}}, \tag{4.20}$$

which can be written as:

$$k_{v,k+I,i} = \frac{\dot{V}_{k+I,i}}{\sqrt{\Delta p_{k+I,i}}}, \tag{4.21}$$

or hydraulic resistance, using formula 3.61, can be found as:

$$r_{k+I,i} = \frac{\Delta p_{k+I,i}}{\dot{V}^2_{k+I,i}}. \tag{4.22}$$

After installing the closing element, the quantities should be measured again, taking the values of k_v, Δp_z, r_z for a complete valve, at the full 100% lift of the closing element and each required position of the throttling one (for a given pre-setting n_i). Naturally, hydraulic resistance r will rise and flow factor k_v will fall. Comparing the results, the distribution of hydraulic resistances and the medium pressure losses in the valve, and thereby—inner authority a_w, can be calculated using the following procedure:

- for a single-regulation valve and a double-regulation valve with one adjustable section of the medium volume flow, it can be written that:

$$r_{reg} = r_z - r_k = \frac{\Delta p_z}{\dot{V}_z^2} - \frac{\Delta p_k}{\dot{V}_k^2}. \tag{4.23}$$

Because in both cases the measurement is performed at a set constant pressure, the following is true:

$$r_{reg} = \Delta p_z \cdot \left(\frac{1}{\dot{V}_z^2} - \frac{1}{\dot{V}_k^2} \right). \tag{4.24}$$

Using formula (4.3), this gives:

$$a_w = \frac{r_{reg}}{r_z} = \frac{r_{reg}}{r_{reg} + r_k} = \frac{\Delta p_z \cdot \left(\frac{1}{\dot{V}_z^2} - \frac{1}{\dot{V}_k^2} \right)}{\frac{\Delta p_z}{\dot{V}_z^2}} = 1 - \left(\frac{\dot{V}_z}{\dot{V}_k} \right)^2. \tag{4.25}$$

The formula can also be derived by means of relations using flow factors:

$$\frac{1}{k_{v,reg}} = \frac{1}{k_{v,z}} - \frac{1}{k_{v,k}} = \frac{\sqrt{\Delta p_z}}{\dot{V}_z} - \frac{\sqrt{\Delta p_k}}{\dot{V}_k}. \tag{4.26}$$

Considering that pressure is constant, the following equation is obtained:

$$\frac{1}{k_{v,reg}} = \sqrt{\Delta p_z} \cdot \left(\frac{1}{\dot{V}_z} - \frac{1}{\dot{V}_k} \right). \tag{4.27}$$

Taking account of formula (4.3) and the condition of the square relation of the pressure drop depending on the flow factor, the following expression is written:

$$a_w = \frac{\frac{1}{k^2_{v,reg}}}{\frac{1}{k^2_{v,z}}} = \left(\frac{k_{v,z}}{k_{v,reg}} \right)^2 = \left(\frac{\frac{\dot{V}_z}{\sqrt{\Delta p_z}}}{\frac{\dot{V}_z - \dot{V}_k}{\sqrt{\Delta p_z}}} \right)^2 = 1 - \left(\frac{\dot{V}_z}{\dot{V}_k} \right)^2. \tag{4.28}$$

Volume flows can be substituted with mass flows because the density of the analysed medium (water in the liquid state) practically does not change.

Calculating a thermostatic valve authority according to Standard EN 215:2004 [5], it is also necessary to know the flow factors for set values of the proportional range and full opening, or the pressure drops Δp_1 and Δp_2 corresponding thereto, or the medium volume/mass flow values, as in Fig. 2.16. These data are usually provided by manufacturers in the form of nomograms. An example nomogram for the single-regulation *Herz TS-E* valve of a linear figure is presented in Fig. 4.4. It shows the lines and points needed to read out the required values of relevant quantities.

As it can be seen, the formulae derived above are identical in form to formula (8.1) transformed to relation (8.2), which is included in the standard concerning the radiator thermostatic valves [5]. However, the two relations [(4.28) and (4.2)] have a different physical sense. The formula included in the standard offers no information about the degree of the original closing characteristic deformation, i.e. on inner authority a_w. As calculations proceed, it can be seen that the value of the quantity calculated in this manner changes with a change in the closing element position, even though it is related each time to the same maximum travel range of the element. Using the nomogram, it can also be noticed that a rise in the opening degree is identical with the line shift to the right and, thereby, a rise in the Δp_2 value until it becomes equal to Δp_1 at full opening. Therefore, if authority a is calculated according to the standard, the obtained value is $a = 0$. As previously stated, in a radiator valve it is impossible for a pressure loss not to occur on the closing element. For this reason, the described formula cannot be treated as a relation defining the valve inner authority a_w, as presented for example in [14, 25, 26], because this would suggest that at the plug full lift there is no pressure loss on it at all. More importantly, this would also lead to incorrect determination of the medium free flow surface area and underestimation thereof in the process of the valve design (cf. valve sizing example in Sects. 4.3 and 5.1). The quantity indicates (indirectly) how much the

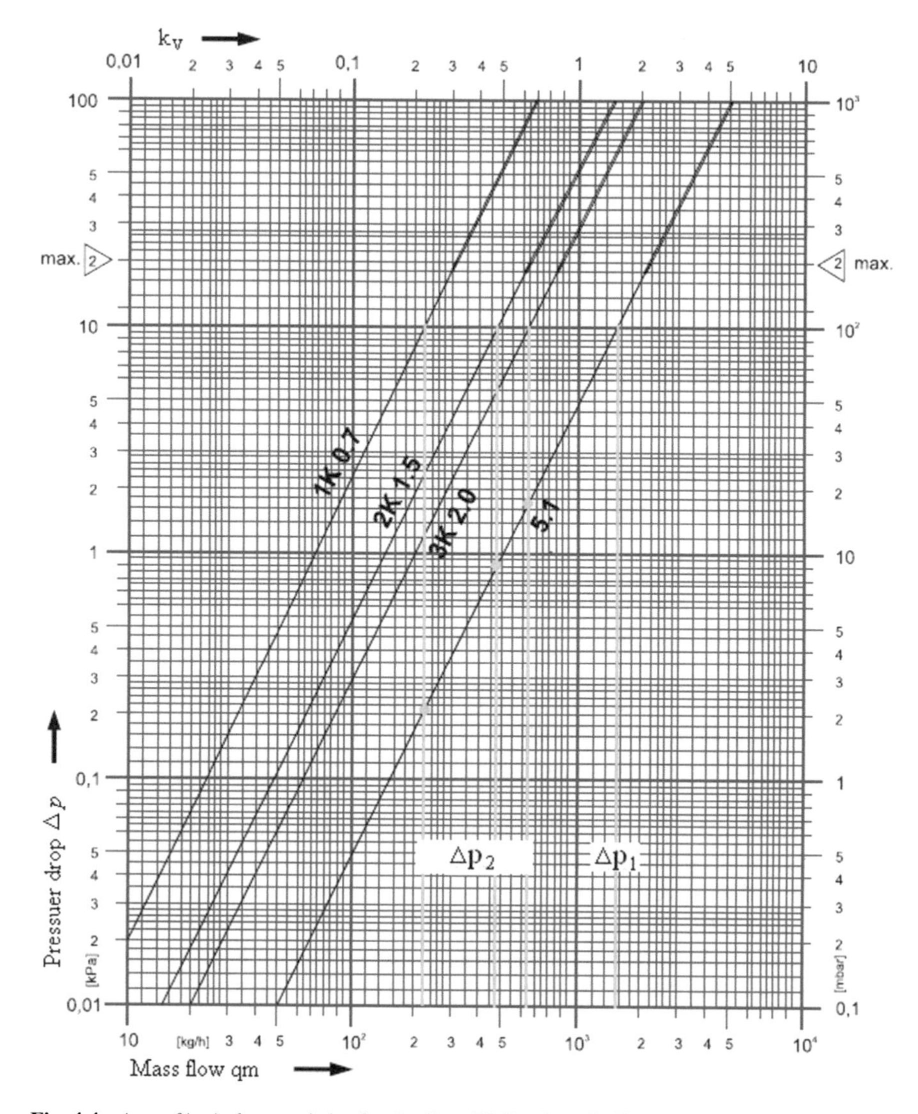

Fig. 4.4 $\Delta p = f(q_m)$ characteristics for the *Herz TS-E* valve of a linear figure [4]

closing element hydraulic resistance for a given opening degree is higher compared to resistance at full opening. As the opening degree gets smaller, the quantity value rises and, consequently, so does hydraulic resistance compared to the position at full opening. For this reason, the parameter should be referred to as the regulated section authority and not inner authority The standard notation relates the medium mass/volume flow for a set opening degree h_x/h_{100} defined by a given value of X_p to the medium maximum volume flow $\dot{V}_{100}$ (the maximum value of X_p), at a given differential pressure on the valve. It thus makes it possible to calculate how much

Table 4.1 Authority of the regulated section for the *Herz TS-E* thermostatic valve of a linear figure

X_p value	1 K	2 K	3 K
Pressure drop readout, formula 2.2	$a = 1 - \frac{0.19}{10} = 0.98$	$a = 1 - \frac{0.9}{10} = 0.91$	$a = 1 - \frac{1.65}{10} = 0.84$
Volume flow readout, formula 2.3	$a = 1 - \left(\frac{220}{1560}\right)^2 = 0.98$	$a = 1 - \left(\frac{460}{1560}\right)^2 = 0.91$	$a = 1 - \left(\frac{620}{1560}\right)^2 = 0.84$
Using factors k_v, formula 2.3	$a = 1 - \left(\frac{0.7}{5.1}\right)^2 = 0.98$	$a = 1 - \left(\frac{1.5}{5.1}\right)^2 = 0.91$	$a = 1 - \left(\frac{2.0}{5.1}\right)^2 = 0.85$

the medium volume flow can possibly rise compared to that determined by a given value of X_p if the closing element opens fully to $X_{\mathrm{p,\,max}}$. The volume of that growth depends on the shape of the original closing characteristic and on the actual value of inner authority a_{w}. The greater the inner authority value, the smaller the characteristic deformation and the greater the increments in the medium volume flow can be (cf. Fig. 4.1). Using the regulated section authority a, subsequent relative volume flow values of $\dot{V}_x/\dot{V}_{100}$ are obtained for subsequent values of X_p, i.e. the shape of the valve closing characteristic already deformed due to inner authority a_{w} is reconstructed. As described herein, for a valve not installed in the pipework, the illustrations of changes in the medium volume flow and in the flow factor are identical.

Table 4.1 presents the results of calculations of authority a for three values *of* X_p by means of the three methods described earlier, according to formulae (2.2) and (2.3) and the readouts of the required values from the nomogram. The number data, if not provided by the manufacturer, can also be read directly from the nomogram. This is because according to the flow factor definition, it is the medium volume flow expressed in m^3/h, at the pressure loss on the valve of 1 bar = 100 kPa.

Analysing the calculation results listed in Table 4.1, it can be seen that the higher the value of authority a for a given proportional range X_{p} (i.e. for the closing element given opening degree h_{x}/h_{100}), the higher the increments in the medium volume flow can be as the valve opens. For $X_{\mathrm{p}} = 2$ for example, the valve under consideration is characterized by the regulated section authority at the level of about $a = 0.9135$, the flow factor value of $k_{\mathrm{v}} = 1.5$, where the maximum value of the parameter totals $k_{\mathrm{v}} = 5.1$. This means that for full opening the flow will reach 5.1/1.5 = 340% of the original value, rising by (5.1 − 1.5)/1.5 = 240%, and the proportion between the available volume flow values at opening and closing is about 29.4%/70.6%. The value of the volume flow increment depending on authority a, if the parameter value is given and there are no relevant values of flow factor k_{v}, can be expressed by transforming relation (2.3) for the regulated section authority a, which gives:

$$\frac{k_{vs}}{k_{vN}} = \frac{1}{\sqrt{1-a}}. \tag{4.29}$$

After substitution, the following result is obtained: $1/(1 - 0.9135)^{1/2} = 340\%$. The proportion between the volume flows can be written in a similar way:

$$\frac{k_{vN}}{k_{vs}} = \sqrt{1-a}, \tag{4.30}$$

which, for the considered plug position h_N (i.e. the set value of proportional range X_p), gives the volume flow value of $(1 - 0.9135)^{1/2} = 29.4\%$ of the total volume flow, and this means that the proportion between the volume flow values at closing and opening is 29.4%/70.6%. If for the same opening degree and proportional range X_p the valve was characterized by the flow factor value at the level of $k_v = 3.0$ for example, the possible increment would total $(5.1 - 3.0)/3.0 = 70\%$, the proportion between the available volume flows at closing and opening due to the set opening degree—about 58.8%/41.2%; the authority value would be $a \approx 0.654$—smaller compared to the previous case. For the recommended limit values of $a = 0.3$ and $a = 0.7$, transforming relation (2.3) for a, the following proportions between the medium flow values at the valve closing and opening are obtained compared to the given original value: about 83.7%/16.3% and about 55%/45%. Naturally, the calculated values concern the situation where $a_z = 1$. As previously mentioned, an increment in the volume flow when the valve opens to a level exceeding the design value is unfavourable and it would be best if the thermoregulator operated without such a rise. Looking at the curves in Fig. 4.1, it can be noticed that such an effect is achieved if the closing characteristic is strongly non-linear, very convex upwards, i.e. for example at small values of inner authority a_w. This conclusion, however, is not right. Such a characteristic, combined with the regulated object (radiator) static characteristic with a similar shape, will give—from the point of view of regulation—a very unfavourable, strongly non-linear shape of the final regulation curve. Therefore, the postulate presented above should be satisfied in a different manner. The element of the first stage of regulation is used for this purpose, defining the preliminary value of hydraulic resistance. Using the quantity of the *regulated section authority*, it can be noticed that low values thereof are favourable and that it is not related qualitatively to the *inner authority* parameter, for which it is favourable to reach high values. The common inaccuracy in theoretical studies and in the adopted nomenclature results from the fact that the parameter is implemented to satisfy the needs of thermostatic valves directly from the manual regulation technique, without introducing the necessary distinctions or divisions.

- For a double-regulation valve with two adjustable sections of the medium flow, individual hydraulic resistances can be determined from the following expression:

$$r_{reg} \equiv r_{II} = r_z - r_{k+I,i} = \frac{\Delta p_z}{\dot{V}_z^2} - \frac{\Delta p_{k+I,i}}{\dot{V}_{k+I,i}^2}, \tag{4.31}$$

which, taking account of the constant pressure value, results in:

$$r_{reg} = \Delta p_z \cdot \left(\frac{1}{\dot{V}_z^2} - \frac{1}{\dot{V}_{k+I,i}^2} \right), \tag{4.32}$$

Consequently, inner authority is described by the following relation:

$$a_w = \frac{r_{reg}}{r_z} = \frac{r_{II}}{r_{II} + r_I + r_k} = \frac{\Delta p_z \cdot \left(\frac{1}{\dot{V}_z^2} - \frac{1}{\dot{V}_{k+I,i}^2} \right)}{\frac{\Delta p_z}{\dot{V}_z^2}} = 1 - \left(\frac{\dot{V}_z}{\dot{V}_{k+I,i}} \right)^2. \tag{4.33}$$

An increase in the throttling element hydraulic resistance, i.e. a decrease in the opening degree (e.g. the i-th pre-setting n_i), causes a natural drop in the medium volume flow $V_{k+I,i}$. According to the formula presented above, this will also involve a drop in the valve inner authority a_w. For the radiator double-regulation thermostatic valves, manufacturers also provide relevant nomograms or tables. These, however, differ from the nomograms (tables) made for single-regulation valves because the lines illustrating flow-dependent pressure losses are plotted not for the closing element different opening degrees corresponding to a given proportional range X_p, but for the set opening degrees of the throttling element, which correspond to a given value of pre-setting n_i and the closing element constant opening degrees h_x/h_{100}, corresponding usually to the proportional range value of $X_p = 2$ K. Since this is not the full available travel range ($h_{100} \neq h_{max}$) of the element, as previously explained, inner authority a_w cannot be determined in the way presented above. Results are necessary of the measurement performed for the closing element full, maximum travel range h_{max}.

Figures 4.5 and 4.6, respectively, present nomograms for the *Herz TS-90-V* double-regulation valve of an angular figure (one adjustable cross-section of the medium flow), and for the *Danfoss RA-N15* double-regulation valve (a twin model to replace the *RTD-N15* variant withdrawn from production, described in details in the book) of a linear figure (two adjustable cross-sections of the medium flow). As indicated in Sect. 2.2, describing the types of solutions of the double-regulation concept, in a valve with a double hanging plug for the set constant value of X_p (i.e. for the plug constant lift above the seat), the valve nominal throttling cannot depend on the pre-setting because the two elements make up a single regulated section. This can be the case only at a variable value of X_p, or if the valve seat does not create a plane parallel to that created by the regulating element. Such is the situation in the *Herz TS-90-V* valve, where the seat is bevelled to make it possible for the closing element placed inside to move even if the throttling element is fully closed, as illustrated by Figs. 2.28 and 2.30.

The charts presented in Figs. 4.5 and 4.6 require a brief comment. The data concerning the flow factor and the pressure loss curves for individual values of pre-setting n_i for the *Herz TS-90-V* valve, as presented in Fig. 4.5, are each time, according to the manufacturer, related to the closing element position corresponding to proportional range $X_p = 2$ K. The situation is a bit different for the *Danfoss RA-N15* valve (and for the twin *RTD-N15* valve replaced by the *RA-N15* variant), the data for which are presented in Fig. 4.6. As specified by the manufacturer, the data for individual values of pre-setting n_i, despite the fact that in this valve type the closing element operates independently of the throttling part, are given for different values of X_p, namely: "*At setting N the k_v value is stated according to EN 215, at*

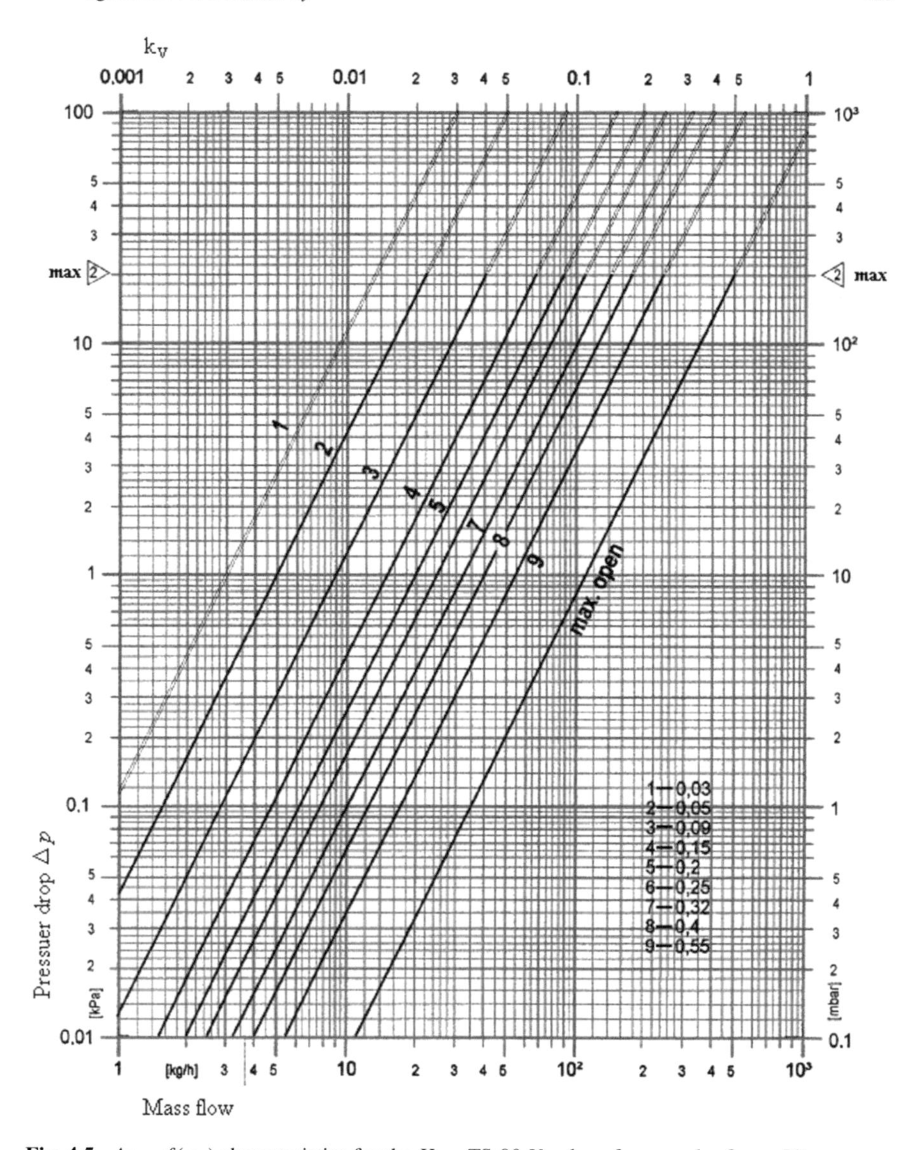

Fig. 4.5 $\Delta p = f(q_m)$ characteristics for the *Herz TS-90-V* valve of an angular figure [4]

$X_P = 2$ K i.e. the valve is closed at 2 °C higher room temperature. At lower settings the X_P value is reduced to 0.5 K of the setting value 1." This is not really due to the deliberately set during the testing different values of the closing plug lift h_x over the seat and, consequently, different values of X_p, but rather due to the specificity of operation of the radiator thermoregulator equipped with such a valve and to the impact of the method of solving the double-regulation principle on the valve regulation characteristics. For increasingly lower values of pre-setting n_i, due to the drop in

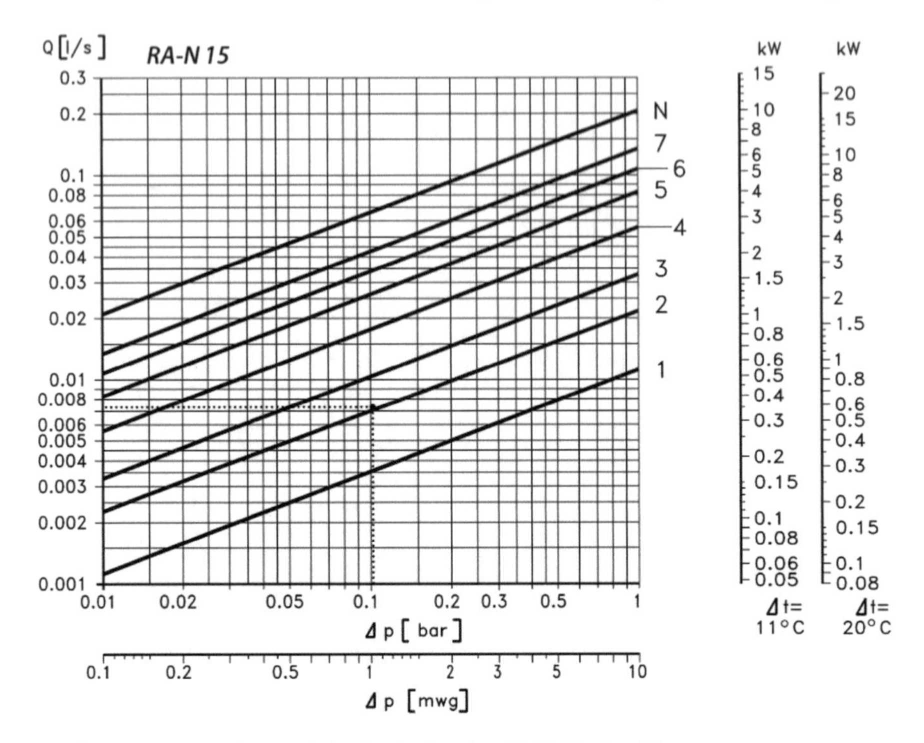

Fig. 4.6 $\Delta p = f(q_m)$ characteristics for the *Danfoss RA-N 15* valve [3]

inner authority a_w, the deformation of the original regulation characteristic becomes bigger and bigger. Due to that, the range of noticeable changes in the flow factor and the medium volume flow values depending on the closing plug opening degree h_x/h_{100} will narrow down to smaller and smaller values of the opening. This is discussed in detail in Sect. 4.1 and illustrated by Figs. 4.1 and 5.18, for example. Figure 5.18 is based on the data of experimental verification performed for the *Danfoss RTD-N15* valve. The specific value of the plug lift h_x above the seat for a given value of proportional range X_p, if the valve co-operates with a thermostatic head, as explained earlier and expressed by formula 2.1, depends on the value of the thermostatic head amplification factor k_m. For the valve under consideration, at the parameter typical value at the level of $k_m = 0.25$ mm/K and the proportional range value of $X_p = 2$ K, the relative lift totals about $h_x/h_{ma} \equiv h_x/h_{100} = 0.294$ (cf. computational example 16 in Chap. 6). According to Fig. 5.18, for pre-setting $n_i = 1.0$, the valve will then reach about 98% of the flow factor maximum value. The considered four-time reduction in the proportional range and in the relative opening degree of the closing plug to $X_p = 0.5$ K and $h_x/h_{100} = 0.0735$, respectively, will make the flow factor fall, but only to about 75% of the maximum value, i.e. only by about 1.3 times. For a valve installed in the pipework, due to a further deformation of the regulation characteristic, the difference will be even smaller. It may therefore turn out that a given value of the medium volume flow locates the working point in such a region of the valve regulation characteristic that the curve becomes almost horizontal and the flow factor/the volume flow value is almost constant in a wide range of changes in the proportional range X_p and the plug lift h_x above the seat. The flow factor/the volume flow can therefore take almost identical values for the proportional range values of X_p included in the interval of (0.5; 2) K. This means that the differences in the values of the medium volume flow, the radiator heat output and the room temperature (which is the target parameter to which the head regulating the valve responds) for even very different values of the closing plug opening degree h_x/h_{100} and proportional range X_p will be so small that the valve-head system will actually set its working point around the lower limit value of the interval of proportional range X_p, e.g. (0.5; 2) K, for which the medium volume flow changes so slightly. This is the cause of the situation described in the manufacturer's information mentioned above.

4.1.2.2 Indirect Method of the Regulation Valve Inner Authority Determination

The data provided by valve manufacturers are usually insufficient for the earlier-presented direct determination of the valve inner authority a_w because the required values taken for a valve without the closing element are not specified. Therefore, the relevant relations can only be applied if own measurements are performed. However, it is also possible to use the indirect method proposed below. So that the valve inner authority a_w and its variability can be calculated depending on a reduction in the maximum range of travel $h_{100} = h_{max}$ to the closing element different value h_{100} and/or pre-setting n_i, it is necessary to know the closing characteristic shape and the mathematical description of the characteristic original form for which the element is designed. The former is provided by the manufacturer in the form of relevant charts or tables. The higher the number of measuring points, the more accurate the valve closing characteristic and, thereby, the more precise the valve inner authority a_w. Based on that and using appropriate mathematical formulae, it is possible to define the distribution of hydraulic resistances occurring in the valve and, consequently, the valve inner authority a_w, for any reduction in the maximum range of the closing element travel h_{max} and for any value of pre-setting n_i.

The following relations are derived in previous subsections to define the valve inner authority a_w based on the hydraulic resistances arising inside the valve:

- for valves with one adjustable current-regulation section of the medium flow, for full opening:

$$a_{w,\min} = \frac{r_{reg,100}}{r_{z,100}} = \frac{r_{reg,100}}{r_{reg,100} + r_k}.$$

So that $r_{reg,\,100}$ can be determined from this equation to define the quantity variability necessary to define changes in a_w, the value of $a_{w,\,min}$ is required. The essence of the indirect method of the parameter determination is that hydraulic resistances are calculated instead of being determined based on the measurement of relevant quantities. A comparison is made between the original regulation characteristic the valve is designed for and the measured characteristic, the shape of which results from the impact of inner authority a_w.

For a valve, as a local obstacle, the pressure loss relation can be written in the form presented in formula (3.58):

$$\Delta p_z = r_z \cdot \dot{V}_z^2.$$

Consequently, the pressure total loss in a valve with one adjustable section of the medium flow, fully open (100% opening) and with any other position (x% of full opening) of the current-regulation element will be, respectively:

$$\Delta p_{z,100} = \dot{V}_{100}^2 \cdot r_{z,100} = \dot{V}_{100}^2 \cdot \left(r_{reg,100} + r_k\right), \tag{4.34}$$

$$\Delta p_{z,x} = \dot{V}_x^2 \cdot r_{z,x} = \dot{V}_x^2 \cdot \left(r_{reg,x} + r_k\right). \tag{4.35}$$

Because the pressure on the valve is constant, the medium volume flows, according to the definition of flow factor k_v, can be replaced with the following flow factors corresponding to them:

$$\Delta p_{z,100} = k_{vs}^2 \cdot r_{z,100} = k_{vs}^2 \cdot \left(r_{reg,100} + r_k\right), \tag{4.36}$$

$$\Delta p_{z,x} = k_{v,x}^2 \cdot r_{z,x} = k_{v,x}^2 \cdot \left(r_{reg,x} + r_k\right). \tag{4.37}$$

Dividing the two equations by each other and taking account of the condition that $\Delta p_{z,\,100} = \Delta p_{z,\,x}$, the following is obtained:

$$\begin{aligned} \frac{k_{v,x}}{k_{vs}} &= \sqrt{\frac{r_{reg,100} + r_k}{r_{reg,x} + r_k}} = \sqrt{\frac{r_{z,100}}{r_{reg,x} + r_{z,100} - r_{reg,100}}} = \sqrt{\frac{1}{\frac{r_{reg,x}}{r_{z,100}} + 1 - \frac{r_{reg,100}}{r_{z,100}}}} \\ &= \sqrt{\frac{1}{\frac{r_{reg,100}}{r_{z,100}} \cdot \frac{r_{reg,x}}{r_{reg,100}} + 1 - \frac{r_{reg,100}}{r_{z,100}}}} = \sqrt{\frac{1}{a_{w,\min} \cdot \frac{r_{reg,x}}{r_{reg,100}} + 1 - a_{w,\min}}}. \end{aligned} \tag{4.38}$$

Considering also, as a consequence of formula (3.61), that:

$$\frac{r_{reg,x}}{r_{reg,100}} = \frac{1}{\left(\frac{k_{v,reg,x}}{k_{v,reg,100}}\right)^2}, \tag{4.39}$$

the result is:

$$\frac{k_{v,x}}{k_{vs}} = \sqrt{\frac{1}{1 - a_{w,\min} + \frac{a_{w,\min}}{\left(\frac{k_{v,reg,x}}{k_{v,reg,100}}\right)^2}}}, \tag{4.40}$$

or, after transformations:

$$\frac{k_{v,x}}{k_{vs}} = \frac{1}{\sqrt{1 + a_{w,\min}\left[\left(\frac{k_{v,reg,100}}{k_{v,reg,x}}\right)^2 - 1\right]}}. \tag{4.41}$$

Replacing the ratio between the flow factors with the specific relation $k_{\text{v, reg, x}}/k_{\text{v, reg, 100}} = f(h_{\text{x}}/h_{100})$, which describes the shape of the closing characteristic, the following is obtained:

- for the original linear characteristic:

$$\frac{k_{v,x}}{k_{vs}} = \frac{1}{\sqrt{1 + a_{w,\min}\left[\left(\frac{h_{100}}{h_x}\right)^2 - 1\right]}}, \tag{4.42}$$

- for the original equal-percentage characteristic:

$$\frac{k_{v,x}}{k_{vs}} = \frac{1}{\sqrt{1 + a_{w,\min}\left(\frac{1}{[\exp[c(h_x/h_{100} - 1)]]^2} - 1\right)}} = \frac{1}{\sqrt{1 + a_{w,\min}(\exp[2c \cdot (1 - h_x/h_{100})] - 1)}}, \tag{4.43}$$

For valves with two adjustable sections of the medium flow the derivation is similar. The difference is that term $r_{\text{z, 100}}$ includes an additional quantity r_{I} representing hydraulic resistance of the first stage of regulation. In its physical sense, it is like r_{k}, i.e. it is not subject to current regulation. In both cases, depending on the selected value of pre-setting n_{i}, a value can be applied which is different from the minimum value of inner authority a_{w} because inner authority is a function of the pre-setting, as described further below by means of relevant relations.

Having derived the formulae presented above, it is possible to demonstrate the relationship between authority a of the valve adjustable section and the valve inner authority a_{w}. Using the relation describing quantity a, transformed by means of flow factors and written as formula (2.3) included in Standard EN 215:2004 [5], the following equations are obtained:

- for the original linear characteristic:

$$a = 1 - \left(\frac{k_{vN}}{k_{vs}}\right)^2 = 1 - \frac{1}{1 + a_{w,\min}\left[\left(\frac{h_{100}}{h_N}\right)^2 - 1\right]}, \tag{4.44}$$

- for the original equal-percentage characteristic:

$$a = 1 - \left(\frac{k_{vN}}{k_{vs}}\right)^2 = 1 - \frac{1}{1 + a_{w,\min}\left(\frac{1}{[\exp[c\cdot(h_N/h_{100}-1)]]^2} - 1\right)}. \quad (4.45)$$

Transforming the derived relations describing the relative flow factor to define the sought value $a_{w,\min}$, the following notation is obtained:

– for the original linear relation:

$$a_{w,\min} = \frac{\left(\frac{k_{vs}}{k_{v,x}}\right)^2 - 1}{\left(\frac{h_{100}}{h_x}\right)^2 - 1}, \quad (4.46)$$

– for the original equal-percentage relation:

$$a_{w,\min} = \frac{\left(\frac{k_{vs}}{k_{v,x}}\right)^2 - 1}{\frac{1}{[\exp[c\cdot(h_x/h_{100}-1)]]^2} - 1} = \frac{\left(\frac{k_{vs}}{k_{v,x}}\right)^2 - 1}{\exp[2c\cdot(1 - h_x/h_{100})] - 1}. \quad (4.47)$$

The reduction in the closing element maximum range of travel h_{max} in valves with one adjustable section of the medium flow is related to determining the value of pre-setting n_i. Considering that the reciprocating motion is in this case achieved due to a turn of an element on a threaded spindle, and that the thread pitch is constant, the change in the element position h_x is proportional to the change in the value of pre-setting n_i, provided that the latter is scaled linearly. For this reason, the plug relative lift h_x/h_{100} can be replaced with relative pre-setting n_i/n_{max}, where n_i is the i-th pre-setting value for the plug lift h_x, and n_{max} denotes the pre-setting maximum value for the full relative lift h_{100}. This gives as follows:

– for the original linear relation:

$$a_{w,\min} = \frac{\left(\frac{k_{vs}}{k_{v,x}}\right)^2 - 1}{\left(\frac{n_{max}}{n_i}\right)^2 - 1}, \quad (4.48)$$

– for the original equal-percentage relation:

$$a_{w,\min} = \frac{\left(\frac{k_{vs}}{k_{v,x}}\right)^2 - 1}{\exp[2c\cdot(1 - n_i/n_{max})] - 1}. \quad (4.49)$$

It can be noticed easily, especially in the case of the formula for the calculation of the valve inner authority a_w with the original linear characteristic, that if the resultant characteristic, representing changes in the flow factor depending on changes in the values of pre-setting n_i, is linear, $a_w = 1$, according to the qualitative description of the phenomenon.

The minimum value of inner authority a_w, for a set reduction in the maximum range of travel $h_{100} = h_{max}$ to the closing element different value h_{100}, can be calculated for any relative opening degree h_x/h_{100} included in this range. A relevant computational example with a comment thereon is presented in the further part of this subsection.

The above derivation assumes the full, maximum (100% in the absolute scale) available travel range of the closing element: $h_{100} = h_{max}$. Nevertheless, as explained earlier, the notation can be carried directly onto any other range of the element travel h_{100} provided that regulation (closing) will proceed from the value downwards only. In such a case, 100% of the new travel of the element of the regulation second stage will be available then. As the available range of travel h_{max} is reduced, inner authority a_w rises from the initial value, marked as $a_{w,\,min}$, to $a_w = 1$ at full closing cutting off the flow completely, defining thereby the available range of changeability. The changeability depends on the closing curve original shape and on the value of $a_{w,\,min}$. The derived definition of a_w can be applied here. It defines the value and changeability of inner authority a_w, which means that it takes account of non-regulated hydraulic resistances arising inside the valve. Once the $a_{w,\,min}$ initial value is determined, from which the parameter will rise at a reduction in the available range of travel h_{max}, the currently non-regulated resistances are already taken into consideration and the changeable resistance of the regulated part will be related not to the entire valve but to this part only, at the part full absolute opening. Therefore, the relation may be written taking account of this fact, i.e. taking account of the formula defining the valve adjustable section authority *a* described earlier. Because the issue under consideration is the change in the medium volume flow at a constant differential pressure, relative volume flow ratio $\dot{V}_x/\dot{V}_{100}$ can be replaced using relative flow factor ratio $k_{v,\,x}/k_{vs}$. Bearing this in mind, it may be written that:

$$a_{w,x} = a_{w,\min} + \left(1 - a_{w,\min}\right) \cdot a = a_{w,\min} + \left(1 - a_{w,\min}\right) \cdot \left(1 - \left(\frac{k_{v,x}}{k_{vs}}\right)^2\right). \tag{4.50}$$

Substituting formula (4.41) in relation (4.50), the result is as follows:

$$a_{w,x} = a_{w,\min} + \left(1 - a_{w,\min}\right) \cdot \left(1 - \frac{1}{1 + a_{w,\min}\left[\left(\frac{k_{v,reg,100}}{k_{v,reg,x}}\right)^2 - 1\right]}\right). \tag{4.51}$$

– for the original linear characteristic the following expression is then obtained:

$$a_{w,x} = a_{w,\min} + \left(1 - a_{w,\min}\right) \cdot \left(1 - \frac{1}{1 + a_{w,\min}\left[\left(\frac{h_{100}}{h_x}\right)^2 - 1\right]}\right), \tag{4.52}$$

– for the original equal-percentage characteristic:

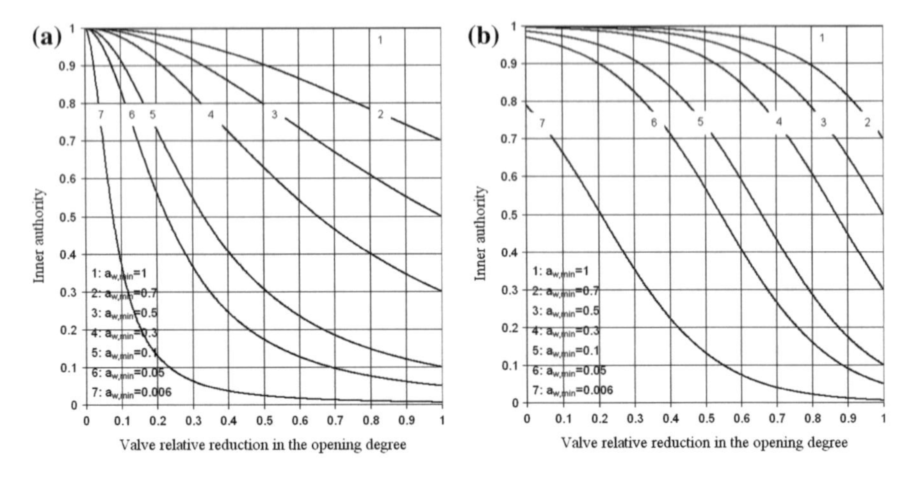

Fig. 4.7 The valve inner authority changeability as a function of a reduction in the closing element opening degree. **a** valve with the original linear characteristic; **b** valve with the original equal-percentage characteristic ($c = 3.22$)

$$a_{w,x} = a_{w,\min} + \left(1 - a_{w,\min}\right) \cdot \left(1 - \frac{1}{1 + a_{w,\min}(\exp[2c \cdot (1 - h_x/h_{100})] - 1)}\right). \tag{4.53}$$

Considering the formula defining relative pre-setting $n_i/n_{\max}$, the following equations are obtained:

- for the original linear characteristic:

$$a_{w,x} = a_{w,\min} + \left(1 - a_{w,\min}\right) \cdot \left(1 - \frac{1}{1 + a_{w,\min}\left[\left(\frac{n_{\max}}{n_i}\right)^2 - 1\right]}\right), \tag{4.54}$$

- for the original equal-percentage characteristic:

$$a_{w,x} = a_{w,\min} + \left(1 - a_{w,\min}\right) \cdot \left(1 - \frac{1}{1 + a_{w,\min}(\exp[2c \cdot (1 - n_i/n_{\max})] - 1)}\right). \tag{4.55}$$

In a valve with one adjustable section of the medium flow where setting initial throttling is not related to the reciprocating motion, the relative opening degree will not represent the relative lift but, for instance, the relative angle of rotation (e.g. the *Goeke Optimal* valve). Figure 4.7 presents curves plotted based on the formulae given above and illustrating changes in the values of inner authority a_w for a valve with one adjustable section of the medium flow depending on the reduction in the closing element opening degree. The curves are plotted for a valve with the original linear and equal-percentage ($c = 3.22$) characteristics, for a few values of $a_{w,\min}$.

The relation presented above and describing changes in inner authority a_w as a function of a reduction in the closing element maximum range of travel $h_{100} =$

$h_{\max}$ can be found using the hydraulic resistance concept in two ways: either by determining the values of individual hydraulic resistances or taking the expression for the ratio between them as the starting point. Formula (4.10) gives the following equation:

$$r_{reg,100} = a_{w,\min} \cdot r_{z,100} = a_{w,\min} \cdot (r_{reg,100} + r_k) = a_{w,\min} \cdot \frac{1}{k_{vs}^2}. \tag{4.56}$$

It may therefore be written that:

$$r_k = r_{z,100} - r_{reg,100} = \frac{1}{k_{vs}^2} - r_{reg,100} = (1 - a_{w,\min}) \cdot \frac{1}{k_{vs}^2}. \tag{4.57}$$

– for the x-th reduction in the closing element range of travel (i-th position of the throttling element), it may be written that:

$$r_{reg,x} + r_k = \frac{1}{k_{v,x}^2}. \tag{4.58}$$

The result is the following equation:

$$r_{reg,x} = \frac{1}{k_{v,x}^2} - r_k = \frac{1}{k_{v,x}^2} - (1 - a_{w,\min}) \cdot \frac{1}{k_{vs}^2}. \tag{4.59}$$

The formulae presented above enable determination of the values of hydraulic resistances arising inside the valve. Their substitution in the relation describing inner authority a_w results in the following expression:

$$a_{w,x} = \frac{r_{reg,x}}{r_{z,x}} = \frac{r_{reg,x}}{r_{reg,x} + r_k} = \frac{\frac{1}{k_{v,x}^2} - (1 - a_{w,\min}) \cdot \frac{1}{k_{vs}^2}}{\frac{1}{k_{v,x}^2} - (1 - a_{w,\min}) \cdot \frac{1}{k_{vs}^2} + (1 - a_{w,\min}) \cdot \frac{1}{k_{vs}^2}} \tag{4.60}$$

$$= \frac{\frac{1}{k_{v,x}^2} - (1 - a_{w,\min}) \cdot \frac{1}{k_{vs}^2}}{\frac{1}{k_{v,x}^2}} = 1 - (1 - a_{w,\min}) \cdot \frac{k_{v,x}^2}{k_{vs}^2}$$
$$= 1 + (a_{w,\min} - 1) \cdot \left(\frac{k_{v,x}}{k_{vs}}\right)^2. \tag{4.61}$$

Using the ratio between hydraulic resistances, the following can be written:

$$\frac{r_{reg,100}}{r_k} = \frac{a_{w,\min}}{1 - a_{w,\min}}. \tag{4.62}$$

Equation (4.62) defines the distribution of hydraulic resistances inside the valve. In order to determine their values, it is necessary to know the medium maximum volume flow rate for a given value of differential pressure. This information is provided by the valve maximum flow factor k_{vs}, which, by definition, describes the medium volume flow in m^3/h at the pressure drop of 1 bar. Taking this into account, and according to earlier derivations, it may be written that:

– for the full range of the adjustable section travel (both elements of regulation):

$$r_{reg,100} + r_k = \frac{1}{k_{vs}^2}. \tag{4.63}$$

This gives the following equation:

$$r_k = \frac{1}{k_{vs}^2} - r_{reg,100}. \tag{4.64}$$

Substituting the relations in Formula (4.62), the following expressions are obtained:

$$r_{reg,100} = \frac{\frac{a_{w,\min}}{1-a_{w,\min}}}{k_{vs}^2 \cdot \left(1 + \frac{a_{w,\min}}{1-a_{w,\min}}\right)}, \tag{4.65}$$

$$r_k = \frac{1}{k_{vs}^2} - \frac{\frac{a_{w,\min}}{1-a_{w,\min}}}{k_{vs}^2 \cdot \left(1 + \frac{a_{w,\min}}{1-a_{w,\min}}\right)}. \tag{4.66}$$

– for the x-th reduction in the closing element range of travel (i-th position of the throttling element), the following relation is obtained:

$$r_{reg,x} + r_k = \frac{1}{k_{v,x}^2}. \tag{4.67}$$

This gives the following equation:

$$r_{reg,x} = \frac{1}{k_{v,x}^2} - r_k. \tag{4.68}$$

Dividing be each other Formulae (4.36) and (4.37) describing total hydraulic resistances and pressure losses, respectively, or Formulae (4.63) and (4.68) that, **respectively**, use the valve flow factors for the 100% opening and the x-th reduction in the regulating part range of travel, the following expression is obtained:

$$\frac{k_{v,x}}{k_{vs}} = \sqrt{\frac{r_{reg,100} + r_k}{r_{reg,x} + r_k}}. \tag{4.69}$$

Formula (4.69) gives the sought value of $r_{reg,\,x}$, which is:

$$r_{reg,x} = \frac{r_{reg,100} + r_k}{\left(\frac{k_{v,x}}{k_{vs}}\right)^2} - r_k. \tag{4.70}$$

Replacing hydraulic resistances with the derived Formulae (4.65) and (4.66), the result is as follows:

$$r_{reg,x} = \frac{\frac{1}{k_{vs}^2}}{\left(\frac{k_{v,x}}{k_{vs}}\right)^2} - \frac{1}{k_{vs}^2} + \frac{\frac{a_{w,\min}}{1-a_{w,\min}}}{k_{vs}^2 \cdot \left(1 + \frac{a_{w,\min}}{1-a_{w,\min}}\right)} = \frac{1}{k_{vs}^2} \cdot \left(\frac{1}{\left(\frac{k_{v,x}}{k_{vs}}\right)^2} + \frac{\frac{a_{w,\min}}{1-a_{w,\min}}}{1 + \frac{a_{w,\min}}{1-a_{w,\min}}} - 1\right). \quad (4.71)$$

If substituted in relation (4.11) describing the sought inner authority, the following equation is obtained:

$$a_{w,x} = \frac{r_{reg,x}}{r_{z,x}} = \frac{r_{reg,x}}{r_{reg,x} + r_k}$$

$$= \frac{\frac{1}{k_{vs}^2} \cdot \left(\frac{1}{\left(\frac{k_{v,x}}{k_{vs}}\right)^2} + \frac{\frac{a_{w,\min}}{1-a_{w,\min}}}{1+\frac{a_{w,\min}}{1-a_{w,\min}}} - 1\right)}{\frac{1}{k_{vs}^2} \cdot \left(\frac{1}{\left(\frac{k_{v,x}}{k_{vs}}\right)^2} + \frac{\frac{a_{w,\min}}{1-a_{w,\min}}}{1+\frac{a_{w,\min}}{1-a_{w,\min}}} - 1\right) + \frac{1}{k_{vs}^2} - \frac{\frac{a_{w,\min}}{1-a_{w,\min}}}{k_{vs}^2 \cdot \left(1+\frac{a_{w,\min}}{1-a_{w,\min}}\right)}}$$

$$= \frac{\frac{1}{k_{vs}^2} \cdot \left(\frac{1}{\left(\frac{k_{v,x}}{k_{vs}}\right)^2} + \frac{\frac{a_{w,\min}}{1-a_{w,\min}}}{1+\frac{a_{w,\min}}{1-a_{w,\min}}} - 1\right)}{\frac{1}{k_{vs}^2} \cdot \left(\frac{1}{\left(\frac{k_{v,x}}{k_{vs}}\right)^2} + \frac{\frac{a_{w,\min}}{1-a_{w,\min}}}{1+\frac{a_{w,\min}}{1-a_{w,\min}}} - 1\right) + \frac{1}{k_{vs}^2} - \frac{\frac{a_{w,\min}}{1-a_{w,\min}}}{k_{vs}^2 \cdot \left(1+\frac{a_{w,\min}}{1-a_{w,\min}}\right)}}$$

$$= \frac{\frac{1}{\left(\frac{k_{v,x}}{k_{vs}}\right)^2} + \frac{\frac{a_{w,\min}}{1-a_{w,\min}}}{1+\frac{a_{w,\min}}{1-a_{w,\min}}} - 1}{\frac{1}{\left(\frac{k_{v,x}}{k_{vs}}\right)^2}} = 1 + \left(\frac{k_{v,x}}{k_{vs}}\right)^2 \cdot \frac{\frac{a_{w,\min}}{1-a_{w,\min}}}{1 + \frac{a_{w,\min}}{1-a_{w,\min}}} - \left(\frac{k_{v,x}}{k_{vs}}\right)^2$$

$$= 1 + \left(\frac{k_{v,x}}{k_{vs}}\right)^2 \cdot \left(\frac{\frac{a_{w,\min}}{1-a_{w,\min}}}{1 + \frac{a_{w,\min}}{1-a_{w,\min}}} - 1\right) = 1 + \left(\frac{k_{v,x}}{k_{vs}}\right)^2 \cdot \left(\frac{\frac{a_{w,\min}}{1-a_{w,\min}} - 1 - \frac{a_{w,\min}}{1-a_{w,\min}}}{1 + \frac{a_{w,\min}}{1-a_{w,\min}}}\right)$$

$$= 1 + \left(\frac{k_{v,x}}{k_{vs}}\right)^2 \cdot \left(\frac{-1}{\frac{1-a_{w,\min}+a_{w,\min}}{1-a_{w,\min}}}\right) = 1 + \left(a_{w,\min} - 1\right) \cdot \left(\frac{k_{v,x}}{k_{vs}}\right)^2$$

$$= a_{w,\min} + \left(1 - a_{w,\min}\right) \cdot \left(1 - \left(\frac{k_{v,x}}{k_{vs}}\right)^2\right). \quad (4.72)$$

The ultimate relation is thus identical with Formula (4.50) obtained from the first derivation. Replacement of the relative flow factor with function $k_{\text{v, reg, x}}/k_{\text{v, reg, 100}} = f(h_x/h_{100})$, defining the shape of the closing characteristic of the entire valve (not of the closing element only), gives Formulae (4.52)–(4.55) already mentioned above.

The presented relations need an additional comment. The ultimate notations of Formulae (4.51)–(4.53), where the valve relative flow factors are replaced by relevant mathematical descriptions of the closing characteristic functions taking account of the pre-setting, may give results a bit different from those produced by the initial relations. Because the formulae are identical, the difference does not result from their different form. The cause is the changeability of the calculated quantity $a_{\text{w, min}}$, if it occurs and if it is not taken into account in the term related to the assumed closing function related to relative pre-setting $n_i/n_{\max}$. The initial notations [cf. (4.50)] do not make use of the original closing function description. Instead, they use flow factors and, due to that, they are not burdened with the need to create the function of the $a_{w,\,min}$ changeability for this term. But the downside of the initial relations is the fact that it is impossible to define the value of $a_{w,\,x}$ without knowing the value of $k_{v,\,x}$. Therefore, in order to apply the formulae to calculate $a_{w,\,x}$, it is necessary, prior to the calculations, to perform measurements of the flow factor changeability depending on the pre-setting, which is not needed in the case of the ultimate relations. Still, in both cases it is necessary to define the initial value of $a_{w,\,min}$. Calculating hydraulic resistances $r_{reg,\,100}$, $r_{reg,\,x}$ and r_k, it is favourable to use the arithmetic mean of authority a_w for all available values of the pre-setting.

- for a valve with two adjustable sections of the medium flow, as described earlier, the closing element range of travel h_{100} is constant, and inner authority varies from the value of $a_{w,\,max}$, achieved at $r_{I,\,100}$ for the throttling element full opening. Using the hydraulic resistance concept, according to the derivations presented earlier (4.15), the following can be written:

$$a_{w,100} \equiv a_{w,\max} = \frac{r_{II,100}}{r_{z,100}} = \frac{r_{II,100}}{r_{reg,100} + r_{I,100} + r_k} = \frac{r_{II,100}}{r_{II,100} + r_{I,100} + r_k}.$$

Consequently, the following equations are true:

$$r_{II,100} = a_{w,\max} \cdot (r_{II,100} + r_{I,100} + r_k) = a_{w,\max} \cdot \frac{1}{k_{vs}^2}, \tag{4.73}$$

and:

$$r_{I,100} + r_k = r_{z,100} - r_{II,100} = \frac{1}{k_{vs}^2} - r_{II,100} = (1 - a_{w,\max}) \cdot \frac{1}{k_{vs}^2}. \tag{4.74}$$

– for the throttling element i-th pre-setting, and expressing the non-regulated hydraulic resistances in a single term as $r_{I,i} + r_k = r_{k+I,\,i}$, the following formula is obtained:

$$r_{II,100} + r_{k+I,i} = \frac{1}{k_{v,x}^2}. \tag{4.75}$$

This gives:

$$r_{k+I,i} = \frac{1}{k_{v,x}^2} - r_{II,100} = \frac{1}{k_{v,x}^2} - a_{w,\max} \cdot \frac{1}{k_{vs}^2}. \tag{4.76}$$

If substituted in relation (4.16) describing inner authority, the following expression is obtained:

$$a_{w,x} = \frac{r_{II,100}}{r_{II,100} + r_{k+I,i}} = \frac{a_{w,\max} \cdot \frac{1}{k_{vs}^2}}{a_{w,\max} \cdot \frac{1}{k_{vs}^2} + \frac{1}{k_{v,x}^2} - a_{w,\max} \cdot \frac{1}{k_{vs}^2}} = a_{w,\max} \cdot \left(\frac{k_{v,x}}{k_{vs}}\right)^2. \tag{4.77}$$

Starting with relation (4.62), which defines the distribution of resistances arising inside the valve, the result is as follows:

$$\frac{r_{II,100}}{r_{k+I,100}} = \frac{a_{w,\max}}{1 - a_{w,\max}}. \tag{4.78}$$

Using the flow factor concept, it may be written that:

– for the throttling element full opening:

$$r_{II,100} + r_{k+I,100} = \frac{1}{k_{vs}^2}. \tag{4.79}$$

This leads to the following formula:

$$r_{k+I,100} = \frac{1}{k_{vs}^2} - r_{II,100}. \tag{4.80}$$

– for the throttling element *i*-th pre-setting, like in Formula (4.75), it may be written that:

$$r_{II,100} + r_{k+I,i} = \frac{1}{k_{v,x}^2}.$$

As written in (4.76), this gives:

$$r_{k+I,i} = \frac{1}{k_{v,x}^2} - r_{II,100}.$$

The relations presented above lead to formulae which are functionally identical to those derived for a valve with one adjustable section of the medium flow, except that in resistances which are not subject to current regulation the r_{k+I} sum appears. Following the same procedure as in the previous case, it may be written that:

– for the throttling element full opening:

$$r_{II,100} = \frac{\frac{a_{w,\max}}{1-a_{w,\max}}}{k_{vs}^2 \cdot \left(1 + \frac{a_{w,\max}}{1-a_{w,\max}}\right)}, \tag{4.81}$$

$$r_{k+I,100} = \frac{1}{k_{vs}^2} - \frac{\frac{a_{w,\max}}{1-a_{w,\max}}}{k_{vs}^2 \cdot \left(1 + \frac{a_{w,\max}}{1-a_{w,\max}}\right)}. \tag{4.82}$$

– for the throttling element i-th pre-setting:

$$r_{II,100} = \frac{\frac{a_{w,\max}}{1-a_{w,\max}}}{k_{vs}^2 \cdot \left(1 + \frac{a_{w,\max}}{1-a_{w,\max}}\right)},$$

$$r_{k+I,i} = \frac{1}{k_{v,x}^2} - \frac{\frac{a_{w,\max}}{1-a_{w,\max}}}{k_{vs}^2 \cdot \left(1 + \frac{a_{w,\max}}{1-a_{w,\max}}\right)}. \tag{4.83}$$

Dividing by each other expressions (4.79) and (4.75) describing the valve total hydraulic resistances for the 100% opening and for the throttling element i-th pre-setting, respectively, the following formula is obtained:

$$\frac{k_{v,x}}{k_{vs}} = \sqrt{\frac{r_{II,100} + r_k + r_{I,100}}{r_{II,100} + r_k + r_{I,i}}} = \sqrt{\frac{r_{II,100} + r_{k+I,100}}{r_{II,100} + r_{k+I,i}}}, \tag{4.84}$$

which means that the sought value of $r_{k+I,i}$ is:

$$r_{k+I,i} = \frac{r_{II,100} + r_{k+I,100}}{\left(\frac{k_{v,x}}{k_{vs}}\right)^2} - r_{II,100}. \tag{4.85}$$

Substituting the formulae presented above in the relation describing inner authority (4.16), the following expression is obtained:

$$a_{w,x} = \frac{r_{II,100}}{r_{z,x}} = \frac{r_{II,100}}{r_{reg,100} + r_{k+I,i}} = \frac{\frac{\frac{a_{w,\max}}{1-a_{w,\max}}}{k_{vs}^2 \cdot \left(1+\frac{a_{w,\max}}{1-a_{w,\max}}\right)}}{\frac{\frac{a_{w,\max}}{1-a_{w,\max}}}{k_{vs}^2 \cdot \left(1+\frac{a_{w,\max}}{1-a_{w,\max}}\right)} + \frac{1}{k_{v,x}^2} - \frac{\frac{a_{w,\max}}{1-a_{w,\max}}}{k_{vs}^2 \cdot \left(1+\frac{a_{w,\max}}{1-a_{w,\max}}\right)}}$$

$$= \frac{\frac{\frac{a_{w,\max}}{1-a_{w,\max}}}{\left(1+\frac{a_{w,\max}}{1-a_{w,\max}}\right)}}{\frac{\frac{a_{w,\max}}{1-a_{w,\max}}}{\left(1+\frac{a_{w,\max}}{1-a_{w,\max}}\right)} + \left(\frac{k_{vs}}{k_{v,x}}\right)^2 - \frac{\frac{a_{w,\max}}{1-a_{w,\max}}}{\left(1+\frac{a_{w,\max}}{1-a_{w,\max}}\right)}} = \left(\frac{k_{v,x}}{k_{vs}}\right)^2 \cdot \frac{\frac{a_{w,\max}}{1-a_{w,\max}}}{\left(1 + \frac{a_{w,\max}}{1-a_{w,\max}}\right)}$$

$$= a_{w,\max} \cdot \left(\frac{k_{v,x}}{k_{vs}}\right)^2.$$

It should be remembered that flow factors $k_{v,x}$ and k_{vs} depending on pre-setting n_i have to be related to the position of the closing element full available range of travel $h_{100} = h_{\max}$. Even if flow factors $k_{v,x}$ and k_{vs} are related to the closing element identical but intermediate lift, the results will be incorrect, as confirmed by the experimental analysis conducted in Sect. (5.3). This situation can be illustrated using a computational example. Assume that a given valve with two adjustable sections of the medium flow is characterized by the following data (units omitted here):

- $r_{\text{reg100}} = 5$, $r_{\text{k + I, 1}} = 5$ and $r_{\text{k + I, 2}} = 11.67$
- the original closing characteristic is linear.

At full opening and pre-setting $n_i = 1$, the valve will be characterized by the flow factor value at the level of:

$$k_{vs} = [k_{v100}]_{n_i=1} = \sqrt{\frac{\Delta p_z}{r_{reg,100} + r_{k+I,1}}} = \sqrt{\frac{1}{5+5}} = 0.316$$

and the inner authority value at the level of:

$$a_{w,\max} = \frac{r_{reg100}}{r_{reg100} + r_{k+I,1}} = \frac{5}{5+5} = 0.5.$$

For the closing element intermediate opening of $h_x/h_{100} = 1/3 \approx 33.33\%$ for example, and for the same pre-setting n_{i}, the valve will obviously be characterized by the same value of inner authority a_{w} and a different value of flow factor k_{k} because the closing element hydraulic resistance will change. Due to the assumption that the original closing characteristic is linear, it may be written, in accordance with the relations mentioned earlier, that:

$$\frac{r_{reg,x}}{r_{reg,100}} = \frac{1}{\left(\frac{k_{v,reg,x}}{k_{v,reg,100}}\right)^2} = \frac{1}{\left(\frac{h_x}{h_{100}}\right)^2} \Rightarrow r_{reg,x} = \frac{r_{reg,100}}{\left(\frac{h_x}{h_{100}}\right)^2} \Rightarrow r_{reg,33,33} = \frac{5}{(1/3)^2} = 45.$$

Therefore, the flow factor will total:

$$k_{v,x} = \left[k_{v,33,33}\right]_{n_i=1} = \sqrt{\frac{\Delta p_z}{r_{reg,33,33} + r_{k+I,1}}} = \sqrt{\frac{1}{45+5}} = 0.1414.$$

If pre-setting $n_i = 2$ is selected, the following flow factors will be obtained respectively for the closing element full and intermediate, set, opening degree:

$$k_{vs} = [k_{v100}]_{n_i=2} = \sqrt{\frac{\Delta p_z}{r_{reg100} + r_{k+I,2}}} = \sqrt{\frac{1}{5+11.67}} = 0.245,$$

$$k_{v,x} = [k_{v,33,33}]_{n_i=2} = \sqrt{\frac{\Delta p_z}{r_{reg33,33} + r_{k+I,2}}} = \sqrt{\frac{1}{45+11.67}} = 0.1328$$

and the inner authority values at the level of:

$$a_{w,x} = \frac{r_{reg100}}{r_{reg100} + r_{k+I,1}} = \frac{5}{5+11.67} = 0.3.$$

In order to get the final results, the obtained values should be substituted in the relation describing inner authority a_{w} as a function of pre-setting n_{i}. The calculations performed for the closing element full (100% in the absolute scale) and set intermediate (33.33% in the absolute scale) opening degree, will give the following results, respectively:

$$a_{w,x} = a_{w,\max} \cdot \left(\frac{k_{v,x}}{k_{vs}}\right)^2 = a_{w,\max} \cdot \left(\frac{[k_{v100}]_{n_i=2}}{[k_{v100}]_{n_i=1}}\right)^2 = 0.5 \cdot \left(\frac{0.245}{0.316}\right)^2 = 0.3,$$

$$a_{w,x} = a_{w,\max} \cdot \left(\frac{k_{v,x}}{k_{vs}}\right)^2 = a_{w,\max} \cdot \left(\frac{\left[k_{v,33,33}\right]_{n_i=2}}{\left[k_{v,33,33}\right]_{n_i=1}}\right)^2 = 0.5 \cdot \left(\frac{0.1328}{0.1414}\right)^2 = 0.44.$$

In the first case, the obtained value agrees with the required one, but in the second—it does not. This proves that, while calculating inner authority a_w for a valve with two adjustable sections of the medium flow, substituting the flow factor values for the closing element intermediate positions is incorrect.

If the valve relative flow factor is to be written taking account of a specific mathematical description of the throttling characteristic so that the throttling element relative opening degree can be applied, the minimum value of the throttling element authority $a_{w,\mathrm{I},\min}$, that will deform the element original regulation characteristic, has to be established. In this case therefore, from the point of view of the throttling element, the other hydraulic resistances in the valve will be the resistance of the valve body and of the closing element. For the throttling element full opening, the authority of this part will have the minimum value; consequently, the value of the closing element authority will be the maximum one. This leads to the following formula describing the changeability of the valve closing element inner authority as a function of the throttling element opening degree:

$$a_{w,x} = a_{w,\max} \cdot \left(\frac{1}{1 + a_{w,I,\min}\left[\left(\frac{k_{vs}}{k_{v,x}}\right)^2 - 1\right]}\right). \tag{4.86}$$

Substituting relevant mathematical descriptions, the result is as follows:

- for a valve with the original linear throttling characteristic:

$$a_{w,x} = a_{w,\max} \cdot \left(\frac{1}{1 + a_{w,I,\min}\left[\left(\frac{n_{\max}}{n_i}\right)^2 - 1\right]}\right), \tag{4.87}$$

- for a valve with the original equal-percentage throttling characteristic:

$$a_{w,x} = a_{w,\max} \cdot \left(\frac{1}{1 + a_{w,I,\min}\left[\exp[2c \cdot (1 - n_i/n_{\max})] - 1\right]}\right). \tag{4.88}$$

The value of the throttling element authority $a_{w,\mathrm{I},\min}$ can be determined indirectly, in the same manner as in the case of the closing element authority, i.e. by transforming the relation describing the relative flow factor as a function of the opening degree to express the sought quantity. The calculations can be performed for any position of the closing element. However, due to the recommendation stipulating that heating circuits with thermostats should be designed for proportional range $X_p = 2$ K [35], the throttling element authority $a_{w,\,I,\,min}$ has to be determined for the closing element position $h_x/h_{100} = h_x/h_{max}$, which corresponds at least to this proportional range value.

If the mathematical description of the analysed element original regulation characteristic is unknown, in order to determine the shape of the initial curve, an analysis and a measurement can be performed of the changeability of the medium flow surface area A_x created between the element and the seat depending on the opening degree (e.g. the plug lift). If the plug moves inside the seat hole, measurements should be carried out of the closing plug geometry (e.g. diameter) changeability along the plug height. Based on that, the changeability of the medium flow surface area created between the seat hole perimeter and the plug should be calculated (cf. the *Danfoss MSV-C* valve, Fig. 2.25b). Using these data, it is possible to determine the changeability of the flow factor. If the plug moves not within but over the seat hole, changing the distance from the hole to the plug face (e.g. the *Herz TS-90-V* and the *Danfoss MSV-I* DN20 valves, cf. Figs. 2.26 and 2.28), the surface area of the created side surface should be measured. Determining the sought curve based on such measurements and comparing it to the curve obtained from the measurement of the closing characteristic, it will be found that their shapes are different. This is the effect of the pressure distribution inside the valve, which deforms the original characteristic and which is expressed as *inner authority*.

4.1.3 Regulation Valve Outer Authority

A valve designed to obtain a given regulation characteristic is always installed in a pipework characterized by a certain hydraulic resistance. The resistance reduces the impact of changes in the valve setting on the medium volume flow rate in the circuit. Arising outside the valve, the impact is referred to as the valve *outer authority*. In qualitative terms, the issue concerns the degree of deformation of the regulation characteristic of the entire valve. At the circuit given available pressure, the medium volume flow rate is affected not only by the valve but also by the pipework. The relation defining outer authority a_z may be written in the following way:

$$a_z = \frac{\Delta p_z}{\Delta p_{ob}}. \tag{4.89}$$

It can be seen that the higher the valve pressure loss Δp_z, the higher the value of outer authority a_z and vice versa. The authority minimum value is obtained for the valve regulating element maximum range of travel h_{max}, when the element hydraulic resistance is the smallest.

At the full (100%), as well as at any other $x\%$ range of travel of the valve regulating element, corresponding to a given pre-setting n_i but not to the element any intermediate position, it may be written that:

$$a_{z,100} \equiv a_{z,\min} = \frac{\Delta p_{z,100}}{\Delta p_{z,100} + \Delta p_{str,100}} = \frac{\Delta p_{II,100} + \Delta p_{I,100} + \Delta p_k}{\Delta p_{II,100} + \Delta p_{I,100} + \Delta p_k + \Delta p_{str,100}}, \tag{4.90}$$

$$a_{z,x} = \frac{\Delta p_{z,x}}{\Delta p_{ob}} = \frac{\Delta p_{z,x}}{\Delta p_{z,x} + \Delta p_{str,x}} = \frac{\Delta p_{II,100} + \Delta p_{I,i} + \Delta p_k}{\Delta p_{II,100} + \Delta p_{I,i} + \Delta p_k + \Delta p_{str,x}}, \tag{4.91}$$

or, using hydraulic resistances:

$$a_{z,100} \equiv a_{z,\min} = \frac{r_{z,100}}{r_{z,100} + r_{str}} = \frac{r_{II,100} + r_{I,100} + r_k}{r_{II,100} + r_{I,100} + r_k + r_{str}}, \tag{4.92}$$

$$a_{z,x} = \frac{r_{z,x}}{r_{ob}} = \frac{r_{z,x}}{r_{z,x} + r_{str}} = \frac{r_{II,100} + r_{I,i} + r_k}{r_{II,100} + r_{I,i} + r_k + r_{str}}. \tag{4.93}$$

If there is no element of the regulation first stage, 0 should be substituted for the quantities corresponding to it. In such a case, $a_{z,x} \equiv a_{z,100}$ and the closing element range of travel for the valve totals 100% in the absolute scale. If on the other hand initial throttling is set by reducing the range of the element travel, the 100% should be understood as a relative value related to the current available range and not as the maximum range $h_{\max}$ for no reduction. Designs realizing the double-regulation principle with one adjustable section of the medium flow, where the closing element range of travel is constant (e.g. the historical *Goeke Optimal* designs or the Polish *708*-type valve) are not very common nowadays.

4.1.4 Regulation Valve Total Authority

Because by definition the part responsible for current regulation of the flow in a circuit is the valve closing element, it is essential what regulation characteristic is obtained after it is installed in the target body or circuit. All in all then, the characteristic deformation is caused by two factors defined together as *total authority* a_c. For this reason, hydraulic resistance, or the medium pressure loss in the closing element, should be related to the sum of hydraulic resistances, or the medium pressure losses in the regulated circuit, being the product of the valve inner and outer authority a_w and a_z, respectively, according to the following formula:

$$a_c = a_w \cdot a_z = \frac{\Delta p_{reg}}{\Delta p_z} \cdot \frac{\Delta p_z}{\Delta p_{ob}} = \frac{\Delta p_{reg}}{\Delta p_{ob}} = \frac{\Delta p_{reg}}{\Delta p_z + \Delta p_{str}}, \tag{4.94}$$

Using hydraulic resistances, the result is:

$$a_c = \frac{r_{reg}}{r_{ob}}. \tag{4.95}$$

Taking account of possible pre-settings, relation (4.95) can be expressed as:

$$a_{c,100} = a_{w,100} \cdot a_{z,100}, \tag{4.96}$$

$$a_{c,x} = a_{w,x} \cdot a_{z,x}. \tag{4.97}$$

- In relation to a valve with one adjustable section of the medium flow, considering the relations written in Sects. 4.1.2 and 4.1.3, defining a_w and a_z, the following equations are obtained:

$$a_{c,100} \equiv a_{c,\min} = \frac{\Delta p_{reg,100}}{\Delta p_{z,100} + \Delta p_{str,100}} = \frac{\Delta p_{reg,100}}{\Delta p_{reg,100} + \Delta p_k + \Delta p_{str,100}}, \tag{4.98}$$

$$a_{c,x} = \frac{\Delta p_{reg,x}}{\Delta p_{z,x} + \Delta p_{str,x}} = \frac{\Delta p_{reg,x}}{\Delta p_{reg,x} + \Delta p_k + \Delta p_{str,x}}, \tag{4.99}$$

Using hydraulic resistances, the result is:

$$a_{c,100} \equiv a_{c,\min} = \frac{r_{reg,100}}{r_{z,100} + r_{str}} = \frac{r_{reg,100}}{r_{reg,100} + r_k + r_{str}}, \tag{4.100}$$

$$a_{c,x} = \frac{r_{reg,x}}{r_{z,x} + r_{str}} = \frac{r_{reg,x}}{r_{reg,x} + r_k + r_{str}}. \tag{4.101}$$

- For a valve with two adjustable sections of the medium flow, it may be written that:

$$a_{c,100} \equiv a_{c,\max} = \frac{\Delta p_{II,100}}{\Delta p_{II,100} + \Delta p_{I,100} + \Delta p_k + \Delta p_{str,100}}, \tag{4.102}$$

$$a_{c,x} = \frac{\Delta p_{II,100}}{\Delta p_{II,100} + \Delta p_{I,i} + \Delta p_k + \Delta p_{str,x}}, \tag{4.103}$$

Using hydraulic resistances, the result is:

$$a_{c,100} \equiv a_{c,\max} = \frac{r_{II,100}}{r_{II,100} + r_{I,100} + r_k + r_{str}}, \tag{4.104}$$

$$a_{c,x} = \frac{r_{II,100}}{r_{II,100} + r_{I,i} + r_k + r_{str}}. \tag{4.105}$$

Summing up the descriptions of inner authority a_w, outer authority a_z and total authority a_c resulting therefrom, two situations need to be distinguished that correspond to the two applied and described methods of double regulation. In the case of valves with one adjustable section of the medium flow, a rise in initial throttling involves a rise both in outer and inner authority (a_z and a_w, respectively). Consequently, there is also an increase in total authority a_c. In the case of valves with two adjustable sections of the medium flow, a rise in initial throttling involves a rise in outer authority a_z, but inner authority a_w is decreased. Due to that, total authority a_c is also decreased.

Comparing the relations defining inner authority a_w, outer authority a_z and total authority a_c, it can be seen that, if the appropriate interpretation is unknown, the calculation of a_c can be performed using any number of combinations of the values of a_w and a_z. In every case, however, the relation presents the ratio between pressure losses on the valve current-regulation element and total pressure losses in the circuit. At present, both in practice and in theoretical studies, inner authority a_w and outer authority a_z often relate to the maximum absolute range of travel h_{max} of the adjustable section of the medium flow (of course, if total authority is calculated being split into the two factors, the method based on formulae (2.2), (4.1) and (4.2) does not provide such information whatsoever). As explained earlier, this is only right if current regulation proceeds from this maximum available range of travel h_{max} of the element of the second stage of regulation, i.e. if the valve operates in the circuit with the maximum value of pre-setting n_i corresponding to such an opening degree. As expected, such cases are very rare. In order to satisfy this postulate, in each and every design case a valve should be selected such that at full opening it should be characterized by a flow factor precisely equal to the required value. It can easily be concluded that manufacturers would then have to offer a huge number of valves of a given type, each with a specific flow factor. The initial regulation concept was developed specifically to make it possible for a single valve to balance hydraulic resistances in a wide range of required values.

In relation to the valve authority and control properties, the distribution of hydraulic resistances and the medium pressure losses in a double-regulation valve can be considered in two ways. Because the solution with two adjustable sections of the medium flow deteriorates the closing element regulation characteristic by reducing the share of this element in total resistance, the throttling element resistance can be included in resistance r_{str} of the pipework. According to formula (4.93), this will cause a drop in the valve outer authority a_z and, consequently, a rise in inner authority a_w according to formula (4.11). Considering the obtained calculation results (the final value of a_c), in quantitative terms this will be right. But the resistance can also be included in the hydraulic resistance of the valve part which is not subject to current regulation. This will cause a decrease in inner authority a_w and a rise in outer authority a_z. Both computationally and from the point of view of the phenomenon physics, adopting the second approach is more justified because the described process takes place inside and not beyond the valve. Moreover, the valve characteristics are taken in conditions where there are practically no losses in the network of connected pipes ($a_z = 1$). Apart from that, the valve may be installed in a pipework with any pressure distribution and with different ratios between its own hydraulic resistance and the circuit total resistance. Due to that, the first method would be less clear during calculations. Therefore, all resistances arising inside the valve and generated beyond the element responsible for current regulation should be included in inner authority a_w, and not in the entire pipework resistance r_{str}, despite the fact that the quantitative effects are the same as those covered by total authority a_c.

Taking account of the described phenomena, it is possible to derive relations defining the changes in the medium volume flow depending on the opening degree of the closing element of a valve installed in the pipework. The derivation process runs in the same manner as in the case of determining the relative flow factor of a valve which is not installed in the pipework. The difference is that here the hydraulic resistances of the network of connected pipes are taken into consideration. For a valve, as a local obstacle, the pressure loss relation can be written in the form of formula (3.58):

$$\Delta p_z = r_z \cdot \dot{V}_z^2.$$

In relation to the other part of the pipework, it may be written that:

$$\Delta p_{str} = r_{str} \cdot \dot{V}_z^n. \tag{4.106}$$

Due to the fact that in practice the other part of the pipework includes a great number of local obstacles, which means that the share of the pressure losses arising on them is relatively high compared to the entire network of pipes, it may be assumed that the exponent of the characteristic is $m = 2$. Therefore, the total pressure loss in the circuit for a fully open valve (100% opening) and with any other position (x% of full opening) of the current-regulation element will be, respectively:

$$\Delta p_{ob} = \Delta p_{z,100} + \Delta p_{str,100} = \dot{V}_{100}^2 \cdot \left(r_{z,100} + r_{str}\right), \tag{4.107}$$

$$\Delta p_{ob} = \Delta p_{z,x} + \Delta p_{str,x} = \dot{V}_x^2 \cdot \left(r_{z,x} + r_{str}\right). \tag{4.108}$$

Here, relative volume flow $\dot{V}_{\mathrm{x}}/\dot{V}_{100}$ cannot be replaced by relative flow factor $k_{\mathrm{v,\,x}}/k_{\mathrm{v}100}$, which was the case for the calculations performed for a valve that is not installed in the pipework, because the pressure drop on the valve is not constant. The pressure loss occurs also in the network of connected pipes, and the pressure constancy is a prerequisite for the application of flow factors.

Dividing the two equations by each other, the following relation is obtained that defines the medium relative flow in the circuit regulated by a given valve:

- For valves with one adjustable section of the medium flow

$$\frac{\dot{V}_x}{\dot{V}_{100}} = \sqrt{\frac{r_{z,100} + r_{str}}{r_{z,x} + r_{str}}} = \sqrt{\frac{r_{reg,100} + r_k + r_{str}}{r_{reg,x} + r_k + r_{str}}} = \sqrt{\frac{r_{ob}}{r_{reg,x} + r_{ob} - r_{reg,100}}}$$
$$= \sqrt{\frac{1}{\frac{r_{reg,x}}{r_{ob}} + 1 - \frac{r_{reg,100}}{r_{ob}}}} = \sqrt{\frac{1}{\frac{r_{reg,100}}{r_{ob}} \cdot \frac{r_{reg,x}}{r_{reg,100}} + 1 - \frac{r_{reg,100}}{r_{ob}}}}$$
$$= \sqrt{\frac{1}{a_{c,\min} \cdot \frac{r_{reg,x}}{r_{reg,100}} + 1 - a_{c,\min}}}. \tag{4.109}$$

Considering that:

$$\frac{r_{reg,x}}{r_{reg,100}} = \frac{1}{\left(\frac{k_{v,reg,x}}{k_{v,reg,100}}\right)^2}, \tag{4.110}$$

the following expression is obtained:

$$\frac{\dot{V}_x}{\dot{V}_{100}} = \sqrt{\frac{1}{1 - a_{c,\min} + \frac{a_{c,\min}}{\left(\frac{k_{v,reg,x}}{k_{v,reg,100}}\right)^2}}}. \tag{4.111}$$

Or, after transformations:

$$\frac{\dot{V}_x}{\dot{V}_{100}} = \frac{1}{\sqrt{1 + a_{c,\min}\left[\left(\frac{k_{v,reg,100}}{k_{v,reg,x}}\right)^2 - 1\right]}}. \tag{4.112}$$

In order to establish the picture of changes in the medium volume flow due to changes in the opening degree of a valve installed in the pipework, the flow factors in formula (4.112) should be replaced with the set relation $k_{v,\ reg,\ x}/k_{v,\ reg,\ 100} = f(h_x/h_{100})$, which defines the closing characteristic shape, e.g. one of those discussed earlier.

– For the original linear characteristic the following expression is then obtained:

$$\frac{\dot{V}_x}{\dot{V}_{100}} = \frac{1}{\sqrt{1 + a_{c,\min}\left[\left(\frac{h_{100}}{h_x}\right)^2 - 1\right]}}, \tag{4.113}$$

– For the original equal-percentage characteristic:

$$\frac{\dot{V}_x}{\dot{V}_{100}} = \frac{1}{\sqrt{1 + a_{c,\min}\left(\frac{1}{[\exp[c\cdot(h_x/h_{100}-1)]]^2} - 1\right)}} = \frac{1}{\sqrt{1 + a_{c,\min}(\exp[2c\cdot(1 - h_x/h_{100})] - 1)}}. \tag{4.114}$$

- For valves with two adjustable sections of the medium flow the derivation is similar. The difference is that the sum with r_{ob} includes an additional quantity r_I representing hydraulic resistance of the first stage of regulation. In its physical sense, it is like r_k, i.e. it is not subject to current regulation.

The value of $a_{\mathrm{c},\ 100}$, according to previous formulae, is the product of $a_{\mathrm{w},\ 100}$ and $a_{\mathrm{z}100}$. The relations presented above indicate that the medium volume flow maximum and relative values in a given circuit are affected not only by the valve flow capacity (defined by flow factor k_v and inversely proportional to the square of hydraulic resistance r_z), but also by its total authority a_c. The smaller the authority value, the bigger the characteristic deformation. In the case of an equal-percentage characteristic, this additionally results in a higher value of leakage. Both these phenomena are unfavourable. The charts plotted based on the derived formulae, taking account of the impact of total authority a_c, for a few selected values thereof, are presented in Fig. 4.7.

Replacing the relative lift with the relative pre-setting, the following formulae are obtained:

– for the original linear characteristic:

$$\frac{\dot{V}_x}{\dot{V}_{100}} = \frac{1}{\sqrt{1 + a_{c,\min}\left[\left(\frac{n_{\max}}{n_i}\right)^2 - 1\right]}}, \tag{4.115}$$

– for the original equal-percentage characteristic:

$$\frac{\dot{V}_x}{\dot{V}_{100}} = \frac{1}{\sqrt{1 + a_{c,\min}\left(\frac{1}{[\exp[c\cdot(n_i/n_{\max}-1)]]^2} - 1\right)}} = \frac{1}{\sqrt{1 + a_{c,\min}(\exp[2c\cdot(1 - n_i/n_{\max})] - 1)}}. \tag{4.116}$$

4.2 Determination of the Regulation Valve Pre-setting

The regulation valve pre-setting is now commonly determined by means of a tabular/graphical method. In it, the value of pre-setting n_i is not calculated directly, but it is read from the table or chart provided by the manufacturer. A certain value of flow factor k_v is assigned to each value of pre-setting n_i. The parameter value is calculated and compared to the data from the table/chart, selecting pre-setting n_i characterized by the k_v value which is the closest to the required value but not lower. If a chart is used only, with no specific values of flow factor k_v, the pre-setting value is read by putting the lines of the required pressure loss and the medium volume flow onto the chart. The intersection point marks the value of pre-setting n_i. Such charts are presented in Figs. 4.5, 6.21 and 6.22. The first of them additionally includes the tabular data on flow factor k_v. Although this method is fast and practical, it has certain limitations. It does not enable a precise determination of pre-setting n_i if the required value of flow factor k_v lies between two neighbouring values of pre-setting n_i defined for a given valve, which practically is always the case. The value is then read out in approximation by means of interpolation. Moreover, the method requires the knowledge of the required pressure loss that will arise on the valve. This is the active pressure value (of the pump, for example) minus pressure losses in a given circuit (the circuit is understood here as a separate part of the installation from the pressure source, e.g. the circulation pump, the pressure stabilizer, etc., to the receiver, e.g. the radiator—in a two-pipe installation the number of circuits is thus equal to the number of radiators). If, however, it is the circuit hydraulic resistance that is given and not the pressure loss in it, the method does not make it possible to determine the required value of pre-setting n_i. Additional calculations need to be performed then. They are not included in the method but they are presented in this publication, which also presents two more analytical methods where the sought value of pre-setting n_i is found directly instead of being read from charts or tables.

Among the first researchers to propose an appropriate method based on a fully analytical approach were *Pyrkov* and *Szaflik*. Below, two alternative methods are presented that enable a more precise determination of the pre-setting required value. The first requires the knowledge of the value of the valve authority and the mathematical description of the valve original regulation characteristic. The second method does not need these data.

4.2.1 Determination of the Pre-setting of the Radiator Regulation Valve with One Adjustable Section of the Medium Flow

The inner and total authorities are in this case related to the element which is responsible for current regulation because this element creates one adjustable section of the medium flow together with the throttling element that determines the pre-setting. As a result, the closing element authority is identical with the authority of the throttling element.

Transforming formulae 4.115 and 4.116, taking account of the fact that the plug relative lift h_x/h_{100} is replaced by relative pre-setting n_i/n_{max}, the following expressions defining the sought value of the pre-setting are obtained:

- for the original linear characteristic:

$$n_i = \frac{n_{max}}{\sqrt{\frac{\left(\frac{\dot{V}_{100}}{\dot{V}_x}\right)^2 - 1}{a_{c,min}} + 1}}, \tag{4.117}$$

- for the original equal-percentage characteristic:

$$n_i = n_{max} \cdot \left(1 - \frac{\ln \frac{\left[\left(\frac{\dot{V}_{100}}{\dot{V}_x}\right)^2 - 1 + a_{c,min}\right]}{a_{c,min}}}{2c}\right). \tag{4.118}$$

The equations include total authority a_c being a product of inner and outer authority (a_w and a_z, respectively). For a given position of the throttling element, this is a constant value resulting from the distribution of hydraulic resistances of all elements of the pipework. However, based on the calculations performed for the *Danfoss*

MSV-I valve for example (cf. Sect. 5.4), it can be seen that the found value of inner authority a_w, though oscillating around a certain mean value, is to a certain degree changeable depending on the value of pre-setting n_i. The cause of this phenomenon is also explained earlier from the computational perspective. In order to determine the required value of pre-setting n_i precisely if the calculated value of inner authority a_w is not constant, the authority arithmetic mean should not be used as proposed by *Pyrkov* and *Szaflik* in [25, 26]. Instead, its changeability [$a_w = f(n_i)$], and, consequently, the changeability of total authority [$a_c = f(n_i)$], should be taken into account. If this phenomenon is ignored, the calculations are simplified considerably. They may even be performed by hand, but the results are less accurate. Nowadays, when specialist computer programs intended for hydraulic and thermal balancing of heating installations are commonly available, implementing a more complex but at the same time a more comprehensive mathematical model is no longer an obstacle. Taking account of the changeability of the valve inner authority a_w, the following equations are obtained:

- for the original linear characteristic:

$$n_i = \frac{n_{\max}}{\sqrt{\frac{\left(\frac{\dot{V}_{100}}{\dot{V}_x}\right)^2 - 1}{a(n_i)_{c,\min}} + 1}}, \tag{4.119a}$$

which can be written as:

$$n_i = \frac{n_{\max}}{\sqrt{\frac{\left(\frac{\dot{V}_{100}}{\dot{V}_x}\right)^2 - 1}{a_{z,100} \cdot a(n_i)_{w,\min}} + 1}}, \tag{4.119b}$$

- for the original equal-percentage characteristic:

$$n_i = n_{\max} \cdot \left(1 - \frac{\ln \frac{\left[\left(\frac{\dot{V}_{100}}{\dot{V}_x}\right)^2 - 1 + a(n_i)_{c,\min}\right]}{a(n_i)_{c,\min}}}{2c}\right), \tag{4.120a}$$

which can be written as:

$$n_i = n_{\max} \cdot \left(1 - \frac{\ln \frac{\left[\left(\frac{\dot{V}_{100}}{\dot{V}_x}\right)^2 - 1 + a_{z,100} \cdot a(n_i)_{w,\min}\right]}{a_{z,100} \cdot a(n_i)_{w,\min}}}{2c}\right). \tag{4.120b}$$

Such expressions are confounded equations, and they can be solved iteratively.

The volume flows ratio $\dot{V}_{100}/\dot{V}_x$ and outer authority a_z can be determined after the maximum of the circuit volume flow $\dot{V}_{100}$ is found. This maximum volume flow value is achieved at the valve full opening, and it results from active pressure and from the sum of the valve and the pipework hydraulic resistances. According to the relations derived above, the pipework hydraulic resistance can be expressed as:

$$r_{str} = \frac{\Delta p_{str,x}}{\dot{V}_x^2} = \frac{\Delta p_{str,100}}{\dot{V}_{100}^2}. \tag{4.121}$$

The valve hydraulic resistance at full opening, as well as for any other opening degree, can be calculated using the flow factor, according to formula (3.61):

$$r_{z,100} = \frac{1}{k_{vs}^2}, \tag{4.122}$$

$$r_{z,x} = \frac{1}{k_{v,x}^2}. \tag{4.123}$$

Knowing the pipework hydraulic resistance r_{str} and the valve resistance at full (100%) opening and at any other opening degree (x%), it is possible to calculate the maximum, as well as any other intermediate volume flow of the medium in this particular case, using formulae (4.107) and (4.108) appropriately transformed to express the sought quantity for the pipework part characterized by active pressure Δp_{cz}. This gives the following formula defining the maximum volume flow of the medium:

$$\dot{V}_{100} = \sqrt{\frac{\Delta p_{cz}}{r_{c,100}}}, \tag{4.124a}$$

which can be written as:

$$\dot{V}_{100} = \sqrt{\frac{\Delta p_{cz}}{r_{z,100} + r_{str}}}, \tag{4.124b}$$

or alternatively as:

$$\dot{V}_{100} = \sqrt{\frac{\Delta p_{cz}}{\frac{1}{k_{vs}^2} + r_{str}}}. \tag{4.124c}$$

The medium intermediate volume flow can be calculated from the following relation:

$$\dot{V}_x = \sqrt{\frac{\Delta p_{cz}}{r_{c,x}}}, \tag{4.125a}$$

which can be written as:

$$\dot{V}_x = \sqrt{\frac{\Delta p_{cz}}{r_{z,x} + r_{str}}}, \tag{4.125b}$$

or alternatively as:

$$\dot{V}_x = \sqrt{\frac{\Delta p_{cz}}{\frac{1}{k_{v,x}^2} + r_{str}}}. \tag{4.125c}$$

Flow factor k_{v} can be written as a function of pre-setting n_{i} by substituting a function relating the two parameters and determined for a given valve. The pipework hydraulic resistance r_{str} can be replaced with the previously written quotient of the pressure loss and the squared volume flow of the medium corresponding thereto. The following formula is then obtained:

$$\dot{V}_x = \sqrt{\frac{\Delta p_{cz}}{\frac{1}{(k_v(n_i))^2} + r_{str}}}, \tag{4.126a}$$

which can be written as:

$$\dot{V}_x = \sqrt{\frac{\Delta p_{cz}}{\frac{1}{(k_v(n_i))^2} + \frac{\Delta p_{str,x}}{\dot{V}_x^2}}}. \tag{4.126b}$$

Equation (4.126b) also enables determination of the required pre-setting n_{i} for given parameters of the circuit operation. A transformation of the equation gives the expression defining the first proposed method of determination of the regulation valve pre-setting n_{i}.

Method #1:

$$\dot{V}_x - \sqrt{\frac{\Delta p_{cz}}{\frac{1}{(k_v(n_i))^2} + r_{str}}} = 0, \tag{4.127a}$$

which can be written as:

$$\dot{V}_x - \sqrt{\frac{\Delta p_{cz}}{\frac{1}{(k_v(n_i))^2} + \frac{\Delta p_{str,x}}{\dot{V}_x^2}}} = 0. \tag{4.127b}$$

Considering that function $(k_v(n_i))^2$ most often has a complex mathematical description, it is pointless and sometimes even impossible to transform the equation to express the sought quantity n_{i} directly, and the pre-setting required value has to be found iteratively.

In order to determine the values of n_i from the relations describing it directly [(4.119a, b) and (4.120a, b)], it is necessary to define the pressure drops on individual elements of the pipework, according to the following formula:

$$\Delta p_{str,100} = r_{str} \cdot \dot{V}_{100}^2, \tag{4.128}$$

$$\Delta p_{z,100} = r_{z,100} \cdot \dot{V}_{100}^2 = \frac{1}{k_{vs}^2} \cdot \dot{V}_{100}^2 = \left(\frac{\dot{V}_{100}}{k_{vs}}\right)^2, \tag{4.129}$$

$$\Delta p_{z,x} = r_{z,x} \cdot \dot{V}_x^2 = \frac{1}{k_{vx}^2} \cdot \dot{V}_x^2. \tag{4.130}$$

According to relations (4.92) and (4.93), respectively, the outer authority for the full and for any other range of travel of the valve regulating element can be expressed as:

$$a_{z,100} \equiv a_{z,\min} = \frac{r_{z,100}}{r_{z,100} + r_{str}},$$

$$a_{z,x} = \frac{r_{z,x}}{r_{z,x} + r_{str}},$$

or using pressure drops, like in formulae (4.90) and (4.91), as:

$$a_{z,100} \equiv a_{z,\min} = \frac{\Delta p_{z,100}}{\Delta p_{z,100} + \Delta p_{str,100}},$$

$$a_{z,x} = \frac{\Delta p_{z,x}}{\Delta p_{z,x} + \Delta p_{str,x}}.$$

Substitution of the derived relations (4.121), (4.122), (4.124a, b, c) and (4.125a, b, c) in the initial formulae (4.119a, b) and (4.120a, b) results in the notation defining the second proposed method of determination of the regulation valve pre-setting n_i:

Method #2:

- for the original linear characteristic:

$$n_i = \frac{n_{\max}}{\sqrt{\frac{\frac{\Delta p_{cz}}{\dot{V}_x^2 \cdot (r_{z,100} + r_{str})} - 1}{a_{z,\min} \cdot a(n_i)_{w,\min}} + 1}}, \tag{4.131a}$$

which can be written as:

$$n_i = \frac{n_{\max}}{\sqrt{\frac{\frac{\Delta p_{cz}}{\dot{V}_x^2 \cdot \left(\frac{1}{k_{vs}^2} + \frac{\Delta p_{str,x}}{\dot{V}_x^2}\right)} - 1}{a_{z,\min} \cdot a(n_i)_{w,\min}} + 1}}, \tag{4.131b}$$

or alternatively as:

$$n_i = \frac{n_{\max}}{\sqrt{\frac{\frac{\Delta p_{cz}}{\left(\frac{\dot{V}_x}{k_{vs}}\right)^2 + \Delta p_{str,x}} - 1}{a_{z,\min} \cdot a(n_i)_{w,\min}} + 1}}. \tag{4.131c}$$

- for the original equal-percentage characteristic:

$$n_i = n_{\max} \cdot \left(1 - \frac{\ln \frac{\left[\frac{\Delta p_{cz}}{\dot{V}_x^2 \cdot \left(r_{z,100} + r_{str}\right)} - 1 + a_{z,\min} \cdot a(n_i)_{w,\min}\right]}{a_{z,\min} \cdot a(n_i)_{w,\min}}}{2c}\right), \tag{4.132a}$$

which can be written as:

$$n_i = n_{\max} \cdot \left(1 - \frac{\ln \frac{\left[\frac{\Delta p_{cz}}{\left(\frac{\dot{V}_x}{k_{vs}}\right)^2 + \Delta p_{str,x}} - 1 + a_{z,\min} \cdot a(n_i)_{w,\min}\right]}{a_{z,\min} \cdot a(n_i)_{w,\min}}}{2c}\right). \tag{4.132b}$$

The final forms of the formulae presented above, where hydraulic resistances on the valve and in the pipework are replaced, for the required volume flow, by pressure drops arising on the valve and in the pipework ($\left(\dot{V}_x/k_{vs}\right)^2$ and $\Delta p_{str,\,x}$, respectively), make it possible to find the required value of pre-setting n_i if pressure losses are imposed in the valve and in the pipework, due for example to the need to achieve appropriate values of authority a'. Consequently, if they are used to calculate the value of pre-setting n_i, pressure losses Δp_z and Δp_{str} on the valve and in the pipework, respectively, will agree with those assumed in input data. The situation in this variant is that pipework structures are selected with different values of hydraulic resistance r_{str} that guarantee an identical pressure drop Δp_{str} on them at different required values of the medium volume flow. But if changes in the volume flow need to be determined as a function of pre-setting n_i of a valve installed in the pipework with a specific and not selected value of hydraulic resistance r_{str}, the relation written with pressure drops cannot be applied because the changes are determined only for the required volume flow rate. In order to obtain the necessary relation, the formulae written with the use of hydraulic resistances should be applied. If the values of flow factor k_v determined for specific values of pre-setting n_i are known, formula (4.125c) may be used. If, however, the medium volume flow needs to be established for any value of pre-setting n_i, not just the one defined for a given valve, relation (4.126b), or the formulae that resulted in the notation defining the sought value thereof, i.e. formulae (4.131a, b, c) and (4.132a, b), should be used. After transforming the expressions to define the sought quantity (i.e. $\dot{V}_x$), formulae are obtained that define the volume flow rate as a function of pre-setting n_i of a valve installed in a pipework with a specific hydraulic resistance r_{str}. It may therefore be written that:

- for a valve with the original linear characteristic:

$$\dot{V}_x = \sqrt{\frac{\Delta p_{cz}}{\left(r_{z,100} + r_{str}\right) \cdot \left[1 + a_{z,\min} \cdot a(n_i)_{w,\min} \cdot \left(\left(\frac{n_{\max}}{n_i}\right)^2 - 1\right)\right]}}$$
$$= \sqrt{\frac{\Delta p_{cz}}{\left(\frac{1}{k_{vs}^2} + \frac{\Delta p_{str,100}}{\dot{V}_{100}^2}\right) \cdot \left[1 + a_{z,\min} \cdot a(n_i)_{w,\min} \cdot \left(\left(\frac{n_{\max}}{n_i}\right)^2 - 1\right)\right]}}$$
$$= \sqrt{\frac{\Delta p_{cz}}{\left(\frac{a_{z,\min}}{k_{vs}^2} + \frac{a_{z,\min} \cdot \Delta p_{str,100}}{\dot{V}_{100}^2}\right) \cdot \left[a(n_i)_{w,\min} \cdot \left(\left(\frac{n_{\max}}{n_i}\right)^2 - 1\right) + \frac{1}{a_{z,\min}}\right]}}. \tag{4.133}$$

Term $\Delta p_{str,\,100}$ can be written using the relation defining $a_{z,\,\min}$. The transformation of formula (4.90) will then result in:

$$\Delta p_{str,100} = \frac{\Delta p_{z,100}}{a_{z,\min}} - \Delta p_{z,100}. \tag{4.134}$$

Substituted in formula (4.133), this gives:

$$\dot{V}_x = \sqrt{\frac{\Delta p_{cz}}{\left(\frac{a_{z,\min}}{k_{vs}^2} + \frac{a_{z,\min} \cdot \left(\frac{\Delta p_{z,100}}{a_{z,\min}} - \Delta p_{z,100}\right)}{\dot{V}_{100}^2}\right) \cdot \left[a(n_i)_{w,\min} \cdot \left(\left(\frac{n_{\max}}{n_i}\right)^2 - 1\right) + \frac{1}{a_{z,\min}}\right]}}. \tag{4.135}$$

According to Eq. (4.129), the pressure loss on a fully (100%) open valve for the maximum volume flow $\dot{V}_{100}$ can be written as:

$$\Delta p_{z,100} = \left(\frac{\dot{V}_{100}}{k_{vs}}\right)^2.$$

Using this in formula (4.135), it may be written that:

$$\dot{V}_x = \sqrt{\frac{\Delta p_{cz}}{\left(\frac{a_{z,\min}}{k_{vs}^2} + \frac{a_{z,\min}\cdot\left(\frac{\left(\frac{\dot{V}_{100}}{k_{vs}}\right)^2}{a_{z,\min}} - \left(\frac{\dot{V}_{100}}{k_{vs}}\right)^2\right)}{\dot{V}_{100}^2}\right)\cdot\left[a(n_i)_{w,\min}\cdot\left(\left(\frac{n_{\max}}{n_i}\right)^2 - 1\right) + \frac{1}{a_{z,\min}}\right]}}$$

$$= \sqrt{\frac{\Delta p_{cz}}{\left(\frac{a_{z,\min}}{k_{vs}^2} + \frac{a_{z,\min}\cdot\left(\frac{\dot{V}_{100}}{k_{vs}}\right)^2\cdot\left(\frac{1}{a_{z,\min}} - 1\right)}{\dot{V}_{100}^2}\right)\cdot\left[a(n_i)_{w,\min}\cdot\left(\left(\frac{n_{\max}}{n_i}\right)^2 - 1\right) + \frac{1}{a_{z,\min}}\right]}}$$

$$= \sqrt{\frac{\Delta p_{cz}}{\left(\frac{a_{z,\min}}{k_{vs}^2} + \frac{\left(\frac{\dot{V}_{100}}{k_{vs}}\right)^2\cdot(1 - a_{z,\min})}{\dot{V}_{100}^2}\right)\cdot\left[a(n_i)_{w,\min}\cdot\left(\left(\frac{n_{\max}}{n_i}\right)^2 - 1\right) + \frac{1}{a_{z,\min}}\right]}}$$

$$= \sqrt{\frac{\Delta p_{cz}}{\left(\frac{a_{z,\min}}{k_{vs}^2} + \frac{1 - a_{z,\min}}{k_{vs}^2}\right)\cdot\left[a(n_i)_{w,\min}\cdot\left(\left(\frac{n_{\max}}{n_i}\right)^2 - 1\right) + \frac{1}{a_{z,\min}}\right]}}$$

$$= k_{vs}\cdot\sqrt{\frac{\Delta p_{cz}}{a(n_i)_{w,\min}\cdot\left(\left(\frac{n_{\max}}{n_i}\right)^2 - 1\right) + \frac{1}{a_{z,\min}}}}. \tag{4.136}$$

- for a valve with the original equal-percentage characteristic, after the same transformations, the following expression is obtained:

$$\dot{V}_x = k_{vs}\cdot\sqrt{\frac{\Delta p_{cz}}{a(n_i)_{w,\min}\cdot\left[\exp\left[2c\cdot\left(1 - \frac{n_i}{n_{\max}}\right)\right] - 1\right] + \frac{1}{a_{z,\min}}}}. \tag{4.137}$$

The derivations presented above concern both a single- and a double-plug valve with one adjustable section of the medium flow. However, the quantities in the formulae are related to the former type, such as the *Danfoss MSV-I* valve for example. If the case concerns a valve with a double suspended plug, or a thermostatic valve (e.g. *Herz TS-90-V*), current regulation will be performed through the travel of the element of the second stage of regulation, i.e. the suspended plug. Consequently, it will proceed not from the throttling element maximum lift $h_{max} \equiv n_{max}$, for which the valve is characterized by flow factors k_{vs}, $a_{w,\,min}$ and $a_{z,\,min}$, but from the reduced position of n_i, for which the valve is characterized by factors $k_{v,x}$, $a_{w,x}$ and $a_{z,x,}$. Such a case, instead of relative pre-setting n_i/n_{max}, concerns the closing plug relative lift h_x/h_{100} in the travel range reduced by pre-setting n_i from $h_{100} = h_{max}$ to a new, smaller value of h_{100}.

This can be confirmed using the data of the example valve from the calculations presented earlier (cf. Sect. 4.1.2.2), for which $r_{reg100} = 5$, $r_{k+I,\,1} = 5$ $(i = 1)$ and the original closing characteristic is linear. For the closing element full (maximum) range of travel h_{max}, the valve was characterized by the values of flow factor $k_{vs} =$

0.3162 and inner authority $a_{\mathrm{w, min}} = 0.5$. Assume that the valve operates in a circuit with $r_{\mathrm{str}} = 15$ and $\Delta p_{\mathrm{cz}} = 1$.

In order to verify the above theorem, a comparison can be made for an example value of the absolute opening degree between the medium volume flow values obtained for two cases. One of them will correspond to a reduction in the closing element maximum range of travel h_{max}, for which new values of the required parameters will be calculated according to the conclusion of the theorem. In the other, there will be no such reduction. Assume that the comparison is to be made between volume flows for the plug absolute lift over the seat at the level of 0.2 and for two variants: no reduction in the plug travel range, i.e. for the plug maximum range of travel h_{max}, and a reduction in the range to the level of 0.4 of the maximum travel h_{max}. In the first case, for the assumed value of $n_{\mathrm{i}}/n_{\mathrm{max}} \equiv h_{\mathrm{x}}/h_{100} \equiv h_{\mathrm{x}}/h_{\mathrm{max}} = 0.2$, the result is:

$$a_{z,\min} = \frac{r_{reg}100 + r_{k+I,1}}{r_{reg}100 + r_{k+I,1} + r_{str}} = \frac{5+5}{5+5+15} = 0.4,$$

$$\dot{V}_x = k_{vs} \cdot \sqrt{\frac{\Delta p_{cz}}{a_{w,\min} \cdot \left(\left(\frac{n_{\max}}{n_i}\right)^2 - 1\right) + \frac{1}{a_{z,\min}}}} = 0.3162 \cdot \sqrt{\frac{1}{0,5 \cdot (5^2 - 1) + 1/0.4}} = 0.083.$$

In the second case, for $h_{100} = 0.4 \cdot h_{\mathrm{max}}$, the new values of the required parameters according to the hypothesis will total:

$$k_{v,x} = \sqrt{\frac{\Delta p_{cz}}{r_{reg,40} + r_{k+I,1}}} = \sqrt{\frac{1}{\frac{5}{(0.4)^2} + 5}} = 0.1661,$$

$$a_{w,x} = \frac{r_{reg,40}}{r_{reg,40} + r_{k+I,1}} = \frac{\frac{5}{(0.4)^2}}{\frac{5}{(0.4)^2} + 5} = 0.862,$$

$$a_{z,x} = \frac{r_{reg,40} + r_{k+I,1}}{r_{reg,40} + r_{k+I,1} + r_{str}} = \frac{\frac{5}{(0.4)^2} + 5}{\frac{5}{(0.4)^2} + 5 + 15} = 0.707.$$

Using these parameters, it should be checked if for the absolute opening degree at the level of $h_{\mathrm{x}}/h_{100} = 0.2$, but with no reduction in the plug maximum range of travel to the value of $h_{100} = 0.4 \cdot h_{\mathrm{max}}$, the obtained volume flow of the medium will be the same as in the first case. The relative opening degree will total $h_{\mathrm{x}}/h_{100} = 0.2/0.4 = 0.5$. The volume flow will then total:

$$\dot{V}_x = k_{v,x} \cdot \sqrt{\frac{\Delta p_{cz}}{a_{w,x} \cdot \left(\left(\frac{h_{100}}{h_x}\right)^2 - 1\right) + \frac{1}{a_{z,x}}}} = 0.1661 \cdot \sqrt{\frac{1}{0.862 \cdot (2^2 - 1) + 1/0.707}} = 0.083.$$

The result coincides with the value obtained for the initial parameters, so the theorem conclusion is right.

The situation is a bit different in the case of thermostatic valves. For manual valves, the maximum and other values of flow factor k_{vs} provided by the manufacturer correspond each time to the full range of travel h_{100} of the closing element for a given position of the throttling part (reducing the maximum range of travel $h_{100} = h_{\mathrm{max}}$

to a new relative value of h_{100}), but this is not so in the case of thermostatic valves. As described earlier, the values of flow factor k_v are given for the closing element intermediate positions, corresponding to a given value of proportional range X_p. For this reason, the derived formula cannot be applied here because the values of $a_{w, x}$ and $a_{z, x}$ have to be calculated for the closing element full lift ($X_p = \max$, h_{max}) and for the value of flow factor k_v corresponding thereto, for a given value of pre-setting n_i. The quantity can be determined experimentally, but it can also be calculated based on the mathematical description of the original closing characteristic and the values of the closing element inner authority a_w and travel range h_{100} (equal to h_{max} in this case). By contrast to the procedure for the selection of the required value of pre-setting n_i, in this situation the knowledge of the actual values is absolutely necessary. Moreover, in the case of thermoregulators, establishing changes in the medium volume flow arising due to changes in the valve pre-setting will not give correct results if the process is performed using computational relations which are right for manual valves. As explained in previous sections, the thermostatic head operation is aimed in every situation to ensure that the obtained measured temperature is possibly the closest to the set value. This means that changes in pre-setting and the possible changes in the medium flow being the effect thereof and conditioning the changes in the room temperature will to some extent be eliminated by changes in the current-regulation element opening degree caused due to the impact of the head. Therefore, the calculations should in such a case be performed taking account of static parameters (the nominal and maximum proportional range, the amplification factor) of the thermostatic/electronic head and the valve it regulates.

If these parameters are determined, it is also possible to carry out alternative calculations to find the medium volume flow if the closing element opens more than the set design value, reduced for example by a given value of pre-setting n_i to a different position, as described at the beginning of Sect. 4.1.2. It is stated that in both cases the calculations should be performed using different (not identical) authority values determined for two set positions of the regulation element, and not for the full opening or for only one of the positions under consideration. So that this thesis should be right, though, the results of the volume flow calculations should coincide. Assume as an example the following data for the calculations: the relative opening degree at the level of 0.2, and the opening degree of 0.4, which will then correspond to the value of $h_x/h_{100} = 0.4/0.2 = 200\%$ ($h_{100}/h_x = 0.2/0.4 = 50\%$) of the relative initial opening degree and of $h_x/h_{100} = 0.4/0.4 = 100\%$ of the target opening. The volume flow should be determined for both variants. In the first case, the result is:

$$k_{v,x} = \sqrt{\frac{\Delta p_{cz}}{r_{reg,20} + r_{k+I,1}}} = \sqrt{\frac{1}{\frac{5}{(0.2)^2} + 5}} = 0.0877,$$

$$a_{w,x} = \frac{r_{reg,20}}{r_{reg,20} + r_{k+I,1}} = \frac{\frac{5}{(0.2)^2}}{\frac{5}{(0.2)^2} + 5} = 0.9615,$$

$$a_{z,x} = \frac{r_{reg,20} + r_{k+I,1}}{r_{reg,20} + r_{k+I,1} + r_{str}} = \frac{\frac{5}{(0.2)^2} + 5}{\frac{5}{(0.2)^2} + 5 + 15} = 0.8965.$$

The volume flow will total:

$$\dot{V}_x = k_{v,x} \cdot \sqrt{\frac{\Delta p_{cz}}{a_{w,x} \cdot \left(\left(\frac{h_{100}}{h_x}\right)^2 - 1\right) + \frac{1}{a_{z,x}}}} = 0.0877 \cdot \sqrt{\frac{1}{0.9615 \cdot (0.5^2 - 1) + 1/0.8965}} = 0.139.$$

For the second case:

$$\dot{V}_x = k_{v,x} \cdot \sqrt{\frac{\Delta p_{cz}}{a_{w,x} \cdot \left(\left(\frac{h_{100}}{h_x}\right)^2 - 1\right) + \frac{1}{a_{z,x}}}} = 0.1661 \cdot \sqrt{\frac{1}{0.862 \cdot (1^2 - 1) + 1/0.707}} = 0.139.$$

The results coincide, so the thesis is right. The case is verified experimentally in Chap. 5.

4.2.2 Determination of the Pre-setting of the Radiator Regulation Valve with Two Adjustable Sections of the Medium Flow and of the Thermostatic Valve

As previously mentioned, in the case of a valve with two adjustable sections of the medium flow, the initial hydraulic resistance and throttling are realized not through the medium flow section co-created by the closing element but through the throttling part. Therefore, the changes in the volume flow resulting from changes in pre-setting n_i should be related to the operation of the throttling element. For this reason, the element original (throttling) characteristic and inner authority $a_{w,I}$ are essential. Consequently, writing the relations defining pre-setting n_i of a valve installed in the pipework, total authority a_c has to be related to the throttling element (the first stage of regulation). The quantity can be found for any position of the closing element, both of a manual and a thermostatic valve. However, as stated earlier, manual valves with two adjustable sections of the medium flow are generally not used in the present radiator regulation practice. Especially in a variant where the computational working point would be related to an intermediate position of the closing element. This is the domain of thermostatic valves, both with one and two adjustable sections of the medium flow. The considerations presented in this subsection, even though they are obviously correct for manual valves too, will therefore be related to thermostatic valves with two adjustable sections of the medium flow.

Because it is recommended that radiator thermostatic valves should be selected for the proportional range of $X_p = 2$ K, the device throttling element inner authority $a_{w,I}$ should be calculated, among others, for the current-regulation closing element lift corresponding to the value of $X_p = 2$ K. Due to that, the calculated value of outer authority a_z should also be related to a specific range of X_p because it concerns the throttling element and can be marked, if there is no reduction in the element range of travel, as $[a_{z,\min}]_{Xp}$. In this case, the other quantities characterizing the valve should also be related to a specific proportional range X_p. Considering the above, the relation defining the pre-setting value will take the following form:

- for the original linear characteristic:

$$n_i = \frac{n_{\max}}{\sqrt{\frac{\frac{\Delta p_{cz}}{\left(\frac{\dot{V}_x}{[k_{vs}]_{Xp}}\right)^2 + \Delta p_{str,x}} - 1}{[a_{z,\min}]_{Xp} \cdot [a(n_i)_{w,I,\min}]_{Xp}} + 1}}, \tag{4.138a}$$

which can be written as:

$$n_i = \frac{n_{\max}}{\sqrt{\frac{\frac{\Delta p_{cz}}{\left(\frac{\dot{V}_x}{[k_{vs}]_{Xp}}\right)^2 + \Delta p_{str,x}} - 1}{[a_{z,\min} \cdot a(n_i)_{w,I,\min}]_{Xp}} + 1}}, \tag{4.138b}$$

or as:

$$n_i = \frac{n_{\max}}{\sqrt{\frac{\frac{\Delta p_{cz}}{\dot{V}_x^2 \cdot \left([r_{z,100}]_{Xp} + r_{str}\right)} - 1}{[a_{z,\min} \cdot a(n_i)_{w,I,\min}]_{Xp}} + 1}}. \tag{4.138c}$$

- for the original equal-percentage characteristic:

$$n_i = n_{\max} \cdot \left(1 - \frac{\ln \frac{\left[\frac{\Delta p_{cz}}{\left(\frac{\dot{V}_x}{[k_{vs}]_{Xp}}\right)^2 + \Delta p_{str,x}} - 1 + [a_{z,\min} \cdot a(n_i)_{w,I,\min}]_{Xp}\right]}{[a_{z,\min} \cdot a(n_i)_{w,I,\min}]_{Xp}}}{2c} \right), \tag{4.139a}$$

which can be written as:

$$n_i = n_{\max} \cdot \left(1 - \frac{\ln \frac{\left[\frac{\Delta p_{cz}}{\dot{V}_x^2 \cdot \left([r_{z,100}]_{Xp} + r_{str}\right)} - 1 + [a_{z,\min} \cdot a(n_i)_{w,I,\min}]_{Xp}\right]}{[a_{z,\min} \cdot a(n_i)_{w,I,\min}]_{Xp}}}{2c} \right). \tag{4.139b}$$

Based on the expression using the changeability of flow factor k_v as a function of pre-setting n_i, the relation defining the medium volume flow, like in formula (4.125b), can be expressed as:

$$\dot{V}_x = \sqrt{\frac{\Delta p_{cz}}{r_{z,x} + r_{str}}} = \sqrt{\frac{\Delta p_{cz}}{\frac{1}{\left([k_v(n_i)]_{Xp}\right)^2} + \frac{\Delta p_{str,x}}{\dot{V}_x^2}}}. \tag{4.140}$$

Like before, the relation can be applied here to calculate the required value of pre-setting n_i:

$$\dot{V}_x - \sqrt{\frac{\Delta p_{cz}}{\frac{1}{\left([k_v(n_i)]_{Xp}\right)^2} + r_{str}}} = 0, \tag{4.141a}$$

which can be written as:

$$\dot{V}_x - \sqrt{\frac{\Delta p_{cz}}{\frac{1}{\left([k_v(n_i)]_{Xp}\right)^2} + \frac{\Delta p_{str,x}}{\dot{V}_x^2}}} = 0. \tag{4.141b}$$

The throttling element operation in valves with two adjustable sections of the medium flow differs from the operation of the closing element. There is usually no reciprocating motion of the element in this case. Instead, the element turns around its own axis (cf. Fig. 2.29) and the rotation is not blocked at the minimum or the maximum opening degree, which makes it possible to set the required value of pre-setting n_i regardless of the direction of the shutter rotation.

Like in the case of valves with one adjustable section of the medium flow, it is possible to formulate the relation defining the medium volume flow for current regulation, i.e. for different positions of the closing element. The formula will have the same form, but the h_{max}/h_x ratio should be replaced by relative lift h_{max}/h_x of the closing element responsible here for current regulation, and the authority values should be related to the selected value of pre-setting n_i. If the flow factor for the closing plug maximum lift h_{max} of a valve not installed in the pipework is to be found, and not for the intermediate lift h_N, for which the $k_{v,x}$ value is specified, the only resistances will be those arising inside the valve. Therefore, the quantity to use will be inner authority a_w depending on pre-setting n_i. The starting point may be the equation derived earlier to define the valve relative flow factor as a function of the opening degree and taking account of hydraulic resistances not subject to current regulation. Considering these issues and transforming Eq. (4.41), the following is obtained:

$$[k_{v100}]_{n_i} = \left[k_{v,x}\right]_{Xp} \cdot \sqrt{1 + a_{w,x}\left[\left(\frac{k_{v,reg,100}}{k_{v,reg,x}}\right)^2 - 1\right]}. \tag{4.142}$$

It can be seen that the relation is similar in form to formula (4.136), which defines the sought volume flow for a valve with one adjustable section of the medium flow installed in the pipework. This is natural because de facto the only difference between the two cases is the occurrence of the pipework additional resistances r_{str}. Considering that in this case $a_z = 1$ and, consequently, term $1/a_{\text{z, min}}$ in formula (4.136) equals 1, and, according to the flow factor definition, $\Delta p_{\text{cz}} = 1$, Eq. (4.136) derived for a valve with one adjustable section of the medium flow is reduced to the form as expressed above in (4.142), which is right for a valve with two adjustable sections of the medium flow.

Replacing the ratio between the flow factors with the specific relation $k_{\text{v, reg, x}}/k_{\text{v, reg, 100}} = f(h_{\text{x}}/h_{100})$, which describes the shape of the closing characteristic, the following is obtained:

- for the original linear characteristic:

$$[k_{v100}]_{n_i} = \left[k_{v,x}\right]_{Xp} \cdot \sqrt{1 + a_{w,x}\left[\left(\frac{h_{100}}{h_x}\right)^2 - 1\right]}, \qquad (4.143)$$

- for the original equal-percentage characteristic:

$$[k_{v100}]_{n_i} = \left[k_{v,x}\right]_{Xp} \cdot \sqrt{1 + a_{w,x}(\exp[2c(1 - h_x/h_{100})] - 1)}. \qquad (4.144)$$

In both cases: a valve with one and a valve with two adjustable sections of the medium flow, relations (4.131a, b, c), (4.132a, b), (4.138a, b, c) and (4.139a, b) for the sought value of pre-setting n_{i} use the values of authorities and flow factors related to the maximum value of pre-setting n_{max}. It is of course possible to perform calculations similar to those presented in the example in Sect. 4.2.1 to find new values of the parameters, if the search for the value of pre-setting n_{i} is carried out in a different range of its value. Naturally, the results will be the same as those demonstrated by the example mentioned above, where the medium volume flow is calculated. However, determining the required value of pre-setting n_{i}, it is more favourable to use the full range of its changeability available for the valve because, as a rule, the required value is sought in this maximum range offered by the valve.

The derivations presented above omit the issues related to the changeability of active pressure depending on the medium volume flow in the conditions of quantitative regulation. If necessary, this phenomenon can be taken into account by creating appropriate functions that describe pressure changeability depending on the volume flow value and substituting them for Δp_{cz} in the formulae intended for determination of the value of pre-setting n_{i}. In the same way, if calculations are to be performed for a wide range of the *Reynolds* numbers, including flows with diminishing turbulence as well as transitional flows, if the pipework characteristic demonstrates a non-negligible share in this area, it is possible to take account of the changeability of the coefficient of linear pressure losses (λ) [e.g. using the *Colebrook-White* formula (3.15)] and of local pressure losses (ζ). This is justified especially if the problems are solved by means of computers, which is practically always the case at present. Implementing a full model of the discussed algorithm will enable more comprehensive calculations and more detailed analyses of the effects of quantitative and, consequently, quantitative-qualitative regulation, minimizing the need to make on-line adjustments to correct the regulation process in the operating installation, which is a rather common practice nowadays [15, 25, 26].

4.3 Sizing of the Radiator Regulation Valve

The process of the radiator regulation valve sizing is now substantially non-analytical in character. This is the effect of the fact that in the theory to date there are no appropriate computational relations describing the valve inner authority a_w, which is one of the two decisive parameters that determine the shape of the obtained regulation characteristic of the device. Numerous examples of the valve optimization in terms of the regulation characteristic and other operating parameters can be found in literature [1, 2, 6, 7, 9, 16, 17, 23, 24, 27, 30–34, 36–40]. Two general approaches can be distinguished in these works—they either use relations which do not take account of the impact of the valve inner authority and/or the original characteristic (as the relationships between them have not been realized so far and they cannot be captured quantitatively) or the sizing process is carried out experimentally, by trial and error. There are very few examples where the radiator valves are designed through the analytical approach. In [13], Kołodziejczyk describes the method used in Poland to size the radiator regulation valves. However, the algorithm presented by him does not take account of the impact of the distribution of hydraulic resistances inside the valve, i.e. the valve authority, on the shape of the assumed closing characteristic. It turned out in practice that valves made based on those relations had no originally-assumed characteristics and corrections had to be made during the prototyping process.

As engineering computer programs, CFD codes in particular, have become more popular and increasingly available, the valve design process can be simplified considerably. These tools are used in computer-assisted design to make subsequent prototypes and create the company know-how. It has to be noted, however, that computer programs intended for the CFD analysis require properly trained engineers and, usually, the purchase of relevant licences, which involves considerable expenses. Making subsequent prototypes also means costs related to materials, production processes and remuneration for work. These factors can be optimized through the analytical approach to the valve design because the process can be automated if the mathematical formulae that strictly relate the regulation valve parameters to each other are known. It is then also possible to predict the impact of one of the valve design parameters on the others, which is rather difficult to do using the experimental approach.

An example is presented below of the regulation valve sizing taking quantitative account of inner authority a_w proposed herein. The differences from the method presented by *Kołodziejczyk* are also indicated, together with a presentation of relevant characteristics and a comparison of the calculated shape of the valve plug as the target outcome. Experimental verification is also carried out on a custom-made design of the valve.

Data:

- water installation, double-regulation valve with one adjustable section of the medium volume flow
  - water (liquid) density $\rho = 965$ [kg/m^3],
  - flow factor at full opening $k_{vs} \equiv k_{v100} = 2.15$ [m^3/h],
  - minimum inner authority value (at full opening) $a_{w,\,min} = 0.35$,
  - valve discharge coefficient $\alpha = 0.7$.

Quantitatively, as the case concerns a single-plug valve, $a_{w,\,min}$ is the ratio between the working medium pressure drop in the valve regulating part, for the valve plug maximum range of travel h_{max} and for the plug full lift over the valve seat h_{100}, and the pressure drop on the entire valve. The *Kołodziejczyk* algorithm takes account of this division in the calculation of the liquid free outlet surface area A_{wyl}, but does not take account of its impact on the regulation characteristic. Even if the impact of the other elements of the circuit is not taken into consideration, the assumed value of $a_{w,\,min} = 0.35$, which can be encountered in practice, is close to the lower permissible value of total authority a_c.

According to the definition of k_v, the flow factor value informs about the working medium volume flow expressed in m^3/h, at the pressure drop on the valve of $\Delta p_z = 10^5$ Pa. Inner authority a_w provides information about the distribution of pressure losses inside the valve, on the currently regulated part and on the other elements. It follows from the assumed data that 35,000 Pa fall on the valve regulating element at full opening of $h_{100} \equiv h_{max}$.

Using relation (3.59) and taking account of the data, the required surface area of the flow cross-section at full opening is obtained at the level of:

$$A_{100} = \frac{\dot{V}_{100}}{\alpha \cdot \sqrt{\frac{2\Delta p_{reg,100}}{\rho}}} = \frac{2.15/3600}{0.7 \cdot \sqrt{\frac{2 \cdot 0.35 \cdot 10^5}{965}}} = 100\,\text{mm}^2. \tag{4.145}$$

The above formula provides justification for the issue mentioned earlier, in the description of the valve regulated section authority a, which is a quantity used within Standard EN 215:2004 [5]. If it is assumed that the parameter is the valve inner authority a_w as stated in [13, 14, 25, 26], following the standard recommendations at full opening, the result is—as proven—$a_{w,\,min} = 1$, which means that the entire available pressure (10^5 Pa) supplied to the valve will appear on the regulated section only, in the closing part. Due to that, the liquid flow surface area will be substantially underestimated, totalling in this case:

$$A_{100} = \frac{\dot{V}_{100}}{\alpha \cdot \sqrt{\frac{2\Delta p_{reg,100}}{\rho}}} = \frac{\dot{V}_{100}}{\alpha \cdot \sqrt{\frac{2\Delta p_{z,100}}{\rho}}} = \frac{2.15/3600}{0.7 \cdot \sqrt{\frac{2 \cdot 10^5}{965}}} = 59\,\text{mm}^2. \tag{4.146}$$

The value of inner authority a_w, as one of the input parameters, can be determined only if the valve body and the valve throttling element geometries are known, together with surface area A_{wyl} for a given range of the valve closing element travel (from which regulation will proceed). This creates a problem because surface area A_{wyl} can be determined only if a certain value of inner authority $a_{w,\,min}$ is assumed. For this reason, the problem would have to be solved iteratively. But it can also be done using the direct method of the valve inner authority determination, as presented in Sect. 4.1.2.1.

The ratios of $h_{100} \equiv h_{max}$ and of the plug base diameter d_1 may have many combinations (cf. Fig. 4.8). Nonetheless, the bigger the plug height and the wider the travel range h_{100}, the higher the regulation accuracy. There is yet a certain condition that has to be satisfied if the case concerns a structure where the plug does not move inside the seat, but over it, changing the surface area of the created side surface, as presented in Fig. 3.14a [e.g. the *Herz TS-90-V* valve (Fig. 2.28)]. Here, the regulating element hydraulic resistance is mainly conditioned by the sum of the inlet and outlet surface areas (i.e. the surface area of the cross-section through which the medium flows into the valve seat and of the side area created between the plug and the valve seat where it flows out)—A_{wl} and A_{wyl}, respectively. They can be expressed as: $A_{wl} = \pi \cdot d_0^2/4, A_{wyl} = \pi \cdot d_0 \cdot h_x$. As the plug rises, A_{wyl} gets bigger and the regulating element hydraulic resistance becomes smaller. Once the lift reaches the value at which the side surface area A_{wyl} is equal to that of the valve seat A_{wl}, the hydraulic resistance values in the two sections will be the same. Relevant comparisons indicate that this will occur for the lift value of $h_{100} \equiv h_{max} = 0.25 d_0$. Further opening of the valve will involve a further reduction in the resistance of the liquid outlet cross-section A_{wyl} and some changes in the total resistance, i.e. in the medium volume flow. These changes, however, will be much smaller than before because the decisive resistance will be the one conditioned by the inlet hole in the valve seat, characterized by a smaller value of A_{wl}. The plug further opening will therefore be of little use and a certain part of its total travel will fall on the area of the hydraulic resistance small changeability. The part will get bigger with an increase in the valve plug travel range above the value of $0.25 \cdot d_0$. Consequently, the regulation accuracy and quality will deteriorate. Valves with a design that takes this phenomenon into account, where the plug travel is limited to the above value, are referred to as *full-lift valves*.

But due to their regulating element design and the way in which they change the surface area of the liquid free flow (most often—proportional and linear), they are still not the optimal solution for the radiator regulation valve. A valve regulating element that can ensure also the original linear characteristic and where the plug moves inside the seat is presented in Fig. 3.14b [e.g. the *Danfoss MSV-I* valve (cf. Fig. 2.26)]. The solution discussed in the computational example concerns a case where the plug moves inside the seat hole to enable the liquid flow surface area changeability according to the required curve. Another reason for that is that this is the variant to which the compared *Kołodziejczyk's* calculations are related.

Using the calculated maximum value of surface area A_x (i.e. $A_{100} \equiv A_{wyl}$), a valve seat is selected with the diameter of $d_0 = 12$ mm, whose free-flow cross-section is

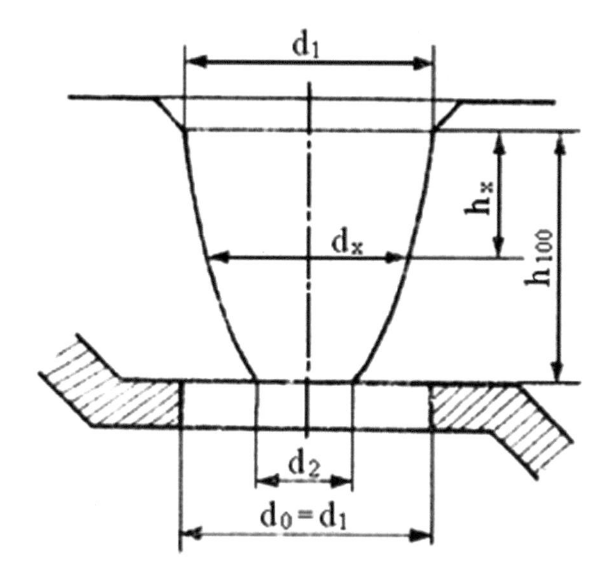

Fig. 4.8 Assumed and sought dimensions of the closing plug [13]

reduced by the full lift of a plug with the face diameter of $d_2 = 4$ mm. This gives the required free-flow surface area:

$$A_{100} = \frac{\pi}{4} \cdot \left(d_0^2 - d_2^2\right) = \frac{\pi}{4} \cdot \left(12^2 - 4^2\right) \cong 100\,\text{mm}^2. \tag{4.147}$$

The plug height $h_{100} \equiv h_{max}$ is assumed as equal to the plug base diameter d_1 and equal to the seat hole diameter d_0. In the case of thermostatic valves, due to the fact that the closing element required maximum available range of travel h_{max} is very small, the plug height $h_{100} \equiv h_{max}$ should be reduced accordingly. The plug example geometry is presented in Fig. 4.8.

In the next step, the shape and the mathematical description of the regulation characteristic have to be determined. Both for manual and thermostatic valves, the most favourable effect is linear regulation. For this reason, the closing characteristic should be selected as symmetrical to the radiator thermal characteristic with respect to the chart diagonal. However, there are some obstacles here that make it impossible to meet this postulate and pose universal assumptions. Firstly, the shape of a given radiator thermal characteristic depends on the thermal load. As a result, it varies throughout the heating season. Secondly, the installation may include different types of radiators with a different design, which are characterized by different shapes of the thermal characteristic, and making universal assumptions thus becomes even more difficult [21, 22]. It is assumed that for a typical radiator and operation under the nominal load ($\phi = 100\%$), the optimal case is the equal-percentage curve with coefficient $c = 3.22$. However, it seems more justified to take as the basis a partial load curve resulting from averaging the entire period of operation in the heating season, e.g. $\phi = 50\%$ because the radiator operates under the nominal load only for

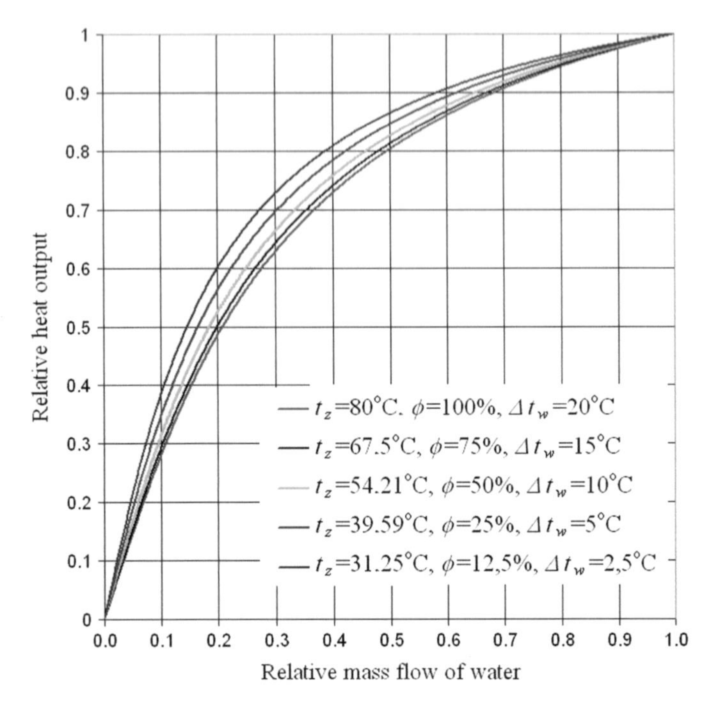

Fig. 4.9 A set of a typical convective radiator static thermal characteristics for different values of the supply temperature at a set range of changes in the working medium mass flow [21, 22]

a very short time. Then, the considerations should be carried out for example for a characteristic with a higher value of coefficient *c* because the thermal characteristic curve of a radiator under a partial load displays a bigger (upward) curvature [21, 22]. Such a case is illustrated in Fig. 4.9.

Moreover, the recommended equal-percentage characteristic cannot be considered for the entire range of the valve plug travel because it does not enable full closing and complete throttling of the medium flow, as illustrated in Fig. 3.13. A combined curve is then required composed of the one ensuring the flow complete cut-off at full closing and of the equal-percentage characteristic. Theoretical studies often recommend that the plug should be designed for linear changes in a certain initial height range (typically—10–30%) and for the equal-percentage characteristic in the remaining range. It should be noted, however, that the condition is appropriate only if the plug travel range has no upper limit, e.g. due to pre-setting. Otherwise, the percentage of the range of the initial, unfavourable, linear characteristic will get bigger in proportion to the reduction in the plug initial available range of travel that results in this case from the method of selecting the pre-setting value. For this reason, it is suggested herein that, most favourably, the plug geometry should be designed for the equal-percentage characteristic in the entire range of travel because in practice full closing will be ensured anyway, and the characteristic at an almost full-closing position will be similar to linear. It is sufficient for the plug base to be connected to a ring, which is often used, for example, to put a gasket on, and which adheres to the seat at full closing. Such design ensures the flow complete cut-off irrespective of the plug shape, as illustrated in Fig. 3.14b.

Another issue that makes it difficult to pose universal requirements as to the shape of the original regulation characteristic is the fact that valves always operate within a given pipework. Characterized by a certain hydraulic resistance, the pipework deforms the characteristic, like the resistance of the valve body (or of the throttling element in valves with two adjustable sections of the medium flow). The deformation caused due to the impact of the pipework depends on the value of the pipework hydraulic resistance, which means that its size may vary.

So that a direct comparison could be made between the calculation results obtained by means of the proposed method and the results presented by Kołodziejczyk in [13], the first variant of shaping the characteristic curvature and the same coefficient c of the equal-percentage part of the valve closing characteristic were selected. It is assumed that the curves are connected to each other at the level of 30% of the plug height, i.e. at $h_x/h_{100} = 0.3$. Using the relations describing the linear and the selected equal-percentage characteristics, *Kołodziejczyk* states that the flow factor variability depending on the opening degree is defined as:

$$\frac{k_{vx}}{k_{v100}} = 0.35 \cdot \frac{h_x}{h_{100}} \quad \text{for} \quad \frac{h_x}{h_{100}} = 0\text{–}0.3, \tag{4.148}$$

$$\frac{k_{vx}}{k_{v100}} = \exp\left[c \cdot \left(\frac{h_x}{h_{100}} - 1\right)\right] \quad \text{for} \quad \frac{h_x}{h_{100}} = 0.3\text{–}1.0. \tag{4.149}$$

In the case of the equal-percentage characteristic at $c = 3.22$, for $h_x/h_{100} = 0.3$, the result is $\dot{V}_x/\dot{V}_{100} = 0.105$, which means that the slope factor for the linear part, as the tangent of the characteristic inclination angle, totals 0.105/0.3 = 0.35, i.e. $k_{v,x} = 0.35h_x$.

At the plug assumed height of $h_{100} \equiv h_{max}$ and the plug base diameter d_1, in order to ensure appropriate changeability of flow factor k_v depending on the regulating element position over the seat, it is necessary to determine the variability of diameter d_x depending on height h_x. This makes it necessary to introduce more equations and assumptions, namely:

- the liquid flow surface area at the valve any partial opening degree:

$$A_x = \frac{\pi}{4} \cdot \left(d_0^2 - d_x^2\right), \tag{4.150}$$

- flow factor k_v is linearly related to the surface area of the valve flow channel opening:

$$\frac{k_{vx}}{k_{v100}} = \frac{A_x}{A_{100}} = \frac{d_0^2 - d_x^2}{d_0^2 - d_2^2}. \tag{4.151}$$

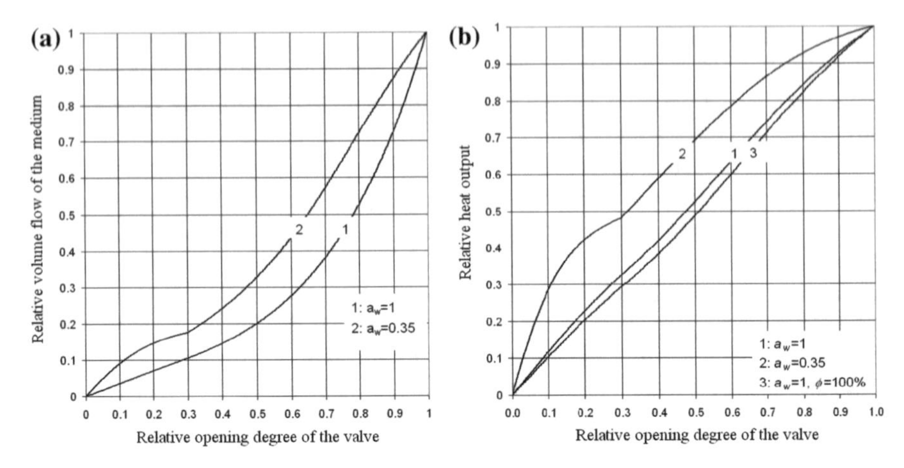

Fig. 4.10 Valve closing characteristics: **a** final regulation curves, **b** 1 and 2: characteristics obtained without (Kołodziejczyk [13]) and with account taken of inner authority, for the radiator thermal load $\phi = 100\%$; 3: characteristic for the radiator load $\phi = 100\%$

After relevant transformations, the following notation is obtained:

$$d_x = d_0 \cdot \sqrt{1 - \frac{k_{vx}}{k_{v100}}\left[1 - \left(\frac{d_2}{d_0}\right)^2\right]}. \tag{4.152}$$

Substitution of appropriate mathematical descriptions of the regulation characteristic function results in the following formulae:

- for the part designed for the linear curve:

$$d_x = d_0 \cdot \sqrt{1 - 0.35\frac{h_x}{h_{100}} \cdot \left[1 - \left(\frac{d_2}{d_0}\right)^2\right]}, \tag{4.153}$$

- for the part designed for the equal-percentage curve:

$$d_x = d_0 \cdot \sqrt{1 - \exp\left[c \cdot \left(\frac{h_x}{h_{100}} - 1\right)\right] \cdot \left[1 - \left(\frac{d_2}{d_0}\right)^2\right]}. \tag{4.154}$$

The required target closing characteristic and the characteristic resulting from *Kołodziejczyk's* proposal, together with the final regulation characteristics created by combining the valve closing characteristic with thermal characteristics of the radiator (assuming that $a_z = 1$), are shown in Fig. 4.10.

As it can be seen in Fig. 4.10b, using relations (4.153) and (4.154), it is only for the practically non-existing case of $a_{\mathrm{w}} = 1$ that the desired curve is obtained, which almost completely coincides with the chart diagonal. However, if the impact of inner authority a_w is taken into account, the obtained curve differs from the required characteristic significantly. The relations proposed in [13] do not take account of the fact that such deformation occurs, assuming that the closing element hydraulic resistance changeability depending on the element opening degree will be the exact hydraulic resistance of the entire valve.

Determining the appropriate relation for the plug curvature along its height consists in finding a function that will give, with respect to the valve, for given opening ratios h_x/h_{100}, the same values of the $k_{v,x}/k_{v100}$ ratio as for the closing part, considering that $a_{w,\min} = 0.35$. This will naturally translate into different values of diameter d_x. The above condition can be written as:

$$\frac{k_{v,x}}{k_{v100}} = \frac{1}{\sqrt{1 + a_{w,\min}\left[\left(\frac{k_{v,reg,100}}{k_{v,reg,x}}\right)^2 - 1\right]}}. \quad (4.155)$$

Replacing the regulating element relative flow factor with the ratios between created surface areas of the liquid flow, the following is obtained:

$$\frac{k_{v,x}}{k_{v100}} = \frac{1}{\sqrt{1 + a_{w,\min}\left[\left(\frac{A_{100,reg}}{A_{x,reg}}\right)^2 - 1\right]}} = \frac{1}{\sqrt{1 + a_{w,\min}\left[\left(\frac{d_0^2 - d_2^2}{d_0^2 - d_x^2}\right)^2 - 1\right]}}. \quad (4.156)$$

After transforming the equation to express the sought quantity d_x, the following is obtained:

$$d_x = d_0 \cdot \sqrt{1 - \frac{\left[1 - \left(\frac{d_2}{d_0}\right)^2\right]}{\sqrt{\frac{1}{a_{w,\min} \cdot \left(\frac{k_{v,x}}{k_{v100}}\right)^2} - \frac{1}{a_{w,\min}} + 1}}}. \quad (4.157)$$

Substitution of appropriate mathematical descriptions of the closing characteristic function that is required in relation to the entire valve results in the following formulae:

- for the plug part designed for the equal-percentage curve, in the range of $h_x/h_{100} = 0.3$–1.0:

$$d_x = d_0 \cdot \sqrt{1 - \frac{\left[1 - \left(\frac{d_2}{d_0}\right)^2\right]}{\sqrt{\frac{1}{a_{w,\min} \cdot \left(\exp\left[2c \cdot \left(\frac{h_x}{h_{100}} - 1\right)\right]\right)} - \frac{1}{a_{w,\min}} + 1}}}, \quad (4.158)$$

- for the plug part designed for the linear curve, in the range of $h_x/h_{100} = 0$–0.3:

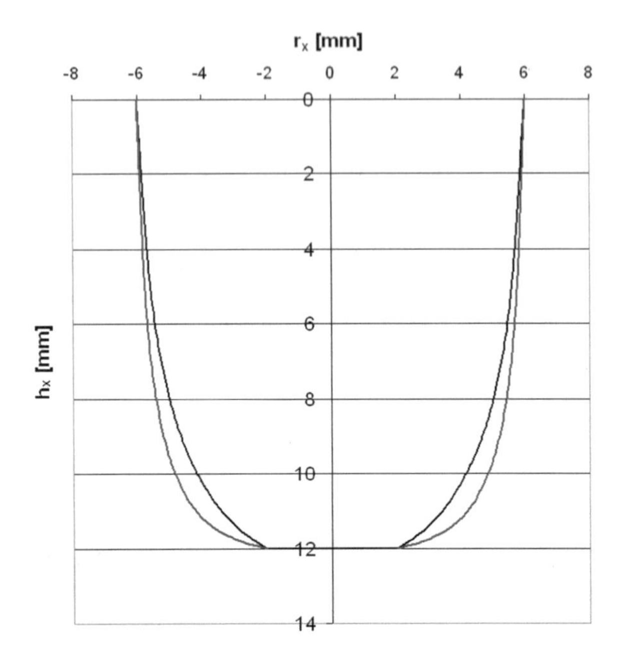

Fig. 4.11 Closing plug contours: black—without taking account of the inner authority impact [13]; red—taking account of the inner authority impact, according to the computational algorithms proposed herein

$$d_x = d_0 \cdot \sqrt{1 - \frac{\left[1 - \left(\frac{d_2}{d_0}\right)^2\right]}{\sqrt{\frac{1}{a_{w,\min} \cdot \left(0.35 \cdot \frac{h_x}{h_{100}}\right)^2} - \frac{1}{a_{w,\min}} + 1}}}. \tag{4.159}$$

Figure 4.11 illustrates the closing plug shape plotted according to the computational algorithm proposed by *Kołodziejczyk* and according to the algorithm put forward herein.

Higher values of diameter d_x can be justified easily, even without any calculations. As discussed above and illustrated in Fig. 4.1, the impact of inner authority a_w (as well as of outer and total authority a_z and a_c, respectively) always involves a rise in the relative volume flow for a set relative degree of the valve opening. The smaller the a_w value, the bigger the increments. Therefore, in order to bring the increased relative volume flow to the initial value, the relative hydraulic resistance of the closing element responsible for regulation should be raised. This means that the liquid flow surface area A_x should be reduced, i.e. the plug diameter value should be increased, for the set value of the valve seat diameter. Illustrating the situation alternatively, it is possible to assume for the plug shape calculations a characteristic with a higher value of coefficient c, which is more convex downwards, knowing that due to the inner authority effect it will be deformed upwards, thus getting closer to the required one. It turns out that higher values of diameter d_x will also be obtained for equal-percentage characteristics with a higher value of coefficient c.

4.4 Concept of a Double-Regulation Valve with a Constant Inner Authority Value and a Constant Range of the Closing Element Travel

It is indicated above that the inner, outer and total authority values are not constant for the valves which are now available on the market. Consequently, the valve regulation characteristic does not have a constant shape, either, because it depends on the valve (total) authority. The smaller it is, the bigger the upward curvature of the regulation characteristic. This is an unfavourable phenomenon because the regulation valve characteristic should be selected for the regulated object characteristic with a specific shape to ensure appropriate final regulation characteristics and, consequently, good quality of the regulation process. But in fact, as the engineering theory and practice have until today not provided correct formulae to calculate the real value of the valve authority (inner, outer and total), and due to the fact that it is impossible to maintain the authority value at an appropriate level because of the so-far applied design solutions, this postulate has not been satisfied.

The changeability of the total authority value of a valve installed in the pipework may have a twofold character. In the case of double-regulation valves with one adjustable section of the medium flow, the parameter value rises with a rise in initial throttling resulting from the valve pre-setting. For valves with two adjustable sections of the medium flow, which are now the most common solution in the radiator regulation technique, the opposite is the case. In both situations, however, the changeability is mainly the effect of a change in the inner authority value. In the case of single-regulation valves, the inner authority value is constant. The changeability of the regulation characteristic shape results only from changes in the outer authority value, for example due to a change in the hydraulic resistance of the connected pipework. From the practical point of view, it is worth analysing the radiator double-regulation valves in the first place. The first limitation concerning the radiator regulation valves results from the fact that as a rule the analysis is only justified for local current-regulation elements (i.e. radiators for example), where the shape of the regulation characteristic is essential. The second limitation concerning double regulation is due to the fact that the double-regulation principle makes it possible to balance the installation circuits thermally and hydraulically, which is not guaranteed by single regulation, which practically impedes proper operation. This is the reason why the latter (single regulation valve) is no longer common.

It is indicated in Sect. 2.2 that the most favourable solution for the radiator valve, especially for a valve intended for co-operation with a thermostatic head, is one that integrates the elements of the regulation first and second stage in one adjustable section of the medium flow, ensures independence of the regulation second-stage element range of travel as a function of the opening degree of the first-stage element and does not limit the possibility of shaping the geometry of the regulation second-stage element to an unfavourable flat profile only. The aim of the integration of the regulation first- and second-stage element is to ensure the highest possible inner authority value. The independence of the closing element range of travel as a function of the throttling element setting serves the purpose of reducing the impact of the inaccuracy of the element making and fitting and of the assembly- and operation-related play on the regulation quality. The freedom of shaping the closing element is to ensure the possibility of free shaping of the regulation characteristic. It is added in the section mentioned above that the *Goeke* historical design of the *Optimal* valve is the closest to meet these requirements. But there is one more condition that needs considering. As described earlier, a change in the valve inner authority results in a change in the shape of the regulation characteristic. Neither a decrease nor an increase in the authority value is desirable from the point of view of co-operation with a regulated object with a specific static characteristic. The best option is to have a constant value of inner authority, ensure that the value is not a function of pre-setting and enable the value regulation regardless of the pre-setting. None of the valves mentioned and described above, whether old or modern, satisfies these postulates.

This section presents an original concept of the regulation valve structural and technical solution that makes it possible to meet the requirements. The analysis is divided into two parts—the first presents a synthetic realization of the regulating assembly that enables the inner authority stabilization, but without ensuring the constancy of the closing element range of travel. The idea is verified experimentally in a further part of this book (cf. Sect. 5.2). Later on, an original concept of the valve design is presented that makes it possible to meet all the postulates discussed above. The solution is patent-protected on the territory of the Republic of Poland.

Analysing relations [(4.10), (4.15), (4.25), (4.33), (4.46), (4.50), (4.77)], it can be concluded that in order to maintain a constant inner authority value, an additional hydraulic resistance should be introduced into the valve that will vary with a change in the hydraulic resistance resulting from changes in pre-setting. In the case of a valve with one adjustable section of the medium flow, the additional hydraulic resistance included in term r_k should be located beyond the current regulation section, i.e. beyond the closing element. If this part hydraulic resistance r_{reg} rises, increasing the numerator and the denominator term in formula (4.25), a parallel rise will occur in the denominator term r_k, keeping a constant value of the quotient. By contrast, in the case of a valve with two adjustable cross-sections of the medium flow, the additional hydraulic resistance should be produced within the current regulation section because if term r_I rises, increasing the denominator in formula (4.33) and (4.15), a parallel rise will occur in term r_{II} in the nominator and denominator term, keeping a constant value of the quotient.

The solution was analysed using a system of two in-series single-regulation valves with one adjustable section of the medium flow, with a known mathematical description of the original closing characteristic. Like in the analysis of the proposed method of the valve inner authority determination (cf. Sect. 5.3), these were the *Danfoss MSV-I* DN20 valves (cf. Fig. 2.26). One of the valves fulfilled the function of the closing element, responsible for current regulation, whereas the other acted as an additional hydraulic resistance, regulated (out of necessity—independently) with a change in the first valve pre-setting. This enabled a simulation of a valve with two adjustable sections of the medium flow making it possible to check the possibility of stabilizing the inner authority value.

In order to find the initial value and the range of changes in the additional regulated hydraulic resistance, the hydraulic resistance two quantities should be established depending on the main valve pre-setting ensuring current regulation. This can be done either by performing measurements or using the experimental-analytical approach discussed in Sect. 4.1.

The computational relations (3.61), (4.11), (4.57) and (4.59) were used to determine the parameters listed in Table 4.2, which presents the data on the regulation characteristic, individual hydraulic resistances and values of the valve inner authority $a_{w,x}$ depending on selected values of pre-setting n_i. It additionally includes the values of the system equivalent hydraulic resistance calculated from the following relation:

$$r_{z,x,zast} = r_{z,x} + r'_{z,x} \tag{4.160}$$

Based on these data and using formulae (4.15) and (3.61) transformed to the following expressions, respectively:

$$r_{I,x} \equiv r'_{z,x} = \frac{r_{reg,x}}{a_{w,100}} - r_{reg,x} - r_k. \tag{4.161}$$

$$k'_{v,x} = \frac{1}{\sqrt{r'_{z,x}}} \tag{4.162}$$

the values of additional hydraulic resistance $r_{I,x} \equiv r'_{z,x}$ and of flow factor k'_{vx} were determined. The values are also presented in Table 4.2. The table also provides the additional valve required value of pre-setting n_i', which corresponds to the calculated values of $r_{I,x} \equiv r'_{z,x}$ and k'_{vx} and which will ensure that the main valve inner authority value is maintained at the required initial level, i.e. $a_{w,100} = 0.3$. The pre-setting value is calculated based on the following polynomial approximating the shape of the valve regulation characteristic [the same polynomial is used in the further part of this book (cf. Chap. 6 presenting computational examples). There, however, the polynomial has a different form due to the smaller number of nodal points—the pre-setting values used to create it, which in that very case was sufficient to ensure adequate accuracy of the calculations].

$$\begin{aligned} k_{v,x}(n_i) = & -0.00193n_i^6 + 0.018378n_i^5 - 0.06921n_i^4 + 0.1552n_i^3 \\ & - 0.4505n_i^2 + 1.6248n_i - 0.008453. \end{aligned} \tag{4.163}$$

Moreover, the table contains data on the system equivalent total hydraulic resistance $r_{z,x,zast}$ and equivalent flow factor $k_{v,x,zast}$.

Analysing the table, it can be seen that from a certain value of the main valve pre-setting n_i, the additional valve pre-setting n_i' cannot be selected to ensure an appropriately high value of its flow factor $k'_{v,x}$ that guarantees that the main valve

Table 4.2 Hydraulic data of the valve/system under analysis

n_i	$k_{v,x}$	$r_{z,x}$	$r_{reg,x}$	$a_{w,x}$,-	r_k	$r_{I,x} \equiv r'_{z,x}$	$k'_{v,x}$	n_i'	$r_{z,x,zast}$	$k_{v,x,zast}$
0.2	0.3	11.11	11	0.99	0.112	25.55	0.198	0.131	36.66	0.165
0.3	0.442	5.12	5.0	0.976	0.112	11.55	0.294	0.196	16.67	0.245
0.4	0.577	3.0	2.89	0.963	0.112	6.63	0.388	0.261	9.63	0.322
0.5	0.7	2.04	1.93	0.946	0.112	4.39	0.477	0.325	6.43	0.394
0.6	0.83	1.452	1.34	0.923	0.112	3.01	0.576	0.4	4.462	0.473
0.8	1.06	0.89	0.778	0.874	0.112	1.703	0.766	0.55	2.593	0.621
1.0	1.269	0.621	0.509	0.82	0.112	1.076	0.964	0.714	1.697	0.768
1.2	1.458	0.47	0.358	0.762	0.112	0.723	1.176	0.91	1.193	0.915
1.4	1.629	0.377	0.265	0.703	0.112	0.506	1.406	1.144	0.883	1.064
1.5	1.7	0.346	0.234	0.676	0.112	0.434	1.52	1.27	0.78	1.132
1.6	1.782	0.315	0.203	0.644	0.112	0.362	1.662	1.443	0.677	1.215
1.8	1.918	0.272	0.16	0.588	0.112	0.261	1.957	1.864	0.533	1.37
2.0	2.039	0.24	0.128	0.533	0.112	0.187	2.312	2.59	0.427	1.53
2.2	2.146	0.217	0.105	0.484	0.112	0.133	2.742	–	–	–
2.4	2.24	0.199	0.087	0.437	0.112	0.091	3.315	–	–	–
2.5	2.3	0.189	0.077	0.407	0.112	0.0677	3.843	–	–	–
2.6	2.322	0.185	0.073	0.38	0.112	0.0583	4.142	–	–	–
2.8	2.394	0.174	0.062	0.356	0.112	0.0327	5.53	–	–	–
3.0	2.457	0.166	0.054	0.325	0.112	0	∞	–	–	–
3.2	2.5	0.16	0.048	0.3	0.112	0	∞	–	–	–

inner authority value is kept at the assumed level of $a_{w,x} = 0.3$. This is because the additional valve does not have a high enough value of flow factor $k'_{v,x}$ (max. $k'_{v,x} = 2.5$), i.e. the value of hydraulic resistance $r'_{z,x}$ is not small enough. The reason for that is simple—for high pre-setting values the main valve is characterized by such a small ratio between the regulated part hydraulic resistance $r_{reg,,x}$ and hydraulic resistance $r_{z,x}$ that, in order to maintain the assumed value of the ratio, i.e. of inner authority $a_{w,x}$, the additional element/valve has to generate a possibly low hydraulic resistance, i.e. it should have a possibly high value of flow factor $k'_{v,x}$. In such a situation, so as not to have to replace the additional valve with another, the upper limit of the main valve pre-setting n_i would have to be reduced, thus increasing the valve hydraulic resistance $r_{reg,x}$. The starting point of the considerations should therefore be the case where the maximum value of the additional valve pre-setting n'_i is selected because then the valve creates the smallest additional hydraulic resistance $r'_{z,x}$ and causes the smallest rise in the system resultant hydraulic resistance $r_{z,x,zast}$. This impact requires selecting an intermediate value of the main valve pre-setting n_i to restore the system inner authority $a'_{w,x}$ to the initial value, i.e. $a_{w,x} = 0.3$. If a fully open additional valve is incorporated into the system and the maximum value of the main valve pre-setting n_i is selected, the inner authority drops, as described by the following formula:

$$a'_{w,x} = \frac{r_{reg,x}}{r_{reg,x} + r_k + r'_{z,x}} \tag{4.164}$$

to the value of:

$$a'_{w,x} = \frac{0.048}{0.048 + 0.112 + 0.16} = 0.15.$$

The required maximum permissible value of the main valve pre-setting n_i can be determined after the required hydraulic resistance $r_{reg,x}$ (i.e. also $r_{z,x}$) or flow factor $k_{v,x}$ is calculated. Transforming formula (4.164) to express the sought value of $r_{reg,x}$, the following is obtained:

$$r_{reg,x} = \frac{r_k + r'_{z,x}}{\frac{1}{a'_{w,x}} - 1} \tag{4.165}$$

This gives:

$$r_{reg,x} = \frac{0.112 + 0.16}{\frac{1}{0.3} - 1} = 0.1166.$$

The valve hydraulic resistance totals:

$$r_{z,x} = r_{reg,x} + r_k = 0.1166 + 0.112 = 0.228$$

And the valve flow factor, according to formula (3.61) thus totals:

$$k_{v,x} = \frac{1}{\sqrt{0.228}} = 2.09.$$

Consequently, according to formula (4.163) and Table 4.2, the required value of the maximum permissible pre-setting is:

$$n_i = 2.095.$$

This should be the parameter maximum value that makes it possible to maintain the inner authority at a level of at least $a_{\mathrm{w,\,x}} = 0.3$. Reducing the value of pre-setting n_{i}, in order to maintain a constant value of parameter $a_{\mathrm{w,\,x}}$, the value of pre-setting n_{i}' should at the same time be reduced to the values calculated earlier and listed in Table 4.2.

Section 5.2 presents results of an experimental analysis conducted for the above considerations.

The solution presented and discussed above is limited by the fact that it is impossible to keep a constant range of the closing element travel because the analysis was performed using a specific regulation valve available on the market, which represents a very common structural and technical solution now in use.

As mentioned in Sect. 2.2, the most favourable solution for the radiator regulation valve, especially for a valve intended for co-operation with a thermostatic head, is one that integrates the elements of the regulation first and second stage (the throttling and the closing element) in one adjustable section of the medium flow, ensures independence of the closing element travel as a function of the throttling element setting and does not limit the possibility of shaping the closing element geometry to a profile that results in a linear original regulation characteristic. Such valves are not available, but the concept is presented by the solution described below. Its geometrical model is shown in Fig. 4.12. The solution makes it possible to generally obtain any regulation characteristic and ensures the characteristic stability as a function of initial throttling, at the closing element constant range of travel. Moreover, it is also possible to obtain two different shapes of the regulation characteristic for a single valve, without having to replace any elements inside. It is a double-regulation valve with two adjustable sections of the medium flow, where the throttling and the closing element create one adjustable section (inlet section) of the regulating device and act independently. Owing to that, it is possible to obtain a relatively high value of the valve inner authority. The second section of the medium flow is created by the throttling element, at the element outlet. The throttling element pre-setting does not limit the range of travel of the closing element.

The liquid flows into the regulating section created by the connection of the throttling element {A} inlet slots {1a} (the element is presented in the figure in its entirety as {A′}, and without the top part as {A}, to show the full geometry of relevant regulating sections) and the closing element slots {3}. It flows out through the outlet slot {1b} of the throttling element {A}. The throttling element {A} tightly fits the valve body {B} to prevent the liquid from flowing around it and force the flow only through the slots {1a, 1b}. Inside the throttling element, downstream the inlet part, the

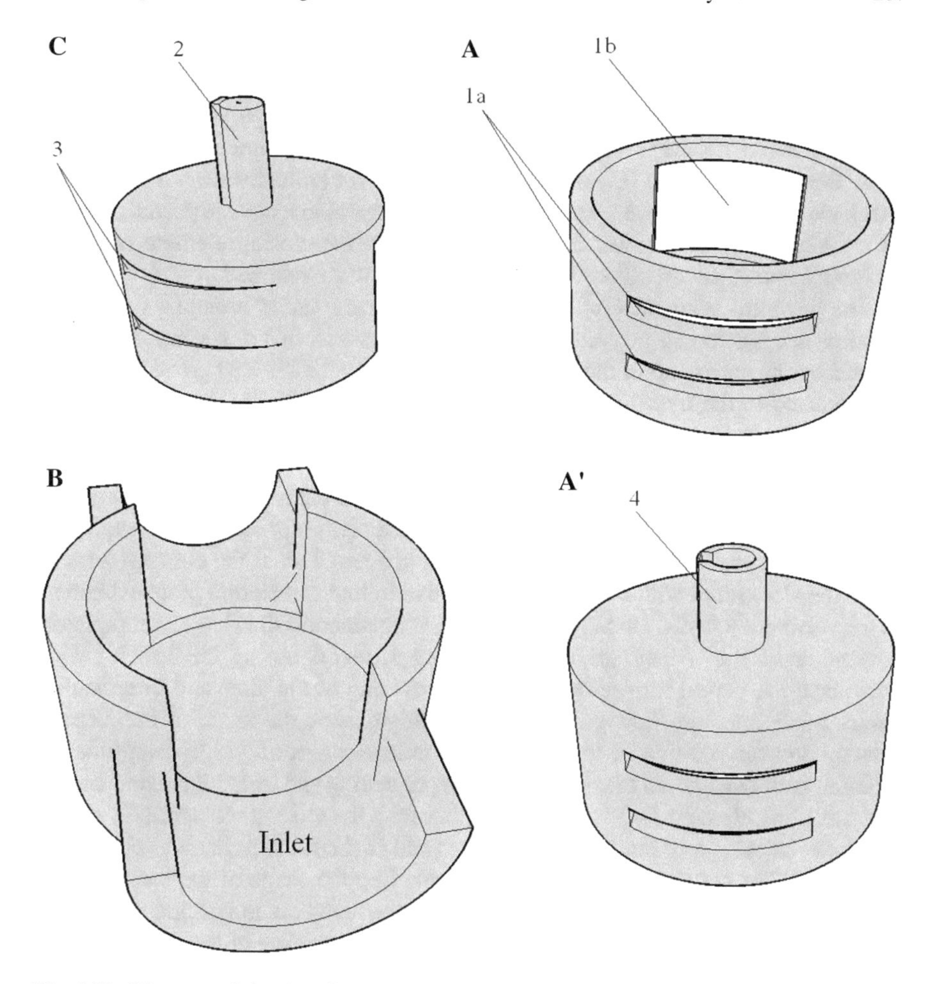

Fig. 4.12 Diagram of the described technical solution of the regulation valve (general view) A—throttling element; B—valve body; C—closing element; 1a—throttling element inlet slot; 1b—throttling element outlet slot; 2—closing element pin; 3—closing element regulating slot

closing element {C} is located, in the form of a properly shaped sector of a cylinder operating through a reciprocating motion and controlled by pressure put on the pin {2}. It has regulating slots {3} and covers the throttling element {3} slots {1a} from the inside, vertically. The shape of the slots {1a} and {3} depends on the required shape of the regulation characteristic. Due to the fact that in the case of the radiator heat output quantitative regulation an equal-percentage characteristic is favourable, the slots {3} are axially asymmetric. Additionally, the shape of the throttling element {A} slots {1a} can be selected to obtain the required regulation characteristic. The possibility of shaping the slots of both elements, not just of one of them, makes it possible to obtain a wide variety of the valve regulation characteristics. Figure 4.12

presents an example shape of the slots. The real shape is determined by the actual distribution of hydraulic resistances inside the valve, which is not known until the specific shape and size of the valve body and of the inner elements are assumed. The same applies to the shape of the throttling element outlet section {1b}.

At the closing element {C} full lift, the inlet slots {1c} remain uncovered, whereas at full closing they are covered completely. Partial opening results in partial covering of the slots. This is the way in which changes in the closing element hydraulic resistance and current regulation of the medium flow are realized.

The throttling element {A} is set by the element rotation around its axis. The rotation is triggered in a typical manner, i.e. by a turn of the setting handwheel located on the upper part of the valve body (cf. e.g. Figs. 2.28–2.31). The handwheel is connected to the throttling element by a bushing {4}. Inside the bushing is the closing element {C} pin {2}, which is controlled through pressure put on it using a manual or an automatic (e.g. thermostatic) head. On the inflow side, the throttling element {A} has slots (or just a single slot) {1a}. If the element position is such that the slots face the valve inflow connecting pipe entirely, water can flow through their entire surface area and the throttling degree is the smallest. If the element turns to make the slots stand with their side to the valve inflow connecting pipe and adhere to the valve body walls, the biggest throttling is obtained. If the slots are completely covered and if they adhere perfectly, assuming perfect tightness, the flow is cut off completely. A partial turn results in partial covering of the slots and intermediate values of the medium flow parameters. For this reason, the angle of the element rotation and the position of the slots to the inflow connecting pipe determine the initial throttling value. In this way, the valve pre-setting and initial throttling are set. The throttling element {A} is turned together with the closing element {C} so that the slots {1a, 1b, 3} of the two elements are still in the same horizontal position to each other. The simultaneous rotation is ensured by the shape of the bushings {2} and {4} of the two elements—where there is a groove cut out in one and a matching wedge formed on the other to eliminate independent rotation of the elements. The grooves/wedges may be located on a part of the height of the bushings {2} and {4} only. Figure 4.12 presents grooves and wedges located along the entire height to illustrate the solution better.

A change in hydraulic resistance in the inlet section, of current regulation {1a-3}, involves a change in the valve inner authority and, consequently, a change in the shape of the regulation (closing) characteristic. In order to compensate for the phenomenon, it is necessary to add, in parallel with the change in the inlet section {1a-3} pre-setting, and vary the value of the additional hydraulic resistance occurring beyond this section. In the solution under consideration, the resistance is created by the throttling element outlet section {1b}—the appropriate groove in the cylinder on the side opposite to the inlet section {1a-3}. As a result, if the cylinder is turned from the opening position towards the valve body wall, a drop occurs on its inlet side in the liquid flow cross-section surface area of the slots {1a-3} and a rise in the hydraulic resistance value. On the outlet side, there is also a drop in the fluid flow surface area of the slot {1b} and a rise in hydraulic resistance. At a position close to full closing, the liquid flow inlet cross-section surface area of the slots

{1a-3} is the smallest. Consequently, the hydraulic resistance is the highest, and so would be the inner authority value. At the same time, however, at such a position the liquid flow outlet cross-section surface area of the slot {1b} is the smallest, so the hydraulic resistance of this part is the biggest. Due to that, the valve inner authority value, resulting among other things from the distribution of individual resistances, is reduced. At certain ratios between the surface areas and between their variability depending on the throttling element {A} opening degree, it is possible to obtain a required, e.g. constant, distribution of their hydraulic resistances and, thereby, the valve constant inner authority value and shape of the regulation characteristic. The condition of the inner authority constancy requires (simplifying) a constant ratio between hydraulic resistances of the medium flow two cross-sections—{1a-3} and {1b}. Moreover, potential changes in the inlet section {1a-3} original regulation characteristic are corrected by the outlet section {1b} appropriate shape and its variability depending on pre-setting (a change in pre-setting involves a change in the shape). The smaller the initial throttling, i.e. the bigger the non-covered surface area of the slots {1a}, the higher the share of the part with a rectangular cross-section. This causes a change in the regulation curve towards a linear characteristic if initial throttling is reduced, irrespective of the change in the valve inner authority. For this reason, the shape of the outlet slot (slots) {1b} does not have to be the same as that of the slot (slots) at the inlet. An example shape of the slot is presented in Fig. 4.12.

Because the cylinder {A} can rotate around its axis in both directions, pre-setting can be selected by turning clockwise or anticlockwise. If the inlet and outlet slots {1a} and {1b} are axially symmetric (e.g. rectangular), the change in throttling does not depend on the direction of the cylinder rotation, but only on the module of the rotation angle value compared to the initial, full opening, position. If the slots are not axially symmetric (e.g. trapezoidal or with curvilinear bases), a change in the direction of the cylinder rotation will result in different throttling because for a given module (absolute value irrespective of direction) of the rotation angle compared to the initial position, the liquid flow cross-section surface areas will be different. For a single valve, this makes it possible, depending on the direction of the cylinder rotation, to set two different shapes of both the throttling characteristic—resulting from different changeability of the created surface areas of the liquid flow in section {1b}—and the closing characteristic—resulting from the change in its original form and the change in the valve inner authority. Figure 4.12 presents an example, axially symmetric, shape of the cylinder outlet slot {1b}. The inlet slots are also axially symmetric.

The surface areas of the inlet and outlet slots {1a} and {1b} depend on the valve required flow capacity. The higher it is, the bigger the surface area of the slots should be. The ratio between the surface area depends on the expected value of the valve inner authority, but also on the size of the closing slot {3} and the value of the valve body hydraulic resistance, i.e. the valve target geometry, because both these factors have an impact on the resultant value of the valve hydraulic resistance and on the distribution of individual resistances within the valve. A coupled analysis is therefore necessary to determine the surface areas, where the geometries of the valve body {B} and of the closing element {C} are designed simultaneously.

Nonetheless, the height of the inlet slot {1a} should remain the same so that the available working range of the closing element {C} travel, which is determined by it, should not change with a change in the valve nominal flow capacity within a given series of types. The valve nominal flow capacity can be changed through changes in the width of the slots {1a}. The solution therefore makes it possible to create easily a series of types of a given valve model characterized by different nominal values. The slot width range is limited because if the slots {1a} are too wide, they will be impossible to cover completely by the wall of the valve body {B} at the cylinder turn. This limitation involves a limitation to the valve maximum flow capacity. In order to remove it, a proposal is made within the presented solution to use two inlet slots {1a} instead of one, with the same height and located one above the other, at an appropriate distance resulting from the closing element maximum range of travel (the distance must be at least equal to the element travel). Depending on the required flow capacity value, more slots can be applied and/or their width may be varied. However, they should not be connected to create a single slot with an equivalent bigger height because this would increase the available range of the closing element travel, which, as described and explained herein, is not favourable. If more slots are used instead of one {1a}, the closing element {C} also has more slots {3}, according to the number of the element inlet slots.

It is recommended within the presented solution that the height of the slots {1a} and {3} and the closing element range of travel should be at the level of 1 mm, which according to the Author is the optimal value from the point of view of co-operation with both types of thermostatic heads now in use—with the liquid and the vapour sensor. Such a relatively small range of travel of the closing element {C} makes it possible to avoid unfavourable phenomena arising from undesirable increments in the medium volume flow to levels higher than the design value due to the thermostatic head operation, as previously described herein. If the valve operates with thermostatic heads which are now available on the market, the range of changeability of the maximum proportional range at the level from about 2 K (for heads with the vapour sensor, with the typical amplification factor values of about 0.4–0.5 mm/K) to about 4 K (for heads with the liquid sensor, with the typical amplification factor values of about 0.2–0.25 mm/K). Anyway, the range should generally be as small as possible.

The structural design and the mutual operating and geometrical relationships of the valve throttling-closing mechanism, which make it possible to obtain the closing element constant range of travel and a constant shape of the regulation (closing) characteristic depending on the throttling element pre-setting, and which ensure freedom of the characteristic shaping, constitute a novel technical solution. The valve can be made using commonly available technical means, materials and technologies.

Pre-setting is carried out in a typical manner—through rotation of the appropriate element. The mechanical connection and the shift of this motion from the valve external manual selector onto the throttling element can be solved in a number of ways. For example, the throttling cylinder {A} may be closed on top like the closing cylinder {C} and may have a bushing connected coaxially to the pin {2} in the axis of its rotation. The pin will then be inside the bushing and the bushing will then protrude outside the valve, into the selector.

References

1. Ali, M.: Knowledge-based optimization model for control valve locations in water distribution networks. J. Water Res. Plann. Manag. **141**(1) (2015)
2. Bianchi, A., Mambretti, S., Pianta, P.: Practical formulas for the dimensioning of air valves. J. Hydraul. Eng. **133**(10), 1177–1180 (2007)
3. Catalogue information of Danfoss
4. Catalogue information of Herz
5. European Standard EN 215:2004: Thermostatic radiator valves—requirements and test methods
6. Farenzena, M., Trierweiler, J.O.: Valve stiction estimation using global optimization. Control Eng. Pract. **20**(4), 379–385 (2012)
7. Garcia, C.: Comparison of friction models applied to a control valve. Control Eng. Pract. **16**(10), 1231–1243 (2008)
8. Gramberg, A.: Die örtliche Regelung der Wasserheizung, Gesundh.-Ing. t. 32, 6/1909
9. Hägglund, T.: A shape-analysis approach for diagnosis of stiction in control valves. Control Eng. Pract. **19**(8), 782–789 (2011)
10. Jablonowski, H.: Termostatyczne zawory grzejnikowe. Pomiar, Regulacja, Montaż, Hydraulika (Thermostatic Radiator Valves. Measurement, Adjustment, Assembly, Hydraulics), Instalator Polski, Warszawa (1995)
11. Jablonowski, H.: Thermostatventil-Praxis. Meβtechnik Regelung Montage Hydraulik, Gentner Verlag, Stuttgart (1994)
12. Koczyk, H. (ed.): Ogrzewnictwo praktyczne. Projektowanie, montaż, eksploatacja (Practical Heating. Design, Installation, Operation), SYSTHERM SERWIS, Poznań (2015)
13. Kołodziejczyk, W.: Armatura regulacyjna w ogrzewaniach wodnych (Control armature in hydronic heating systems). Arkady, Warszawa (1985)
14. Kołodziejczyk, W.: Termostatyczne zawory grzejnikowe w instalacjach centralnego ogrzewania (Thermostatic radiator valves in central heating systems). Centralny Ośrodek Informacji budownictwa, Warszawa (1992)
15. Kozłowski, B.: Równoważenie hydrauliczne obiegów grzejnych i chłodzących (Hydraulic balancing of heating and cooling circuits). Instytut Techniki Budowlanej, Warszawa (2012)
16. Lee, P., Vítkovský, J., Lambert, M., Simpson, A.: Valve design for extracting response functions from hydraulic systems using pseudorandom binary signals. J. Hydraul. Eng. **134**(6), 858–864 (2008)
17. McPherson, D.L.: Air Valve Sizing and Location: A Prospective. In: Proceedings of Pipelines Conference, pp. 905–919 (2009)
18. Mielnicki, J.S.: Centralne ogrzewanie. Regulacja i eksploatacja Cetral heating. Regulation and exploatation), Arkady, Warszawa (1985)
19. Mielnicki, J.S.: Możliwości regulacji wstępnej i eksploatacyjnej za pomocą zaworów grzejnikowych (Possibility of preliminary and exploitation regulation by means of radiator valves), District Heatin, Heating, Vemtilation, (1), 3/-1969, pp. 73–80
20. Mielnicki, J.S.: Własności statyczne ogrzewań wodnych w zakresie warunków termicznych i hydraulicznych (Static properties of water heating systems in terms of thermal and hydraulic conditions), Przegląd Informacyjny—Ciepłownictwo, 1/1971
21. Muniak, D.: Grzejniki w wodnych instalacjach grzewczych. Dobór, konstrukcja i charakterystyki cieplne Radiators in hydronic heating installations. (Structure, selection and thermal characteristics), WNT/PWN, Warszawa (2015)
22. Muniak, D.: Radiators in hydronic heating installations. Structure, selection and thermal characteristics. Springer (2017)
23. Palau-Salvador, G., González-Altozano, P., Balbastre-Peralta, I., Arviza-Valverde, J.: Improvement in a Control Valve Geometry by CFD Techniques. In: Proceedings of Pipelines Conference, pp. 202–215 (2015)
24. Peffer, T., Pritoni, M., Meier, A., Aragon, C., Perry, D.: How people use thermostats in homes: a review. Build. Env. **12**(46), 2529–2541 (2011)

25. Pyrkov, V.: Gidrawliczeskoje regulirowanije sistem otoplenija i ochłażdjenija. Teorija i praktika, Danfoss, Kijów (2005)
26. Pyrkov, V.: Regulacja hydrauliczna systemów ogrzewania i chłodzenia. Teoria i praktyka (Hydraulic regulation of heating and cooling systems. Theory and practice), Systherm Serwis, Poznań (2007)
27. Rahmeyer, W., Driskell, L.: Control valve flow coefficients. J. Transp. Eng. **111**(4), 358–364 (1985)
28. Roos, H.: Hydraulik der Wasserheizung, wydanie 3, Oldenbourg Verlag GmbH, Monachium (1995)
29. Roos, H.: Zagadnienia hydrauliczne w instalacjach ogrzewania wodnego (Hydraulic issues in water heating installations). PNT CIBET, Warszawa (1997)
30. Scali, C., Ghelardoni, C.: An improved qualitative shape analysis technique for automatic detection of valve stiction in flow control loops. Control Eng. Pract. **16**(12), 1501–1508 (2008)
31. Scott, D.: The Importance of Valves to Asset Management and Pipeline Performance. In: Proceedings of Pipelines Conference, pp. 1–8 (2008)
32. Shoukat Choudhury, M.A.A., Thornhill, N.F., Shah, S.L.: Modelling valve stiction. Control Eng. Pract. **13**(5), 641–658 (2005)
33. Suda, M.: Simulation of valve closure after pump failure in pipeline. J. Hydraul. Eng. **117**(3), 392–396 (1991)
34. Weker, P., Mineur, J.M.: A performance index for thermostatic radiator valves. Appl. Energy **6**(3), 203–215 (1980)
35. Wymagania techniczne COBRTI Instal, zeszyt 2: Wytyczne projektowania instalacji centralnego ogrzewania (Technical requirements of COBRTI Instal, book 2: Design guidelines for central heating installations), COBRTI Instal, Warszawa (2001)
36. Xie, L., Cong, Y., Horch, A.: An improved valve stiction simulation model based on ISA standard tests. Control Eng. Pract. **21**(10), 1359–1368 (2013)
37. Xu, B., Fu, L., Di, H.: Dynamic simulation of space heating systems with radiators controlled by TRVs in buildings. Energy Build. **40**(9), 1755–1764 (2008)
38. Xu, B., Huang, A., Fu, L., Di, H.: Simulation and analysis on control effectiveness of TRVs in district heating systems. Energy Build. **43**(5), 1169–1174 (2011)
39. Xu, J., Nie, Z., Liu, T., Guan, X., Zhang, P., Wu, Z., Sun, P., Ke, L.: The Control Valve Sizing and Selection Used in Central Asia-China Gas Pipeline Project. In: Proceedings of ICPTT, pp. 262–269 (2013)
40. Zhang, S., Zhang, J., Liu, L., Zhang, Q.: Simulating the working conditions of pipeline system with control valves. In: Proceedings of ICPTT, pp. 532–541 (2011)

Chapter 5
Experimental Verification of the Proposed Methods of Determination of the Valve Inner Authority, Pre-setting and the Closing Plug Geometry

This chapter presents the results of an experimental analysis of the computational algorithms concerning the proposed methods of determination of the valve authority, pre-setting and the valve plug geometry. Curves illustrating hydraulic characteristics are presented for selected radiator regulation valves and balancing valves. The characteristics were taken by the Author to verify the proposed computational algorithms concerning the valve authority. They are compared to the curves plotted based on the algorithms. A computational and experimental verification is also carried out of the algorithms proposed for the selection of the required value of the regulation valve pre-setting n_i. The considerations concern the valves described earlier in this book.

5.1 Experimental Verification of the Proposed Method of Determination of the Regulation Valve Plug Geometry

The verification is carried out using a modified design of the *Herz GP* manual regulation valve illustrated in Fig. 2.25a and Fig. 5.1. It is a double-regulation valve with one adjustable section of the medium flow, with a modified closing plug. An element was made (cf. Fig. 5.1b) that, when installed in the valve body, was supposed to ensure a linear regulation characteristic if the maximum pre-setting value was selected.

In formula (4.157), and in its target form, i.e. formula (4.159), parameter $a_{w,min}$ is defined using the direct method of the valve inner authority determination, as presented in Sect. 4.1.2.1.

Figure 5.2 presents the required closing characteristic and the curve obtained from measurements of the modified valve. Satisfactory agreement was obtained. The differences appearing along most of the curve are the effect, among other things, of the

D. P. Muniak, *Regulation Fixtures in Hydronic Heating Installations*, Studies in Systems, Decision and Control 187, https://doi.org/10.1007/978-3-030-03128-2_5

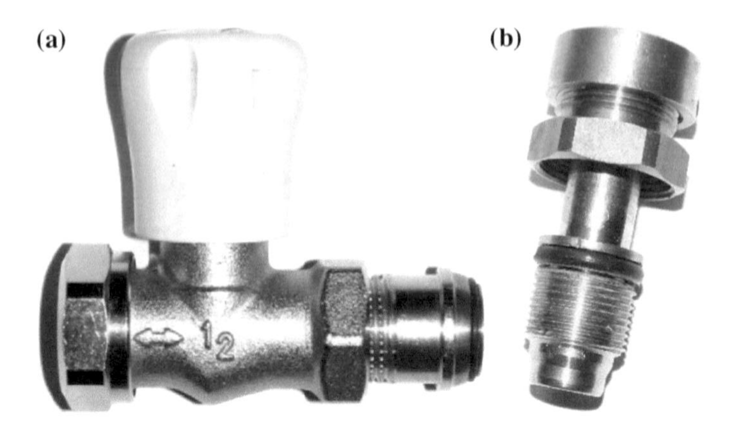

Fig. 5.1 **a** *Herz GP* regulation valve; **b** modified closing plug (own materials)

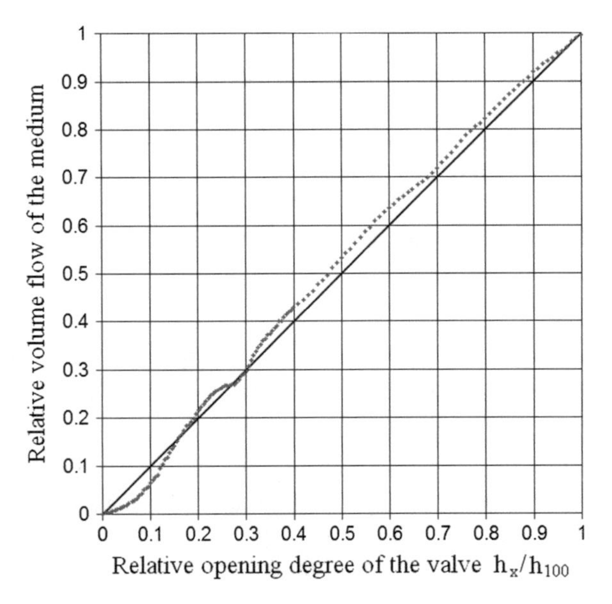

Fig. 5.2 Regulation characteristics: required curve (black continuous line), curve obtained from experiments (red dotted line)

finite accuracy of the valve plug making, as well as of assembly- and operation-related play, which cause the deviation from the original linear regulation characteristic.

Considerable differences, of the order of a few dozen percent and more, can be observed in the lower range of the opening degree, which is due to the plug structure. Flowing through the cylindrical section created between the plug and the valve seat, the liquid flows under the plug base spilling out to the sides through a channel created between the plug base and the seat. At high values of opening degree h_x/h_{100}, the channel has a big cross-section and, consequently, creates low hydraulic resistance. This means that it does not have a significant impact on the hydraulic resistance of the cylindrical section between the plug and the seat, or on the shape of the initial regulation characteristic. At relatively low values of opening degree h_x/h_{100}, in a position close to full closing, the impact is significant because the cross-section of the channel from which the medium flows from underneath the plug base is comparable to and smaller than the cross-section of the cylinder created between the plug side surface and the valve seat. Due to that, an additional and relatively high hydraulic resistance arises, which adds to the hydraulic resistance created by the plug-seat interface, causing deformation of the regulation characteristic. This fact is not taken into account in the presented algorithm. It could be taken into consideration, but that would require an individual approach depending on the structure of the interface between the valve plug and seat. The range of the *Reynolds* numbers in which the testing is carried out is also of some importance. The theoretical curve concerns a fully turbulent flow of the medium, which in practice is not always the case, especially in the lower range of the opening degree. Some additional factors that cause the observed deviations are related to the flow structure and the liquid velocity field inside the valve, e.g. the flow contraction, the jet local separation from the wall, the jet local swirling, etc. These phenomena are partially taken into account by the valve discharge coefficient α.

5.2 Experimental Verification of the Possibility of the Regulation Valve Inner Authority Stabilization

This section is devoted to experimental verification of the thesis advanced in Sect. 4.4. For selected values of the valve pre-setting $n_i = 1.0$ and $n_i = 2.0$, Fig. 5.3 presents the results of an experimental analysis verifying the proposed methodology and the data determined based on it and listed in Table 4.2. The basic characteristic is the one obtained for no reduction in the range of the valve closing element travel, i.e. for $n_i = 3.2$, for which $a_{w,x} \equiv a_{w,100} = 0.3$. The other two curves are obtained at intermediate values of pre-setting n_i and values of pre-setting n_i' calculated according to the presented algorithms.

Satisfactory agreement was obtained. The deviations between the curves plotted based on experimental data, resulting from the proposed computational algorithms, are, among other things, the effect of assuming inner authority a_w as a mean computational value, found by means of the herein-presented indirect method from the entire available range of the plug travel, and not the computational value for the analysed pre-setting n_i' selected on the additional valve (not to be confused with the value of authority $a_{w,x}$ in the case of a reduction in the closing plug range of travel to this pre-setting value). Moreover, it is also an effect of the finite accuracy of the pre-setting selection on valves, their assembly- and operation-related play (resulting in different flow factor values at closing and opening, despite selecting the same pre-setting),

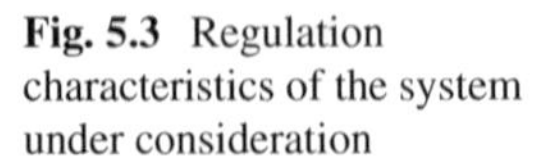
Fig. 5.3 Regulation characteristics of the system under consideration

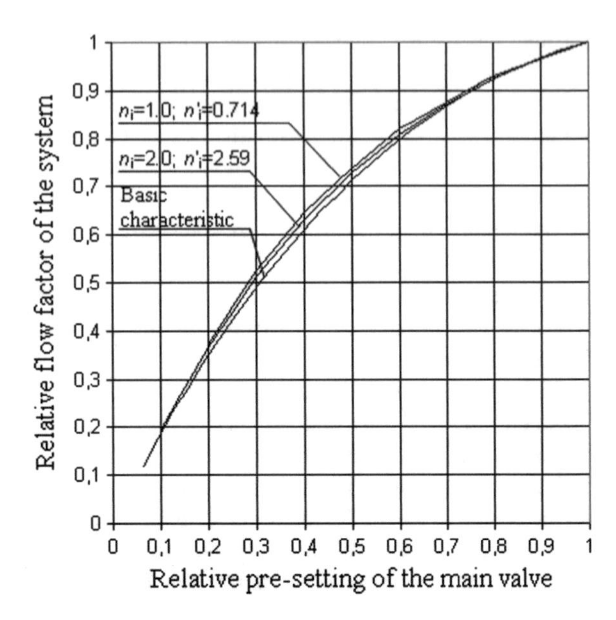

additional hydraulic resistances arising on elements connecting the valves, measuring errors, etc.). It is of course possible to achieve better convergence correcting the pre-setting on a given valve or taking account of the computational variability in the valve inner authority, as previously described in this book. In terms of the methodology presentation, the obtained agreement is sufficient.

5.3 Experimental Verification of the Proposed Method of Determination of the Regulation Valve Inner Authority

The regulation characteristics of the valves under consideration were taken on a purpose-built measuring stand. Its schematic diagram and structure are presented in Fig. 5.4. The stand is composed of the following main elements:

- a set of five glass rotameters, Class 2.5 with the measuring range of 0.2–1012 dm^3/h of water in the temperature of 20 °C,
- water differential manometer, Class 1.6, for the pressure difference range of 100 mbar,
- water differential manometer, Class 2, for the pressure difference range of 4000 mbar,
- *Herz 4002 42* pressure difference stabilizer, with the range of 0.05–0.3 bar,
- *Herz 4004* pressure relief valves, with flow factor $k_{vs} = 2.2$ m^3/h,
- *Herz HT* multi-layered pipes,

- *IMP PUMPS GHN 25/60* circulating pump,
- ball cut-off valves, water filter, water tank, connectors, screws, three-way pipes, rubber pipes, air vent, etc.

The setting of the valves closing elements was selected using a micrometre screw with a pitch of 0.01 mm. The testing was performed in a room with temperature variations of less ±1.5 °C. For most of the testing duration time, the variations did not exceed ±1.0 °C. The changes in the temperature of water were included in the same range. Temperature affects the values read from the rotameter scale, but considering that the variations and deviations from the benchmark value (20 °C) are so small, the impact may be omitted. Temperature also affects the calculated value of the flow factor because water density in the generalized formula defining the quantity is a function of temperature. The factor value, as previously described herein, should be calculated for water with a temperature from the range of 5–40 °C. This condition was satisfied during the testing. However, heating installations usually operate at higher temperatures, of up to 80 °C. For such a temperature level, water density will be slightly lower. Consequently, the flow factor value will be different for the same values of the medium volume flow and of the pressure drop on the valve. Still, the difference will be so small that it may be omitted.

The stand structure ensured the analysed element operation at the outer authority value close to one ($a_z \approx 1$), i.e. without the element characteristic deformation due

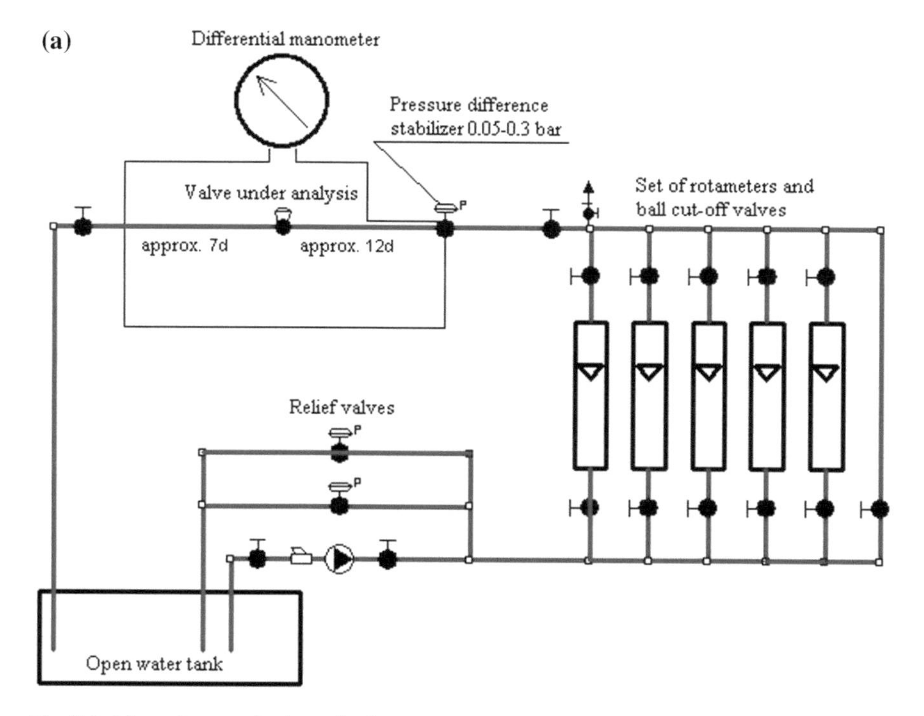

Fig. 5.4 Measuring stand schematic diagram (**a**) and the stand view (**b**)

Fig. 5.4 (continued)

to the effect of external elements. From the practical point of view, a detailed estimation of hydraulic resistances of short sections of straight pipes connected to an analysed valve between the points of pressure intake to the manometer and intended for stabilization of the liquid velocity field is not necessary. The pipes are made of a material characterized by low roughness (copper) and their nominal diameter is relatively large. Consequently, they create a hydraulic resistance which is many times lower compared to the resistance of the analysed valves in their full opening positions, which causes the regulation characteristic smaller deformation than the deformation resulting from the measuring error.

The measurements were performed in two variants. In one of them, a constant differential pressure value was set on the valve each time, for every value of pre-setting n_i. In the other, different pressure was set for each pre-setting n_i so that for every pre-setting value under consideration the medium volume flow values changed in the same range. Therefore, as the pre-setting value was reduced, i.e. as initial throttling was increased, the differential pressure value was raised appropriately. Such a measurement method is justified because it relates directly to the calculations performed in the second variant of each computational example presented in Chap. 6 and to the thesis about the changeability of inner authority a_w of a valve with one adjustable section of the medium flow as a function of pre-setting n_i .

An analysis was also conducted of closing characteristics of valves with two adjustable sections of the medium flow depending on pre-setting n_i. Moreover, the alternative relations presented herein and those proposed by *Pyrkov* and *Szaflik* that are used to determine the value of inner authority a_w of this type of valves were verified. The differences between them were indicated—the issue is described and discussed at the end of Sect. 4.1.2. The verification was carried out using a valve with one adjustable section of the medium flow, with a known mathematical description of the original closing characteristic, with an element connected thereto with a regulated hydraulic resistance, in the form of the same valve, which made up the second adjustable section of the medium flow. This made it possible to simulate a valve with two regulated hydraulic resistances—a solution commonly used in the radiator thermostatic valves. A comprehensive analysis and verification of the proposed computational relations were carried out for this type of valves on such a model because precise mathematical descriptions of original closing and throttling characteristics in the case of the double-regulation valves with two adjustable sections of the medium flow presented herein are unknown. Manufacturers do not provide such descriptions, or they offer approximate relations which, considering how much they differ from reality, cannot be used to verify the proposed relations. The mathematical description of the original closing characteristic can also be determined by means of the method described at the end of Sect. 4.1.2, but in some cases this would require destruction of the tested valve.

For valves where a precise mathematical description of the closing characteristic is unknown, curves plotted based on experimental data would not be smoothed out. Nor would their shape be interpolated. This was only done in the case of the *Danfoss MSV-I* DN20 valve, based on which the presented computational algorithms were verified comprehensively.

5.3.1 Valves with One Adjustable Section of the Medium Flow

- manual double-regulation valve *Danfoss MSV-I* DN20 (cf. Fig. 2.26) with one adjustable section of the medium flow, with the value of pre-setting n_i realized by a reduction in the regulating element (single plug) maximum range of travel h_{max}. The original throttling/closing characteristic, specified by the manufacturer and being the effect of the changeability of the surface area of the free flow cross-section depending on the opening degree, is linear. Table 5.1 presents the flow factor values depending on pre-setting n_i.

Table 5.1 Throttling/closing characteristic data of the *Danfoss MSV-I* DN20 valve ([1] and own testing results)

Number of pre-setting n_i	The valve flow factor k_{vx} [m^3/(h bar$^{0.5}$)]	Number of pre-setting n_i	The valve flow factor k_{vx} [m^3/(h bar$^{0.5}$)]
0.2	0.3	1.6	1.782
0.3	0.442	1.8	1.918
0.4	0.577	2.0	2.0 (2.039)
0.5	0.7	2.2	2.146
0.6	0.83	2.4	2.24
0.8	1.06	2.5	2.3
1.0	1.3 (1.269)	2.6	2.322
1.2	1.458	2.8	2.394
1.4	1.629	3.0	2.5
1.5	1.7	3.2	2.5

Measurements were performed in a few cases and variants, setting different values of differential pressure on the valve, different pre-settings n_i and taking closing characteristics. The measurement results were compared to the results obtained from the herein-presented methods of determination of the valve inner authority a_w.

Variant 1

The testing concerned the valve closing characteristic shape for different pre-settings n_i and the differential pressure value selected so that the changes in the medium volume flow should be included in the same range. The results are compared to the results obtained from the proposed computational algorithms.

(a) differential pressure on the valve: $\Delta p_\mathrm{z} \approx \Delta p_\mathrm{cz} = 0.05$ bar, pre-setting $n_\mathrm{i} = 3.2 = n_\mathrm{max}$. For this pre-setting, the valve inner authority calculated according to the proposed relation (4.48) will total (as to the mean value) $a_\mathrm{w,min} = 0.3$, and the determined closing characteristic will be described by a curve corresponding to this value. The medium maximum volume flow, calculated using formula (4.124c) (assuming that $r_\mathrm{str} \approx 0$), totals:

$$\dot{V}_{100} = \sqrt{\frac{\Delta p_{cz}}{\frac{1}{k_{vs}^2} + r_{str}}} = \sqrt{\frac{0.05}{\frac{1}{2.5^2} + 0}} = 0.56\,\mathrm{m}^3/\mathrm{h} = 0.155\,\mathrm{dm}^3/\mathrm{s}.$$

(b) pre-setting $n_\mathrm{i} = 2.0$. For this position of the valve plug, the valve inner authority, according to the proposed computational relations (4.48) and (4.54), will total $a_\mathrm{w,x} = 0.523$, and the determined closing characteristic will be described by a curve corresponding to this value. In order to ensure the same volume flow of the medium as in case *a*, the differential pressure value on the valve should for this pre-setting n_i be set at the level determined from formula (4.125c):

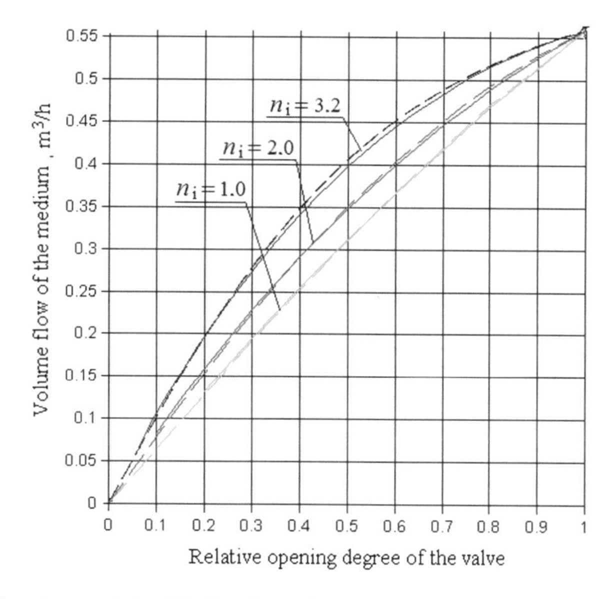

Fig. 5.5 Closing characteristic of the *Danfoss MSV-I* DN20 valve for a few pre-settings: continuous lines—experimental data; dashed lines—data obtained by means of the computational algorithms presented herein; black dashed line—closing characteristics for all pre-setting values assuming the inner authority value as proposed by *Pyrkov* and *Szaflik* in [6, 7]

$$\dot{V}_x = \sqrt{\frac{\Delta p_{cz}}{\frac{1}{k_{v,x}^2} + r_{str}}} \Rightarrow \Delta p_{cz} = \frac{\dot{V}_x^2}{k_{v,x}^2} = \frac{0.56^2}{2.039^2} = 0.0754\,\text{bar}.$$

(c) pre-setting $n_i = 1.0$. For this position of the valve plug, the valve inner authority, according to the proposed computational relations (4.48) and (4.54), will total $a_{w,x} = 0.814$, and the determined closing characteristic will be described by a curve corresponding to this value. In order to ensure the same volume flow of the medium as in case *a*, the differential pressure value on the valve should for this pre-setting n_i be set at the level determined from formula (4.125c):

$$\dot{V}_x = \sqrt{\frac{\Delta p_{cz}}{\frac{1}{k_{v,x}^2} + r_{str}}} \Rightarrow \Delta p_{cz} = \frac{\dot{V}_x^2}{k_{v,x}^2} = \frac{0.56^2}{1.269^2} = 0.195\,\text{bar}.$$

The results are compared in Fig. 5.5. The chart is scaled in the system of the valve relative opening degree for every pre-setting and the absolute volume flow.

The experimental testing results are illustrated by continuous lines, whereas the characteristics determined by means of the computational algorithms proposed herein, defined by relation (4.136),—by dashed lines. According to the thesis advanced in this book and described in the computational example from Sect. 4.2.1, factor k_{vs} in the relation each time denotes the regulating element maximum available travel range for a given pre-setting n_i, which at the same time determines the full relative range of travel in term n_{max}/n_i and h_{100}/h_x. Consequently, $a_{w,min}$ and $a_{z,min}$ denote the authority values for the maximum relative lift, also determined by this value of pre-setting n_i, and they will differ in value for different pre-settings n_i, i.e. for different reductions in the regulating element maximum range of travel h_{max}. In this way, after the authority values had been found according to formulae (4.48) and (4.54), calculations were carried out for the valve under analysis. Good agreement can be seen with the experimental data. The differences are the effect of the issues already mentioned in this book, i.e. the finite accuracy of the measurement, the inaccuracy of the valve regulating element making and the operation-related play, which cause some changes in the original linear closing characteristic adopted for the calculations performed according to the proposed methods. In addition, the black dashed line illustrates all three closing characteristics obtained assuming the inner authority value a_w as proposed by *Pyrkov* and *Szaflik* in [6, 7]. It can be seen that the postulate of the constancy of the inner authority value as a function of pre-setting is incorrect for this valve type.

Variant 2

The testing concerned the valve closing characteristic shape for different pre-settings n_i and constant differential pressure. The results are compared to the results obtained from the computational algorithms proposed herein.

(a) differential pressure on the valve: $\Delta p_z \approx \Delta p_{cz} = 0.1$ bar, pre-setting value $n_i = 3.2 = n_{max}$;
(b) differential pressure on the valve: $\Delta p_z \approx \Delta p_{cz} = 0.1$ bar, pre-setting value $n_i = 2.0$;
(c) differential pressure on the valve: $\Delta p_z \approx \Delta p_{cz} = 0.1$ bar, pre-setting value $n_i = 1.0$.

The values of the valve inner authority are calculated in Variant 1. The chart in Fig. 5.6, scaled in relative co-ordinates in every case, illustrates the experimental testing results (continuous lines) and the characteristics determined by means of the herein-proposed computational algorithms (4.48), (4.54) and (4.114) (dashed lines). The parameters are calculated according to the procedure discussed in the previous variant. The testing concerned the valve closing characteristics in a typical situation, i.e. downward regulation, from the position of the valve plug determined by a given value of pre-setting n_i.

Figures 5.7 and 5.8 present characteristics for a valve operating with a constant pre-setting value $n_i = 1.0$ and $n_i = 2.0$, respectively, and for the process of the valve further opening. These parts of the curves are plotted using computational relations (4.48), (4.54) and (4.42). A comparison is also made between the curves plotted assuming a constant value of inner authority a_w as proposed by P*yrkov* and *Szaflik* and the curve obtained experimentally. This also verifies the thesis advanced in Sect. 4.1.2 and described in the computational example from Sect. 4.2.1.

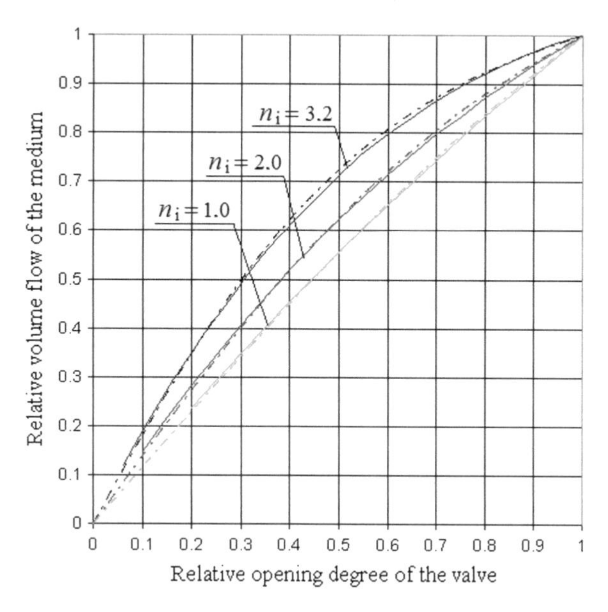

Fig. 5.6 Closing characteristic (relative values) of the *Danfoss MSV-I* DN20 valve for a few pre-settings: continuous lines—experimental data, dashed lines—data obtained by means of the computational algorithms presented herein; black dashed line—closing characteristics for all pre-setting values according to *Pyrkov* and *Szaflik* in [6, 7]

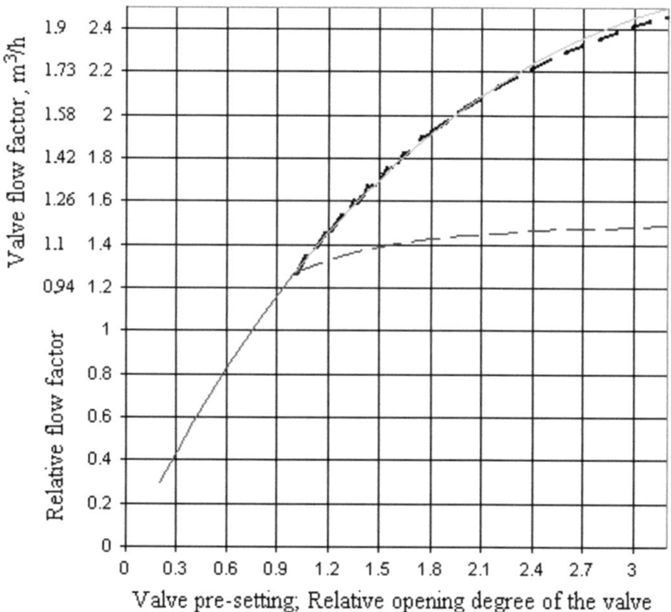

Fig. 5.7 Closing characteristic of the *Danfoss MSV-I DN20* valve for $n_i = 1.0$ and $n_i = 3.2$: continuous lines—experimental data, black dashed line—data obtained by means of the herein presented computational algorithms; blue dashed line—data obtained assuming the inner authority value as proposed by *Pyrkov* and *Szaflik* in [6, 7]

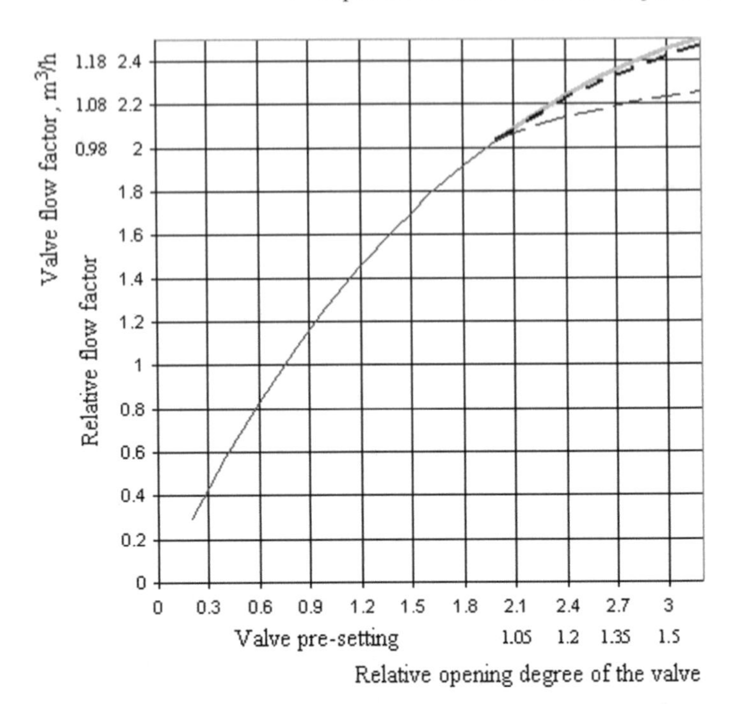

Fig. 5.8 Closing characteristic of the *Danfoss MSV-I* DN20 valve for $n_i = 2.0$ and $n_i = 3.2$: continuous lines—experimental data, black dashed line—data obtained by means of the herein presented computational algorithms; blue dashed line—data obtained assuming the inner authority value as proposed by *Pyrkov* and *Szaflik* in [6, 7]

For pre-setting $n_i = 1.0$, which in this case means $n_{max} = 100\%$ of the relative opening degree h_x/h_{100}, each further opening exceeds 100%. For example, pre-setting $n_i = 1.2$ means that $n_i/n_{max} = h_x/h_{100} = 120\%$, pre-setting $n_i = 1.8$—that $n_i/n_{max} = h_x/h_{100} = 180\%$, etc. Consequently, also the valve flow factor for this pre-setting is a reference value, maximum in the relative scale ($k_{vs} = 1.3$ m^3/h), and the calculated quantity is $k_{v,x}$. The valve inner authority for the closing element range of travel corresponding to this pre-setting, being a reference value, determined according to relations (4.48) and (4.54), totals $a_{w,x} = 0.814$. Calculating the valve flow factor k_v according to formula (4.42), these values should therefore be substituted in it. The situation is the same in the case of pre-setting $n_i = 2.0$, for which flow factor k_{vs}, to which the calculated values of $k_{v,x}$ are related, is $k_{vs} = 2.0$ m^3/h; the inner authority value totals $a_{w,x} = 0.523$. Because at $a_z = 1$ the valve regulation characteristic is generally identical to the curve illustrating changes in the medium volume flow, the only differences being the absolute values (resulting from the ratio between the difference in pressures on the valve), the y-axis is described by both quantities. So that the medium volume flow values can be calculated according to formula (4.136), the same values of the discussed parameters should be used as in the case of formula (4.42). The curves obtained from the two relations have an identical shape.

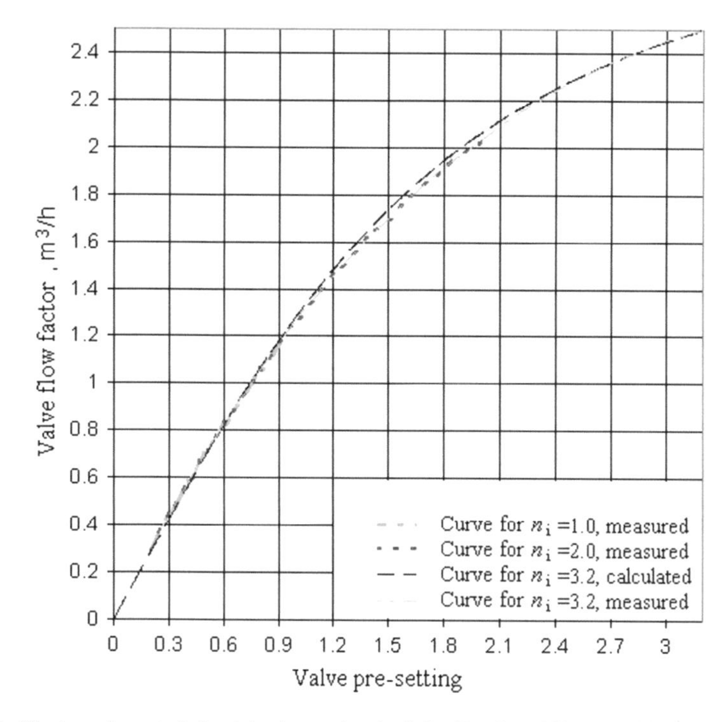

Fig. 5.9 Closing characteristics (absolute values) of the *Danfoss MSV-I DN20* valve for a few pre-settings

Figure 5.9, scaled in absolute values for each case under consideration, illustrates characteristics for the same situation, i.e. for regulation exceeding the set position of the valve regulating element for which the circuit operating parameters are established. As it can be seen, the closing characteristics for intermediate values of pre-settings n_i, though described by the above values of inner authority a_w, when determined for the range of absolute values, i.e. for the regulating element maximum available range of travel h_{max} (the maximum value of pre-setting) corresponding to the lower value of authority a_w, have the same shape and are a part of the full characteristic. This confirms the thesis presented in Sect. 4.1.2.

- the radiator double-regulation thermostatic *Herz TS-90-V* DN15 valve (cf. Fig. 2.28) with one adjustable section of the medium flow, with the value of pre-setting n_i realized by a reduction in the maximum range of travel h_{max} of the closing element with a closing plug in it.

The testing concerned the valve closing characteristic shape for different pre-setting values. Differential pressure was maintained at a constant level of 0.1 bar. The measurement results are presented in Fig. 5.11. The figure shows that as the surface area of the medium flow cross-section created by the throttling element gets

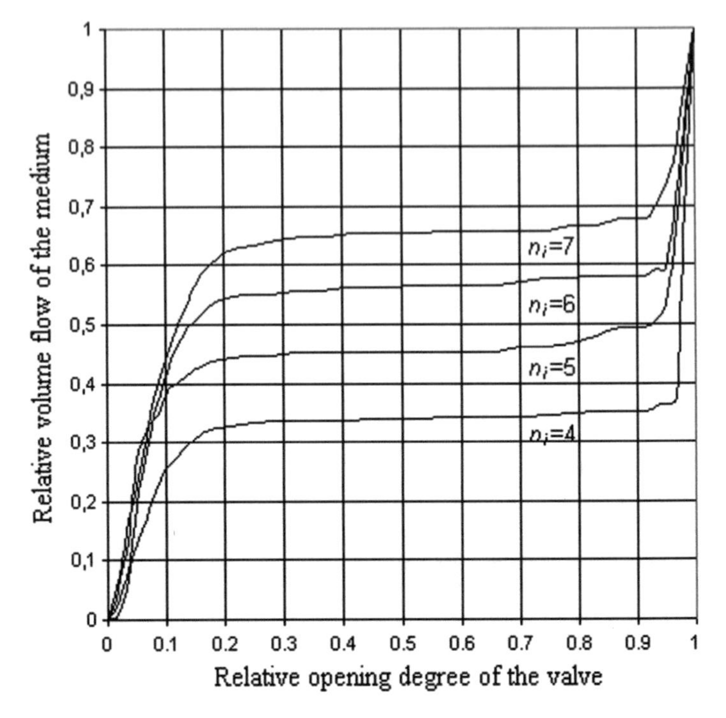

Fig. 5.10 Closing characteristics of the *Herz TS-90-V* valve for selected pre-settings

smaller, the closing characteristics are shifted downwards. This means that the valve inner authority a_w rises (Fig. 5.10).

- Polish double-regulation *708*-type valve with one adjustable section of the medium flow, with the value of pre-setting n_i realized by a change in the opening degree of the throttling plug located inside the closing element. The valve structure is presented in Fig. 2.8b. A set of closing characteristics plotted for a few values of pre-setting n_i is presented in Fig. 5.11. In the original, the pre-setting has a rising numbering depending on throttling. For the purpose of this work, the pre-setting numbering is inverted so that the rise in the pre-setting value should mean a higher opening degree and a drop in throttling, which is the case in the other situations and as has commonly been marked in practice for a long time.

The Fig. 5.11 shows that in this case, as the surface area of the medium flow cross-section created by the throttling element gets smaller, the value of the valve inner authority a_w rises, which is proved by the closing characteristic smaller deformation (i.e. increased downward convexity). Alternatively, the higher the value of pre-setting n_i, i.e. the lower the initial hydraulic resistance, the bigger the characteristic upward deformation and, consequently, the smaller the value of the valve inner authority a_w.

- double-regulation *Goeke Optimal* valve with one adjustable section of the medium flow, with the value of pre-setting n_i realized by means of a change in the sur-

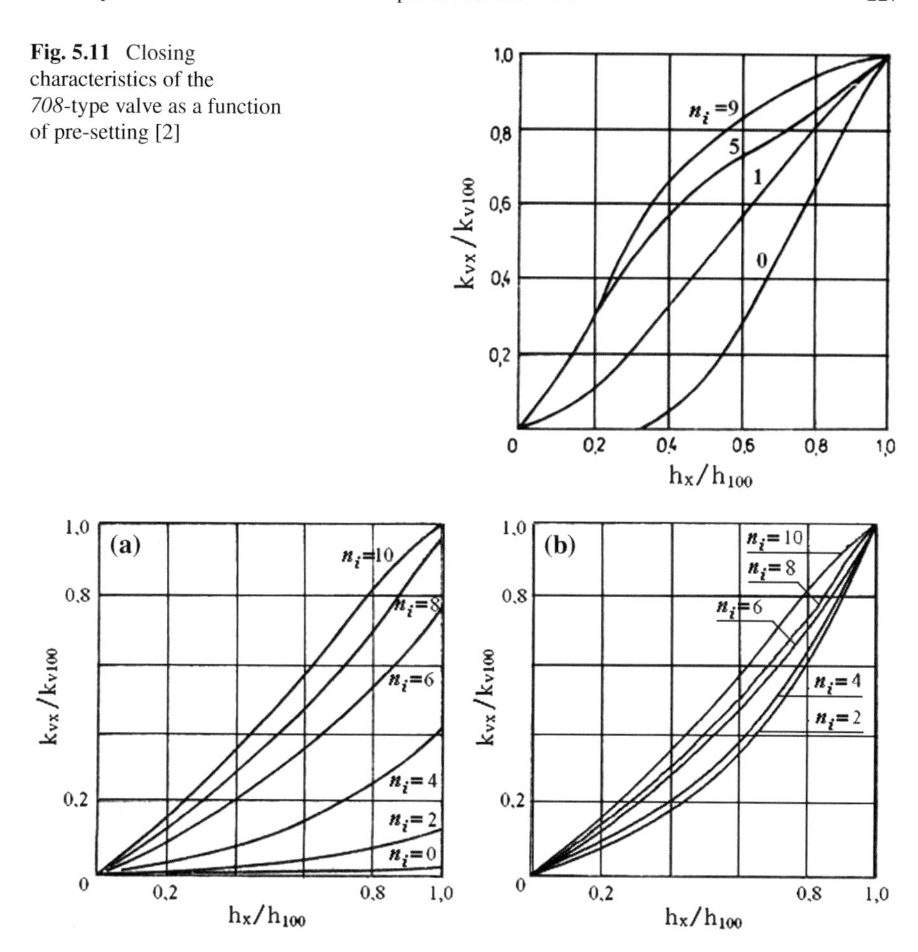

Fig. 5.11 Closing characteristics of the *708*-type valve as a function of pre-setting [2]

Fig. 5.12 **a** Closing characteristics of the *Goeke Optimal* valve as a function of pre-setting [2]. **b** Closing characteristics of the *Goeke Optimal* valve as a function of pre-setting related to the maximum flow factors for a given pre-setting

face area of a single cross-section of the medium flow by revolution of a non-symmetrical curvilinear plug in relation to the flow channel of the valve body. The valve structure is presented in Fig. 2.2c. The closing characteristics as a function of pre-setting are presented in the chart in Fig. 5.12a. From the point of view of the analysis of the problem, it is necessary to provide a chart presenting relative values for characteristics depending on pre-setting n_i, where flow factor k_v is related not to its maximum value occurring for the throttling element full opening (maximum value of pre-setting $n_i = 10$), but to the maximum value of k_v for a given value of pre-setting n_i. This is illustrated by the chart in Fig. 5.12b plotted based on the data from Fig. 5.12a.

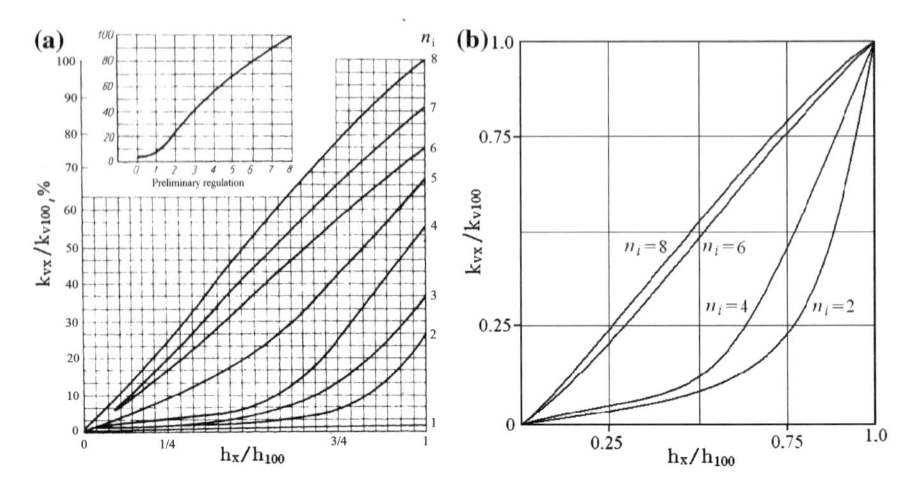

Fig. 5.13 **a** Closing characteristics of the *Pruss* valve as a function of pre-setting [4, 5]. **b** Relative closing characteristics of the *Pruss* valve as a function of pre-setting

It can be noticed that the higher the value of pre-setting n_i, i.e. the lower the initial hydraulic resistance, the bigger the characteristic upward deformation and, consequently, the smaller the value of the valve inner authority a_w.

- double-regulation *Pruss* valve with the value of pre-setting n_i realized by a reduction in the lift of a single plug. The closing characteristics as a function of pre-setting n_i are presented in Fig. 5.13a. Like in the case of the *Goeke Optimal* valve, they are plotted in relation to the maximum flow factor and, for the needs of this analysis, they have to be rescaled in the same way as before. But even from the original set of characteristics it can be seen that the lines are characterized by different curvatures. The higher the pre-setting value, the bigger the characteristic upward deformation. For the pre-setting highest value $n=8$, the line has an upward curvature, whereas already for $n_i=5$ the curvature is downwards, with the deformation degree increasing with a drop in the value of pre-setting n_i. The closing characteristics as a function of pre-setting n_i are presented, in a system of relative co-ordinates, in Fig. 5.13b.

Also in this case it can be noticed that the higher the value of pre-setting n_i, i.e. the lower the initial hydraulic resistance, the bigger the characteristic upward deformation and, consequently, the smaller the value of the valve inner authority a_w.

- Polish double-regulation *M-3176* valve (cf. Fig. 2.1c) with the value of pre-setting n_i realized by a reduction in the maximum range of travel h_{max} of a single regulating plug. The valve closing characteristics as a function of pre-setting are presented in Fig. 5.14.

The plug of the valve under discussion is made according to the linear–equal-percentage characteristic, where 20% of the plug height is shaped for linear changes, and the other part—for equal-percentage changes. Due to that, as initial throttling

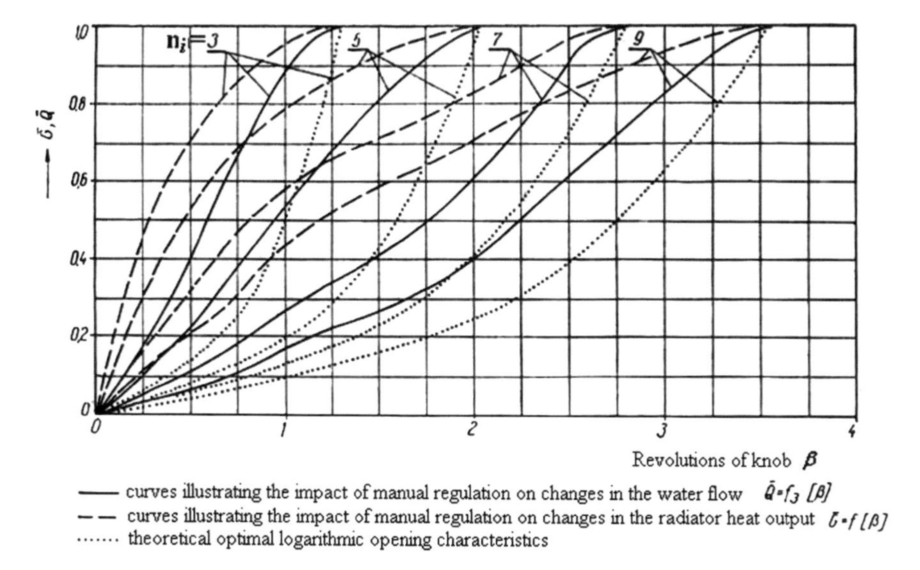

Fig. 5.14 Closing characteristics of the *M-3176* valve as a function of pre-setting (continuous lines) [3]

gets bigger by a reduction in the plug lift over the seat, the share of the original characteristic linear part in the entire range available for a given pre-setting is increased. For example, for pre-setting $n_i = 5$, i.e. for the reduction in the plug maximum range of travel to $h_{100} = 0.5 \cdot h_{max}$, the share of the original linear characteristic will total 2/5 = 40%. For $n_i = 3$, the value will be equal to 2/3 = 66.7%, etc. For this reason, with every decrease in the value of pre-setting n_i, despite the fact that the value of inner authority a_w rises, characteristics with an increasing downward curvature are not obtained because the share of the original linear characteristic gets bigger. And this linear characteristic, deformed by the valve inner authority a_w, is always curved upwards. The closing characteristics will thus have a shape with an inflexion point that separates the parts with opposite curvatures. The higher the value of pre-setting n_i, the smaller the share of the part with an upward curvature will be. Moreover, with a decrease in the value of pre-setting n_i, the deformation degree of both parts will get smaller due to the rise in the value of inner authority a_w. It follows that for low values of pre-setting n_i, if the share of the original linear characteristic is considerable or prevailing, the upward deformation will be smaller than for the same original linear part of the characteristic for low pre-setting values characterized by a lower value of inner authority a_w. The phenomena described above are confirmed by Fig. 5.14, which illustrates the valve closing characteristics.

5.3.2 *Valves with Two Adjustable Sections of the Medium Flow*

- a set of two manual double-regulation *Danfoss MSV-I* DN20 valves (cf. Fig. 2.26) with one adjustable section of the medium flow.

Variant 1

The testing concerned the closing characteristic shape for different pre-settings $n_{\rm i}$ and constant differential pressure. The results are compared to the results obtained from the computational algorithms proposed herein and from the algorithms given by *Pyrkov* and *Szaflik* in [6, 7].

The valve under analysis is characterized by the value of inner authority $a_{\rm w,min} = 0.3$. Considering the flow factor value, at full opening, at the level of $k_{\rm vs} = 2.5\ {\rm m^3/(h\ bar^{0.5})}$, the following is obtained from formulae (4.56) and (4.57):

$$r_{reg,100} = a_{w,\min} \cdot \frac{1}{k_{vs}^2} = 0.3 \cdot \frac{1}{2.5^2} = 0.048 \frac{{\rm h^2 \cdot bar}}{{\rm m^6}},$$

$$r_k = r_{z,100} - r_{reg,100} = \frac{1}{k_{vs}^2} - r_{reg,100} = \frac{1}{2.5^2} - 0.048 = 0.112 \frac{{\rm h^2 \cdot bar}}{{\rm m^6}}.$$

After an additional identical valve is connected, a hydraulic resistance is introduced that arises on the valve body and on the regulating element, which in this case constitutes the regulation first stage. The resistance minimum value, i.e. the value for full opening $n_{\rm i} = 3.2 = n_{\max} \equiv h_{\max}$, totals $r_{\rm I,100} + r_{\rm k} = 0.16\ ({\rm h^2\ bar})/{\rm m^6}$. Considering this, formula (4.15) gives the value of inner authority $a_{\rm w}$ at the level of:

$$a_{w,100} \equiv a_{w,\max} = \frac{r_{reg,100}}{r_{z,100}} = \frac{r_{reg,100}}{r_{reg,100} + r_{I,100} + 2r_k} = \frac{0.048}{0.048 + 0.048 + 2 \cdot 0.112} = 0.15.$$

The throttling element authority $a_{\rm w,I}$ will total:

$$a_{w,I,\min} = \frac{r_{I,100}}{r_{z,100}} = \frac{r_{I,100}}{r_{I,100} + r_{reg,100} + 2r_k} = \frac{0.048}{0.048 + 0.048 + 2 \cdot 0.112} = 0.15.$$

In a similar manner, using formula (4.16), the closing element inner authority $a_{\rm w}$ is calculated for the other values of pre-setting $n_{\rm i}$ under analysis. The value of this parameter is also calculated according to the proposed relations (4.77) and (4.87). This case makes use of the mathematical descriptions of the throttling element characteristic function and of the element value of inner authority $a_{\rm w,I}$ calculated above, as well as the relation proposed by *Pyrkov* and *Szaflik* in [6, 7], using also

Table 5.2 The valve inner authority and flow factor as a function of pre-setting for the closing element full and partial opening

Number of pre-setting n_i	The valve flow factor $k_{vx} = [k_{v100}]\, n_i$, $[m^3/(h\ bar^{0.5})]$	The valve flow factor $k_{vx} = [k_{v,31.25}]n_i$, $[m^3/(h\ bar^{0.5})]$	Inner authority a_{wx}			
			According to (4.15) and (4.16)	According to (4.77)	According to (4.87)	According to [6, 7]
0.2	0.298	0.292	0.0042	0.00425	0.00382	0.01
0.5	0.674	0.613	0.022	0.0217	0.0214	0.044
1.0	1.153	0.91	0.064	0.0633	0.063	0.097
1.5	1.4	1.017	0.095	0.938	0.098	0.121
2.0	1.56	1.07	0.117	0.116	0.121	0.134
2.5	1.69	1.111	0.138	0.138	0.137	0.145
3.0	1.76	1.127	0.15	0.147	0.147	0.149
3.2	1.77	1.131	0.15	0.15	0.15	0.15

the measured values of the flow factor listed in Table 5.2 together with the results of the inner authority calculations.

The value of the closing element inner authority a_w can also be calculated treating the connected valve in its entirety as an element of the first stage of regulation. In such a situation, replacing $r_{I,100}$ with the total hydraulic resistance of the valve body and of the regulating element, a relation is obtained that produces the same value, namely:

$$a_{w,100} \equiv a_{w,\max} = \frac{r_{reg,100}}{r_{z,100}} = \frac{r_{reg,100}}{r_{reg,100} + r_{I,100} + r_k} = \frac{0.048}{0.048 + 0.16 + 0.112} = 0.15.$$

However, the first approach is better-grounded because it distinguishes clearly which of the system elements are characterized by a regulated hydraulic resistance (the body is not one of them) and makes it possible to determine their authority and, consequently, the final regulation characteristic.

The algorithms for determination of the value of inner authority a_w of double-regulation valves with two adjustable sections of the medium flow proposed by *Pyrkov* and *Szaflik* in [6, 7] operate on values of flow factor k_v for the closing element partial opening, at a given value of pre-setting n_i. The computational relation is similar to relation (4.77) derived herein, except that the substituted values of flow factor k_v are related to the closing element partial opening. The proposal made by *Pyrkov* and *Szaflik* concerns valves co-operating with thermostatic heads and therefore the degrees of partial opening are in that case calculated from proportional range X_p. The issue is discussed in detail in Sect. 2.1.2. The calculations presented below are performed according to the method, assuming the opening degree at the level of $h_x/h_{max} \equiv n_i/n_{max} = 1.0/3.2 = 0.3125 = 31.25\%$, using the following formula:

$$a_{w,x} = a_{w,\max} \cdot \left(\frac{k_{v,x}}{k_{vs}}\right)^2_{hx/h100=31.25}.$$

Table 5.3 Valve flow factor depending on pre-setting for the closing element full opening (calculated values—left; measured values—right)

Number of pre-setting n_i	The valve flow factor $k_{vx} = [k_{v100}]n_i$ [m^3/(h bar$^{0.5}$)], according to (4.143)	The valve flow factor $k_{vx} = [k_{v100}]n_i$ [m^3/(h bar$^{0.5}$)]
0.2	0.298	0.298
0.5	0.672	0.674
1.0	1.148	1.153
1.5	1.39	1.4
2.0	1.54	1.56
2.5	1.676	1.69
3.0	1.74	1.76
3.2	1.75	1.77

Analysing the data listed in Table 5.2, it can be seen that the proposed computational relations (4.15), (4.16), (4.77) and (4.87) give similar results of inner authority a_w. Moreover, as indicated by the values of flow factor k_v corresponding to the closing element partial and full opening, it is confirmed that the maximum increments in the volume flow above the value for partial opening are not the same for different values of the throttling element pre-setting, and they get smaller with a decrease in the valve closing element authority a_w. The values can be calculated using the proposed relation (4.143). Table 5.3 presents the calculation results, using the values of authority $a_{w,x}$ found from relations (4.15) and (4.16) and listed in Table 5.2. The data in the table point to good agreement between the computational and experimental results.

Figure 5.15 presents a comparison between the closing characteristics obtained experimentally and the curves plotted according to herein-proposed algorithms (4.15), (4.16) and (4.113) for selected values of pre-setting n_i. Figure 5.16 presents a comparison between the curves plotted using the values of inner authority a_w calculated as proposed by *Pyrkov* and *Szaflik* in [6, 7] and the obtained experimental data, for the same values of pre-setting n_i.

The proposed computational relations for the valve inner authority a_w are confirmed by experimental verification results.

- the radiator double-regulation thermostatic *Herz TS-FV* DN15 valve (cf. Fig. 2.29e) with two adjustable sections of the medium flow, with the value of pre-setting n_i realized by a change in the surface area of the liquid flow cross-section created in the first stage of regulation by a cylinder with orifices on the side surface and with a closing plug moving inside.

The testing concerned the valve closing characteristic shape for different pre-settings n_i. Differential pressure was maintained at a constant level of 0.1 bar. The measurement results are illustrated in Fig. 5.17. As initial throttling is increased, i.e. with a decrease in the value of inner authority a_w, the closing characteristics are increasingly curved upwards.

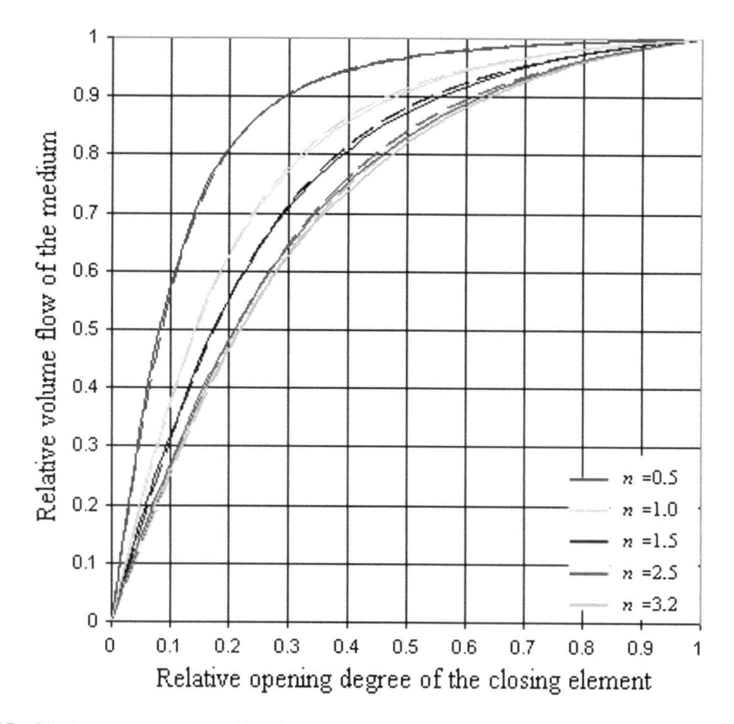

Fig. 5.15 Closing characteristic of the Danfoss valve *MSV-I* DN20 valve co-operating with a throttling element, for selected pre-settings: continuous lines—experimental data, dashed lines—data obtained from the computational algorithms presented in this book

- the radiator double-regulation thermostatic *Danfoss RTD-N15* valve (cf. Fig. 2.29b) with two adjustable sections of the medium flow, with the value of pre-setting n_i realized by a change in the surface area of the medium flow cross-section created in the first stage of regulation by a revolving cylindrical shutter with a groove cut on the side surface.

The testing concerned the valve closing characteristic shape for different pre-settings n_i. Differential pressure was maintained at a constant level of 0.1 bar. The measurement results are illustrated in Fig. 5.18. Like before, as initial throttling is increased, the closing characteristics demonstrate an increasing upward deformation, which is the effect of a decrease in the value of inner authority a_w.

- double-regulation Polish *M-3172* valve with two adjustable sections of the medium flow, with the value of pre-setting n_i realized by a change in the surface area of the medium flow cross-section created in the first stage of regulation by a revolving cylindrical shutter.

The valve structure is shown in Fig. 2.1b, and the relevant characteristics are presented in Fig. 5.19. The data concern the already modified structure (among other things: a conical plug) characterized by improved characteristics compared to the

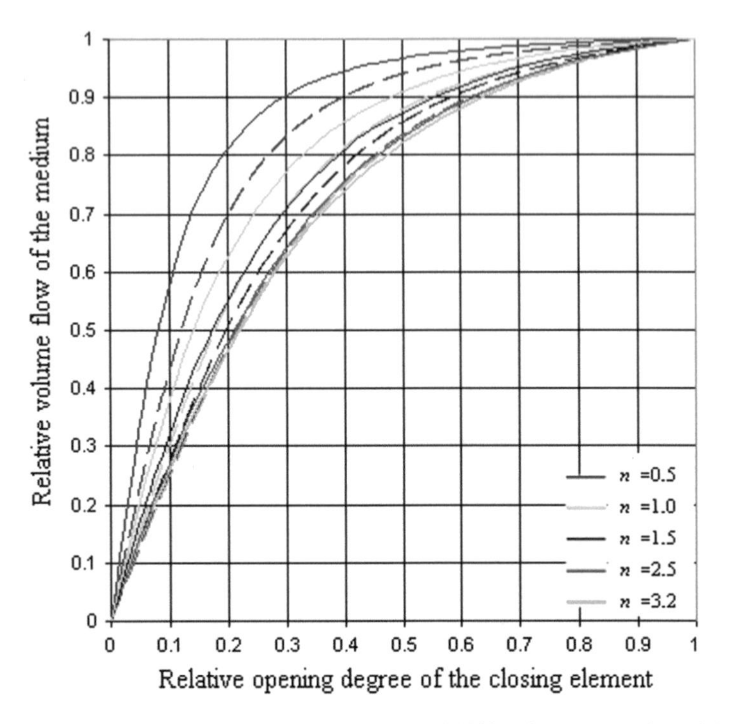

Fig. 5.16 Closing characteristic of the *Danfoss MSV-I* DN20 valve co-operating with a throttling element, for selected pre-settings: continuous lines—experimental data, dashed lines—data obtained using the inner authority calculated as proposed by *Pyrkov* and *Szaflik* in [6, 7]

original design. Pre-settings n_i are defined by the angle of the shutter revolution from the minimum opening position (0°) to full opening (180°). The closing curves are plotted for three pre-setting values: 0°, 90°, 180°, in the system of the relative flow factor depending on the relative opening degree.

The figure indicates that with a drop in the surface area of the cross-section of the medium flow through the throttling element, i.e. with a rise in the element hydraulic resistance, the deformation of the closing element characteristics increases. The curves are deformed upwards, which means that the deformation is the effect of a reduction in the valve inner authority a_w.

Analysing the characteristics, it can also be seen that changes in the valve inner authority a_w as a function of pre-setting n_i are twofold. According to the thesis advanced herein, in the case of valves with one adjustable section of the medium flow, a rise in initial throttling involves a rise in inner authority a_w. Due to that, the closing characteristic deformation compared to the original curve gets smaller. Such valves in the group presented above are elements where initial hydraulic resistance is realized by a reduction in the lift of either a single or a double plug (e.g. the *Danfoss MSV-I, Herz TS-90-V, M-3176*, *Pruss* valves), as well as elements with a curvilinear plug (the *Goeke Optimal* valve). For valves with two adjustable sections of the medium

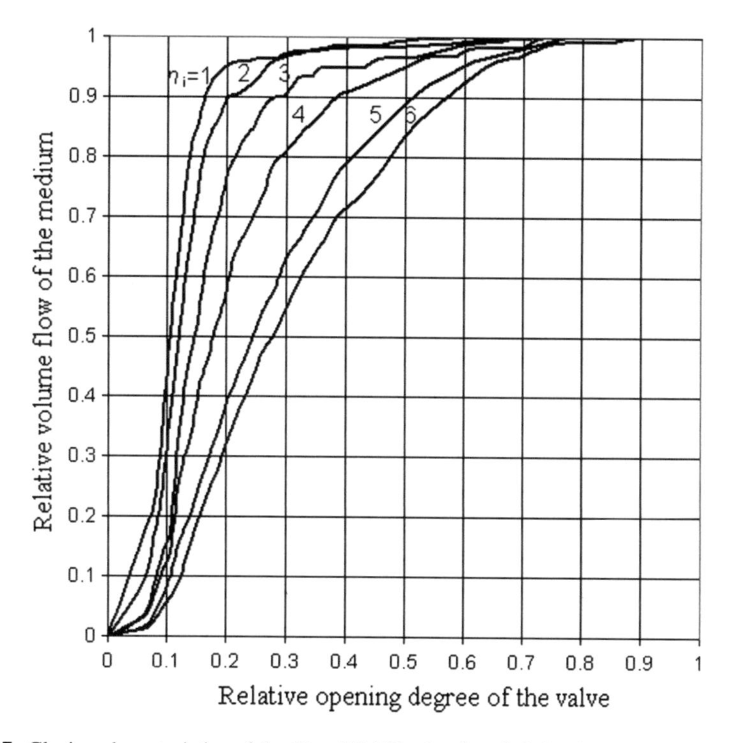

Fig. 5.17 Closing characteristics of the *Herz TS-FV* valve for all defined pre-settings

flow, the opposite is the case. Inner authority a_w decreases with a rise in throttling, which involves increased deformation of the closing characteristic compared to the original curve. This can be observed on the example of the characteristics of the *Danfoss RTD-N 15*, *Herz TS-FV* and *M-3172* valves.

To some extent, a change in the curvature of the curves arises also from a change in the original closing characteristic. This is because a reduction in the regulating element maximum available range of travel h_{max} involves a change in the distribution of the regulated section hydraulic resistances at the inlet (under the valve seat) and at the outlet (over the seat, from underneath the plug). When the valve opens, the resistance at the regulated section outlet changes, but the resistance at the inlet (in the valve seat) remains the same. Because total resistance is the sum of the two quantities, if only one of them changes, the change in total resistance is not proportional to the change in the varying quantity.

The measurements of flow factor k_v, which is the key element in making regulation curves of the valves under consideration, were burdened with a measurement uncertainty determined by the accuracy of the measuring instruments and readout. The real value was included in the range:

$$k_{v,rz} = \left(k_v - \Delta k_{v,\max}\,;\; k_v + \Delta k_{v,\max}\right) = k_v \pm \Delta k_{v,\max}. \tag{5.1}$$

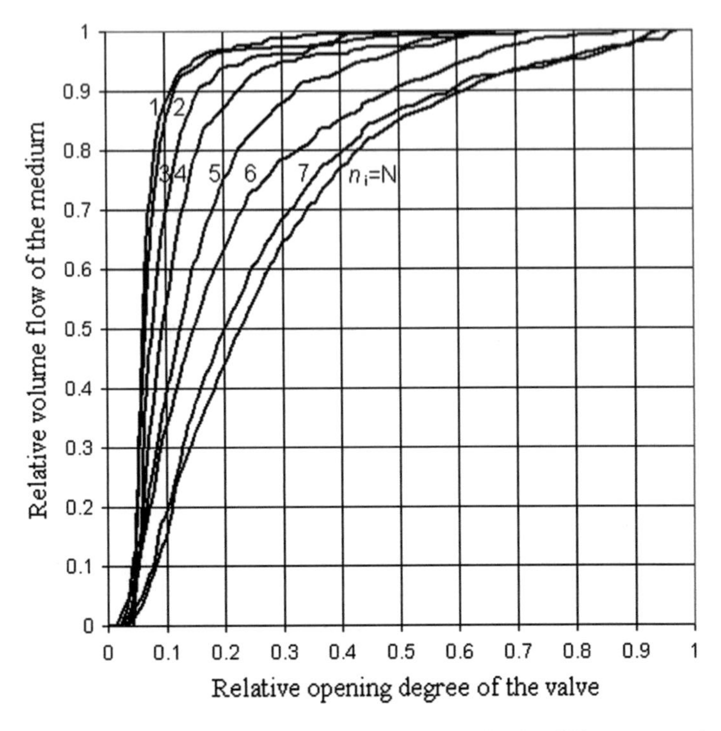

Fig. 5.18 Closing characteristics of the *Danfoss RTD-N* 15 valve for different pre-settings

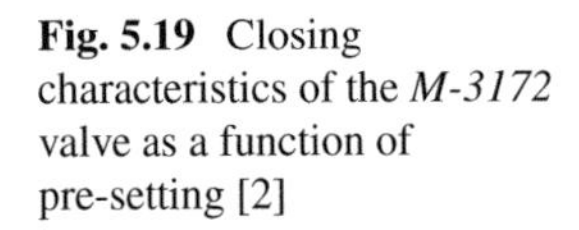

Fig. 5.19 Closing characteristics of the *M-3172* valve as a function of pre-setting [2]

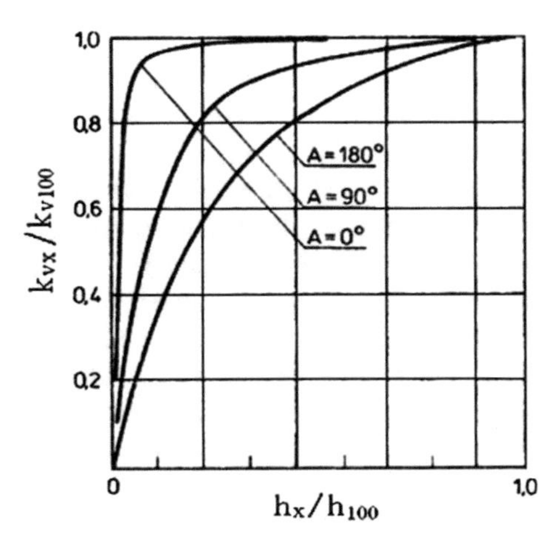

The maximum error in determination of flow factor k_v for a one-off measurement, taking account of the dimension of the flow factor and of the units of the measuring instruments scaling, can be found using the following relation:

$$\Delta k_{v,\text{max}} = 0.001 \cdot \left| \frac{\partial k_v}{\partial \dot{V}} \Delta \dot{V}\text{max} \right| + 0.001 \cdot \left| \frac{\partial k_v}{\partial \Delta p_z} \Delta p_{z,\text{max}} \right|. \quad (5.2)$$

Substituting in the above equation the mathematical description of flow factor k_v, formula (5.2) gives:

$$\Delta k_{v,\text{max}} = 0.001 \cdot \left| \frac{1}{\sqrt{\Delta p_z}} \Delta \dot{V}\text{max} \right| + 0.001 \cdot \left| -\frac{\dot{V}}{2\sqrt{\Delta p_z^3}} \Delta p_{z,\text{max}} \right|. \quad (5.3)$$

The values of $\Delta \dot{V}\text{max}$ and $\Delta p\text{max}$ can be calculated knowing the class and the measuring range of the measuring instruments. All the rotameters were made in Class 2.5 and the differential manometers—in Class 1.6 and Class 2.0. The following is then obtained for the manometers:

$$\Delta p_{z,\text{max}} = 0.016 \cdot 0.1\,\text{bar} = 0.0016\,\text{bar}. \quad (5.4)$$

$$\Delta p_{z,\text{max}} = 0.02 \cdot 0.4\,\text{bar} = 0.008\,\text{bar}. \quad (5.5)$$

For five rotameters with the measuring range of 2, 10, 40, 160 and 800 dm^3/h, at maximum indications, the following values are obtained, respectively:

$$\Delta \dot{V}\text{max} = 0.025 \cdot 2\,\text{dm}^3/\text{h} = 0.05\,\text{dm}^3/\text{h} = 0,00005\,\text{m}^3/\text{h},$$
$$\Delta \dot{V}\text{max} = 0.025 \cdot 10\,\text{dm}^3/\text{h} = 0.25\,\text{dm}^3/\text{h} = 0.00025\,\text{m}^3/\text{h},$$
$$\Delta \dot{V}\text{max} = 0.025 \cdot 40\,\text{dm}^3/\text{h} = 1\,\text{dm}^3/\text{h} = 0.001\,\text{m}^3/\text{h},$$
$$\Delta \dot{V}\text{max} = 0.025 \cdot 160\,\text{dm}^3/\text{h} = 4\,\text{dm}^3/\text{h} = 0.004\,\text{m}^3/\text{h},$$
$$\Delta \dot{V}\text{max} = 0.025 \cdot 800\,\text{dm}^3/\text{h} = 20\,\text{dm}^3/\text{h} = 0.02\,\text{m}^3/\text{h}.$$

According to the principle of the valve flow factor k_v and regulation curves determination, the value of Δp_z is maintained at a constant level. For the first manometer, the maximum value is $\Delta p_z = 0.1$ bar. Depending on the rotameter used for the measurement and for this manometer, formula (5.3) is therefore reduced, respectively, to the following form:

$$\Delta k_{v,max} = 1.58 \times 10^{-4} + 2.53 \times 10^{-5} \dot{V}\,\text{m}^3/\text{h},$$
$$\Delta k_{v,max} = 7.9 \times 10^{-4} + 2.53 \times 10^{-5} \dot{V}\,\text{m}^3/\text{h},$$
$$\Delta k_{v,max} = 3.16 \times 10^{-3} + 2.53 \times 10^{-5} \dot{V}\,\text{m}^3/\text{h},$$
$$\Delta k_{v,max} = 1.264 \times 10^{-2} + 2.53 \times 10^{-5} \dot{V}\,\text{m}^3/\text{h},$$
$$\Delta k_{v,max} = 6.32 \times 10^{-2} + 2.53 \times 10^{-5} \dot{V}\,\text{m}^3/\text{h}.$$

According to the transformation, the value of $\dot{V}$ should be expressed in units of the rotameter scaling, i.e. in dm^3/h. For the first rotameter, for example, replacing $\dot{V}$ with the maximum indication, the result is:

$$\Delta k_{v,max} = 1.58 \times 10^{-4} + 2.53 \times 10^{-5} \times 2 \cong 0.00021\ \text{m}^3/\text{h}.$$

For this value of the water volume flow (2 dm^3/h), the flow factor totals:

$$k_v = \frac{0.002}{\sqrt{0.1}} = 0.00632\ \text{m}^3/\text{h}.$$

The flow factor real value is then as follows:

$$k_{v,rz} = 0.00632\,[\text{m}^3/\text{h}] \pm 0.000210\ \text{m}^3/\text{h},$$

which means that the measuring uncertainty is at the level of ±3.3%. For the other rotameters, at maximum indications, the following values are obtained, respectively:

$$k_{v,rz} = 0.0316\,\text{m}^3/\text{h} \pm 3.3\%,$$
$$k_{v,rz} = 0.1265\,\text{m}^3/\text{h} \pm 3.3\%,$$
$$k_{v,rz} = 0.506\,\text{m}^3/\text{h} \pm 3.3\%,$$
$$k_{v,rz} = 2.53\,\text{m}^3/\text{h} \ \pm 3.3\%.$$

For the measurement performed using the Class 2.0 manometer made for the range of 0.4 bar, after similar calculations for the maximum measuring range, the result is:

For the first rotameter:

$$\Delta k_{v,max} = 0.001 \cdot \left| \frac{1}{\sqrt{0.4}} 0.05 \right| + 0.001 \cdot \left| -\frac{2}{2\sqrt{0.4^3}} 0.008 \right|,$$

$$\Delta k_{v,max} = 7.9 \times 10^{-5} + 3.16 \times 10^{-5} = 11.06 \times 10^{-5}\ \text{m}^3/\text{h}.$$

For this value of the water volume flow (2 dm^3/h), the flow factor totals:

$$k_v = \frac{0.002}{\sqrt{0.4}} = 0.00316\ \text{m}^3/\text{h}.$$

The flow factor real value is then as follows:

$$k_{v,rz} = 0.00316\,\text{m}^3/\text{h} \pm 0.000110\,\text{m}^3/\text{h},$$

which means that the measuring uncertainty is at the level of ±3.5%. For the other rotameters, at maximum indications, the following values are obtained, respectively:

$$k_{v,rz} = 0.0158\,\text{m}^3/\text{h} \pm 3.5\%,$$
$$k_{v,rz} = 0.0632\,\text{m}^3/\text{h} \pm 3.5\%,$$

$$k_{v,rz} = 0.253\,\mathrm{m}^3/\mathrm{h} \pm 3.5\%,$$
$$k_{v,rz} = 1.26\,\mathrm{m}^3/\mathrm{h} \pm 3.5\%.$$

5.4 Experimental and Computational Verification of the Proposed Methods of Determination of the Regulation Valve Pre-setting

Experimental verification of the proposed methods of determination of the value of the regulation valve pre-setting is carried out performing calculations of the quantity for different example parameters and then comparing the results with experimental data. Due to the fact that such calculations are (or at least should be) a common practice in the design of any heating installation, this part is included in the chapter with computational examples to create a coherent whole.

References

1. Catalogue information of Danfoss
2. Kołodziejczyk, W.: Armatura regulacyjna w ogrzewaniach wodnych (Control armature in hydronic heating systems). Arkady, Warszawa (1985)
3. Kołodziejczyk W.: *Zmodernizowana konstrukcja zaworu grzejnikowego z podwójną regulacją* (*Modernized radiator valve structure with double regulation*), District Heatin, Heating, Vemtilation, (3), 10/1971, pp. 298–303
4. Lehrner R.: Entwicklugsstand bei heizungsarmaturen, Sanitär+Heizugstechnik, t.4, /1969, 2/1969
5. Mielnicki J.S.: Możliwości regulacji wstępnej i eksploatacyjnej za pomocą zaworów grzejnikowych (Possibility of preliminary and exploitation regulation by means of radiator valves), District Heatin, Heating, Vemtilation, (1), 3/1969, pp. 73–80
6. Pyrkov, V.: Gidrawliczeskoje regulirowanije sistem otoplenija i ochłażdjenija. Teorija i praktika, Danfoss, Kijów (2005)
7. Pyrkov V.: Regulacja hydrauliczna systemów ogrzewania i chłodzenia. Teoria i praktyka (Hydraulic regulation of heating and cooling systems. Theory and practice), Systherm Serwis, Poznań 2007

Chapter 6
Computational Examples

This chapter presents computational examples related to the issues discussed in the previous chapters. They are arranged chronologically in relation to subsequent chapters and within every issue they are organized from the basic to the most complex tasks.

The computational examples concerning determination of the regulation valve pre-setting n_{i} are developed using the two methods proposed in this book (Method 1 and Method 2), the method proposed by *Pyrkov* and *Szaflik* and the method which is now in common use. The obtained results are discussed and compared. The calculations are performed for the two types of double-regulation valves discussed in this book—with one and with two adjustable sections of the medium flow. They are carried out for selected valves, also discussed herein: the *Danfoss MSV-I* DN20, the *Danfoss RTD-N* 15, the *Herz TS-FV 7523* and the *Herz TS-90-V* valve.

Example 6.1
Determine changes in the room temperature, the final temperatures and the room static thermal characteristics for the following data:

- design thermal load: $\dot{Q}_i = 1000\,\mathrm{W}$,
- heat loss coefficient, calculated according to Standard EN 12831:2003 [3]: $H = 25$ W/K,
- design outdoor temperature: $t_{\mathrm{e}} = -20$ °C,
- design indoor temperature: $t_{\mathrm{e}} = 20$ °C,
- heat disturbance:

(a) heat gains of $\Delta\dot{Q}_i = 100\,\mathrm{W}$,
(b) heat gains of $\Delta\dot{Q}_i = 200\,\mathrm{W}$,
(c) heat losses of $\Delta\dot{Q}_i = 100\,\mathrm{W}$,
(d) heat losses of $\Delta\dot{Q}_i = 200\,\mathrm{W}$.

D. P. Muniak, *Regulation Fixtures in Hydronic Heating Installations*, Studies in Systems, Decision and Control 187, https://doi.org/10.1007/978-3-030-03128-2_6

Solution

The change in temperature and the temperature final values are found from the relation describing the room static thermal characteristic. For this purpose, the amplification factor should be determined. According to the data, the heat loss coefficient totals 25 W/K. The amplification factor is the inverse of the heat loss coefficient. Using formula (2.5), the following value is obtained:

$$k_s = \frac{1}{H} = \frac{1}{25} = 0.04\,\mathrm{K/W}.$$

Expressing the coefficient in percentages and giving the value of the increment in temperature per every 1% of the increment in the heat output, the following value is obtained according to formula (2.6):

$$k_{s,\%} = k_s \cdot 1\% \cdot \dot{Q}_i = 0.04 \cdot 0.01 \cdot 1000 = 0.4\,\mathrm{K/\%}$$

According to formulae (2.7) and (2.8), the changes in the room temperature and the temperature final values total:

(a) for gains at the level of $\Delta\dot{Q}_i = 100\,\mathrm{W}$:

$$\Delta t_i = k_s \cdot \Delta\dot{Q}_i = 0.04 \cdot 100 = 4\ ^\circ\mathrm{C},$$

$$t_{i,k} = t_i + \Delta t_i = 20 + 4 = 24\,^\circ\mathrm{C}.$$

(b) for gains at the level of $\Delta\dot{Q}_i = 200\,\mathrm{W}$:

$$\Delta t_i = k_s \cdot \Delta\dot{Q}_i = 0.04 \cdot 200 = 8\,^\circ\mathrm{C},$$

$$t_{i,k} = t_i + \Delta t_i = 20 + 8 = 28\,^\circ\mathrm{C}.$$

(c) for losses at the level of $\Delta\dot{Q}_i = 100\,\mathrm{W}$:

$$\Delta t_i = k_s \cdot \Delta\dot{Q}_i = 0.04 \cdot (-100) = -4\,^\circ\mathrm{C},$$

$$t_{i,k} = t_i + \Delta t_i = 20 - 4 = 16\,^\circ\mathrm{C}.$$

(d) for losses at the level of $\Delta\dot{Q}_i = 200\,\mathrm{W}$:

$$\Delta t_i = k_s \cdot \Delta\dot{Q}_i = 0.04 \cdot (-200) = -8\,^\circ\mathrm{C},$$

$$t_{i,k} = t_i + \Delta t_i = 20 - 8 = 12\,^\circ\mathrm{C}.$$

The thermal characteristic is presented in Fig. 6.1.
The changes in the room temperature are proportional to the changes in the supplied heat.

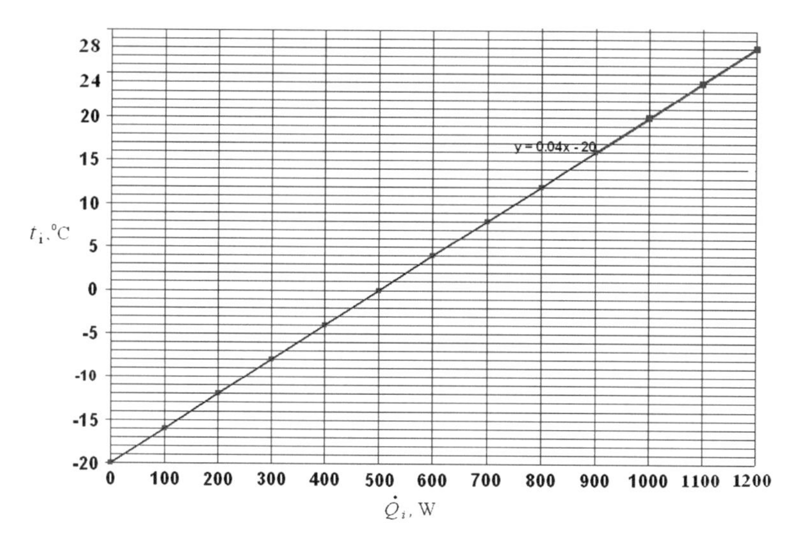

Fig. 6.1 Static thermal characteristic of the room

Example 6.2

Determine changes in the room temperature, the final temperatures and the room static thermal characteristics for the following data:

- design thermal load: $\dot{Q}_\mathrm{i} = 1000\,\mathrm{W}$,
- heat loss coefficient, calculated according to Standard EN 12831:2003: $H = 25$ W/K,
- design outdoor temperature: $t_\mathrm{e} = -20$ °C,
- design indoor temperature: $t_\mathrm{e} = 20$ °C,
- the following are used to regulate the room temperature:

(1) a manual valve,
(2) a thermoregulator with a thermostatic head with design proportional range $X_\mathrm{p} = 2$ K,

- heat disturbance:

(a) heat gains of $\Delta\dot{Q}_i = 100\,\mathrm{W}(10\%)$,
(b) heat gains of $\Delta\dot{Q}_i = 200\,\mathrm{W}(20\%)$,
(c) heat losses of $\Delta\dot{Q}_i = 100\,\mathrm{W}(10\%)$,
(d) heat losses of $\Delta\dot{Q}_i = 200\,\mathrm{W}(20\%)$.

Solution

(1) The room temperature is regulated by a manual valve

In this case, no automatic regulation takes place and the changes in the room temperature will be the function of changes in the amount of supplied heat only. Therefore, the obtained results are the same as in the previous example.

(a) $\Delta t_i = k_s \cdot \Delta \dot{Q}_i = 0.04 \cdot 100 = 4\,°\text{C}, t_{i,k} = t_i + \Delta t_i = 20 + 4 = 24\,°\text{C},$
(b) $\Delta t_i = k_s \cdot \Delta \dot{Q}_i = 0.04 \cdot 200 = 8\,°\text{C}, t_{i,k} = t_i + \Delta t_i = 20 + 8 = 28\,°\text{C},$
(c) $\Delta t_i = k_s \cdot \Delta \dot{Q}_i = 0.04 \cdot (-100) = -4\,°\text{C}, t_{i,k} = t_i + \Delta t_i = 20 - 4 = 16\,°\text{C},$
(d) $\Delta t_i = k_s \cdot \Delta \dot{Q}_i = 0.04 \cdot (-200) = -8\,°\text{C}, t_{i,k} = t_i + \Delta t_i = 20 - 8 = 12\,°\text{C},$

Figure 6.2 presents the system static characteristics as a static characteristic of a typical radiator taking account of the heat gains or losses. It can be observed that the curves illustrating cases (c) and (d), i.e. heat losses, do not originate in the coordinate system zero point and for the minimum outdoor temperature assumed at the level of $t_e = -20$ °C they display a higher-than-zero heat flux required to ensure this temperature inside the room. It follows that for this temperature value the radiator will need to operate. This is due to the fact that according to the conditions in these two cases, unpredicted additional heat losses occurred in the room causing a rise in the design thermal load from 1000 to 1100 and 1200 W, respectively. As mentioned earlier, if the required room temperature is equal to the outdoor temperature level, there is no need to supply the room with heat. However, a non-zero heat output value for a given room temperature simply means that the outdoor temperature is lower. This is also indicated by the curves. As they run to the left, it turns out that they intersect the temperature axis in points $t_e = -24$ °C and $t_e = -28$ °C, respectively. For this reason, a certain amount of heat has to be supplied to keep the room temperature at the level of $t_i = -20$ °C. The curves will show that the difference compared to the outdoor temperature initial value $t_e = -20$ °C will be by as much as the difference between the lower indoor temperature compared to the required value $t_i = 20$ °C.

It is assumed in the example that the radiator heat output is the same irrespective of the room temperature, and that the temperature does not affect the radiator thermal characteristic parameters. But in practice there are factors which are omitted in this analysis that will cause changes in the results. The heat transfer driving force is the difference between the temperatures of the radiator and the environment. The smaller the difference, the weaker the force and the smaller the heat output. The moment that an object in the room generates heat gains making the room temperature rise, the difference between the temperatures of the room and of the air in it gets smaller and due to that the radiator heat output decreases. Consequently, the resultant total output of the two heat sources is smaller than the sum of the radiator initial heat output and the heat output of the additional object. Moreover, the difference in the temperatures has an impact on one of the radiator static thermal characteristic parameters, i.e. the value of the surface film conductance from the radiator external surface to the environment. Its value also drops with a decrease in the difference between the temperatures of the radiator wall and the environment. This causes a further drop in the radiator heat output. As a result, the effect of the two factors is that the room temperature will not rise as calculated. The smaller the initial difference between the temperatures of the radiator and the environment, the stronger the impact of the reduction in the temperature increment. For this reason, low-temperature installations, e.g. surface (especially underfloor) heating systems, will be more effective in stabilizing the room temperature using radiator thermoregulators. Assuming for example that the underfloor radiator and the room temperatures for the design conditions are (typically) 28 °C (as the mean

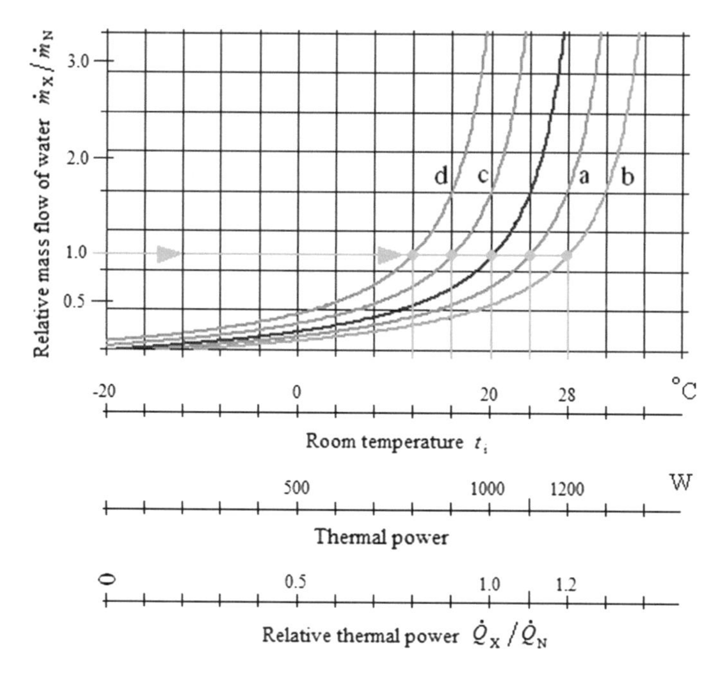

Fig. 6.2 Changes in the room temperature if there is no automatic regulation: (a, b, c, d) curves illustrating the resultant thermal characteristic of heat sources in the room

value resulting from the radiator supply and return temperatures) and 20 °C, respectively, the heat transfer driving force in the form of the temperature difference totals 8 °C, and the emitted unit heat output is about 88 W/m^2 (cf. [8, 9]). If more additional heat sources appear in the room, raising the temperature to e.g. 24 °C, the heat transfer driving force will fall to 4 °C, i.e. by 50% compared to the initial value. Due to the parallel drop in the surface film conductance, the radiator heat output will be more than halved—it will drop to 41 W/m^2, by 53% compared to the initial value (cf. [8, 9]). This heat output, to which the heat output of additional heat sources in the room is added, conditions the temperature for the new state of thermal balance. The balance will be characterized by a lower temperature value than the one resulting from the sum of the radiator initial heat output and the output of the additional heat sources. If the situation concerned a typical convector radiator, with the typical design operating temperature of 70 °C (as the mean value resulting from the supply and return temperatures), in a case like this a drop in the temperature difference would occur from 50 to 46 °C, i.e. only by 8% compared to the initial value. The unit heat output would then drop from about 500 W/m^2 to about 448 W/m^2, i.e. by about 10% only compared to the initial value (cf. [8, 9]). In this case therefore, although the resultant temperature will still be lower than the value resulting from the sum of the heat output of the radiator and of the additional objects generating heat gains, it will still be closer to the temperature value resulting from the sum compared to an underfloor radiator. A complete analysis of the phenomenon should also take account of the change in the radiator temperature that occurs at a change in the temperature of the room. A rise in the room temperature, causing a drop in the temperature difference between the radiator and the environment and a drop in the surface film conductance from the radiator wall to the environment, naturally involves a drop in the heat output. This results in a rise in the mean temperature of the radiator because the water flowing through it, at set values of the mass flow and supply temperature, cools less and its temperature is closer to the temperature at the supply. The effect, however, is not big and can be omitted in practice, especially in the case of high-temperature installations with classical convector radiators.

The described phenomenon of a considerable reduction in the radiator heat output at a rise in the temperature of the environment is referred to as self-regulation and in practice concerns underfloor radiators mainly because these devices are characterized by low operating temperatures. Nonetheless, it can be noticed easily that the effect under discussion increases as the temperature of the radiator surface gets closer to the temperature of the room. Figure 6.3 presents an example of such a situation and the self-regulation phenomenon, assuming a constant temperature value of the radiator surface.

(2) Automatic regulation of temperature is carried out by means of a thermoregulator with a thermostatic head with design proportional range $X_{\text{p}} = 2$ K:

In this case, the changes in the room temperature are conditioned not only by changes in the supplied heat but also by the operation of the radiator thermoregulator. According to its static characteristic, the thermoregulator reduces the temperature variations, which can be defined using formula (2.10). Amplification factors should then be expressed as percentages. For the room, according to formulae (2.5) and (2.6), they total:

$$k_s = \frac{1}{H} = \frac{1}{25} = 0.04\,\text{K/W},$$

$$k_{s,\%} = k_s \cdot 1\% \cdot \dot{Q}_i = 0.04 \cdot 0.01 \cdot 1000 = 0.4\,\text{K}/\%.$$

For the thermoregulator, according to formula (2.9), the percentage amplification factor totals:

$$k_{m,\%} = 100\% \cdot 1/X_{\text{p}} = 100\% \cdot 1/2\,\text{K} = 50\%/\text{K}.$$

Substituting the data in formula (2.10), the following is obtained:

(a)

$$\Delta t_i = \frac{1}{1 + k_{s,\%} \cdot k_{m,\%}} \cdot k_{s,\%} \cdot \Delta\dot{Q}_i = \frac{1}{1 + 0.4 \cdot 50} \cdot 0.4 \cdot 10 = 0.19\,°\text{C},$$

$$t_{i,k} = t_i + \Delta t_i = 20 + 0.19 = 20.19\,°\text{C},$$

(b)

$$\Delta t_i = \frac{1}{1 + k_{s,\%} \cdot k_{m,\%}} \cdot k_{s,\%} \cdot \Delta\dot{Q}_i = \frac{1}{1 + 0.4 \cdot 50} \cdot 0.4 \cdot 20 = 0.38\,°\text{C},$$

$$t_{i,k} = t_i + \Delta t_i = 20 + 0.38 = 20.38\,°\text{C},$$

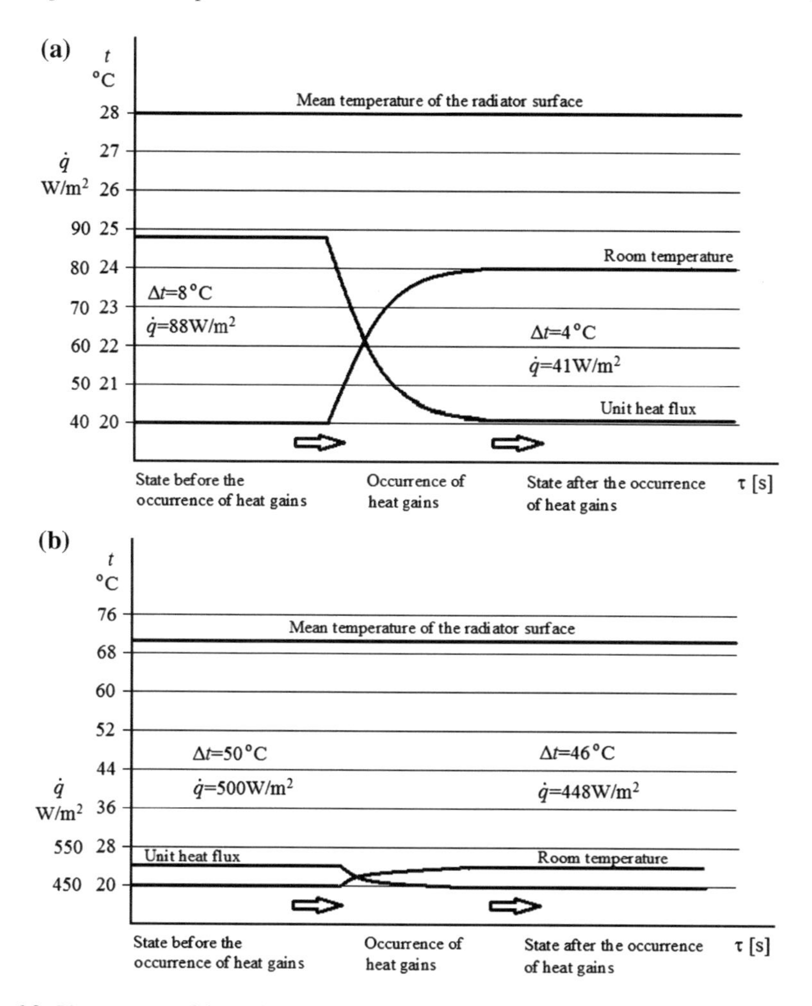

Fig. 6.3 Phenomenon of the radiator heat output self-regulation: **a** underfloor radiator; **b** convector radiator

(c)
$$\Delta t_i = \frac{1}{1 + k_{s,\%} \cdot k_{m,\%}} \cdot k_{s,\%} \cdot \Delta \dot{Q}_i = \frac{1}{1 + 0.4 \cdot 50} \cdot 0.4 \cdot (-10) = -0.19\,°\mathrm{C},$$
$$t_{i,k} = t_i + \Delta t_i = 20 - 0.19 = 19.81\,°\mathrm{C},$$

(d)
$$\Delta t_i = \frac{1}{1 + k_{s,\%} \cdot k_{m,\%}} \cdot k_{s,\%} \cdot \Delta \dot{Q}_i = \frac{1}{1 + 0.4 \cdot 50} \cdot 0.4 \cdot (-20) = -0.38\,°\mathrm{C},$$
$$t_{i,k} = t_i + \Delta t_i = 20 - 0.38 = 19.62\,°\mathrm{C}.$$

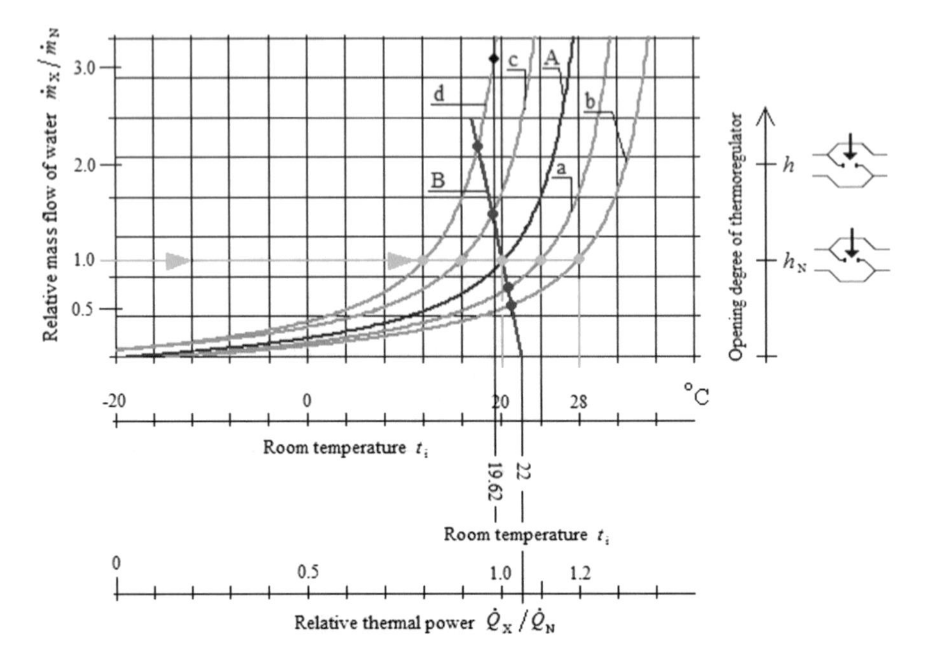

Fig. 6.4 Changes in the room temperature at the operation of a radiator thermoregulator with a proportional effect: a, b, c, d—curves illustrating the resultant thermal characteristic of heat sources in the room; A: curve illustrating the radiator thermal characteristic; B: static characteristic of the radiator thermoregulator

Figure 6.4 presents the system static characteristics as a static characteristic of a typical radiator taking account of the heat gains or losses and considering the operation of a radiator thermoregulator with a proportional regulation type.

The chart should be treated as conventional because it offers a correct illustration of the problem in terms of quality but not in terms of quantity, as described earlier. It can be seen that the temperatures of the new states of balance do not agree with the calculated values. The static characteristic in the chart is plotted with its own, additional axis representing the valve plug position over the seat (the valve opening degree) h_X. Using such an approach, it is a straight line. However, as already mentioned, it cannot be put directly onto the medium mass flow axis, which is plotted in the same chart for the radiator thermal characteristic, because the changes in the medium mass flow $\dot{m}_X/\dot{m}_N$ do not depend on the valve opening degree h_X/h_N linearly. The relation between the two parameters is usually strongly non-linear, which results from the shape of the original regulation characteristic of the valve and from the valve total authority a_c. Due to the impact of the latter, for the same thermoregulator its static and flow characteristic (cf. Fig. 2.15a) may vary considerably, depending on hydraulic resistance (r_{str}) of the pipework the valve operates in. The issue is analysed in detail in Chap. 4. Moreover, the chart presents characteristics of the thermoregulator for the device both closing and opening. In the case of opening, this means that proportional range X_p is bigger than the value assumed for the calculations of temperature difference Δt_i, and, consequently, that amplification factor $k_{m,\%}$, taken into account in the calculations, is smaller. This is another reason why the values of calculated parameters cannot be read from this chart directly.

A comparison should also be made between the situations when the thermoregulator has to increase and decrease the medium flow through the radiator if additional heat gains or losses arise in the room. It should be remembered that though the thermoregulator capability to reduce the medium mass flow is in fact unlimited (to zero), the possibility of raising it above the design value may be substantially restricted. This is the effect of a few factors:

- the relative opening degree h_N/h_{100} of the valve closing plug for a given proportional range X_p, which informs how much the element can open above the value,
- the value of the valve total authority a_c, which informs about the degree of the valve original regulation characteristic deformation and about the valve regulating capacity,
- the location of the initial working point on the radiator static thermal characteristic, informing about how much the radiator heat output can be increased at a rise in the medium mass flow.

The valve opening degree depends on the closing plug available range of travel h_{100}, the thermostatic head amplification factor k_m and the selected proportional range X_p. For example, taking typical values of these parameters, i.e. $h_{max} = 1.3$ mm, $k_m = 0.22$ mm/K and $X_p = 2$ K, the relative opening degree h_N/h_{100} totals $(0.22 \cdot 2)/1.3 = 34\%$. There are then large reserves of the plug available travel upwards and the valve can open significantly above the initial value of h_N. This does not necessarily mean considerable increments in the medium mass flow because the increments are most often very small due to very small values of the valve total authority a_c. Another factor limiting increments in the radiator heat output at a rise in the medium mass flow is the shape of the radiator static thermal characteristic. Significant increments in the heat output at a given rise in the medium flow are only noticeable in its initial part, becoming much smaller later and decreasing further as regions of higher ranges of the mass flow values are entered.

Due to all that, the thermoregulator is unable to prevent drops in the room temperature if additional heat losses arise. The situation can be analysed on the example under consideration. Additional heat losses in (d) total 200 W. This means that the losses have to be compensated for by the radiator, whose heat output in this case needs to reach the level of 1200 W to maintain the room temperature at the level of $t_i = 20$ °C. If the system includes a thermoregulator, due to its specificity, the required heat output will be a bit lower because the temperature resulting from the operation of the thermoregulator trying to maintain the initial set value totals $t_i = 19.62$ °C, and not $t_i = 20$ °C. Comparing this value to the radiator static characteristic {d} in Fig. 6.4, illustrating such a situation (as previously stated—the impact of the points of intersection with the thermoregulator characteristic should be ignored), it can be seen that the required mass flow of the medium will need to be more than three times bigger (by about 200%) than originally. This cannot be achieved in practice because for typical values of the valve total authority a_c, typical shapes of regulation characteristics and relative opening degrees for design conditions, the increments are usually included in the range from several to a few dozen per cent. Significant increments in the radiator heat output could occur if the working point was located on the characteristic in the range of its bigger curvature, i.e. in its initial part. This would practically mean that the radiator would have to be designed for a bigger drop in the water temperature in it (cf. Fig. 2.22), i.e. the dimensions of the radiator would have to be increased.

Example 6.3

Determine changes in the room temperature, the final temperatures and the room static thermal characteristics for the following data:

- design thermal load: $\dot{Q}_i = 1000\,\mathrm{W}$,
- heat loss coefficient, calculated according to Standard EN 12831:2003: $H =$ 25 W/K,
- design outdoor temperature: $t_e = -20\,°\mathrm{C}$,
- design indoor temperature: $t_e = 20\,°\mathrm{C}$,
- the following are used for automatic regulation of the room temperature:

(1) a thermoregulator with a thermostatic head with design proportional range $X_p = 2$ K,
(2) a thermoregulator with a thermostatic head with design proportional range $X_p = 1$ K,

- heat disturbance:

(a) heat gains of $\Delta\dot{Q}_i = 500\,\mathrm{W}(50\%)$,
(b) heat gains of $\Delta\dot{Q}_i = 1000\,\mathrm{W}(100\%)$,
(c) heat gains of $\Delta\dot{Q}_i = 1500\,\mathrm{W}(150\%)$,

Solution

The room static thermal characteristic parameters are the same as in the previous example and so are the parameters of the first thermoregulator. Consequently, the amplification factor values are identical and total:

$$k_{s,\%} = k_s \cdot 1\% \cdot \dot{Q}_i = 0.04 \cdot 0.01 \cdot 1000 = 0.4\,\mathrm{K}/\%,$$

$$k_{m,\%} = 100\% \cdot 1/X_p = 100\% \cdot 1/2\mathrm{K} = 50\%/\mathrm{K}.$$

The proportional range of the second thermoregulator is different, and, according to formula (2.9), its amplification factor totals:

$$k_{m,\%} = 100\% \cdot 1/X_p = 100\% \cdot 1/1\mathrm{K} = 100\%/\mathrm{K}.$$

According to formulae (2.10) and (2.8), the changes in the room temperature and the final temperature values total:

(1) for the thermoregulator with a thermostatic head with the design proportional range of 2 K:

(a)
$$\Delta t_i = \frac{1}{1 + k_{s,\%} \cdot k_{m,\%}} \cdot k_{s,\%} \cdot \Delta\dot{Q}_i = \frac{1}{1 + 0.4 \cdot 50} \cdot 0.4 \cdot 50 = 0.95\,^\circ\text{C},$$
$$t_{i,k} = t_i + \Delta t_i = 20 + 0.95 = 20.95\,^\circ\text{C},$$

(b)
$$\Delta t_i = \frac{1}{1 + k_{s,\%} \cdot k_{m,\%}} \cdot k_{s,\%} \cdot \Delta\dot{Q}_i = \frac{1}{1 + 0.4 \cdot 50} \cdot 0.4 \cdot 100 = 1.9\,^\circ\text{C},$$
$$t_{i,k} = t_i + \Delta t_i = 20 + 1.9 = 21.9\,^\circ\text{C},$$

(c)
$$\Delta t_i = \frac{1}{1 + k_{s,\%} \cdot k_{m,\%}} \cdot k_{s,\%} \cdot \Delta\dot{Q}_i = \frac{1}{1 + 0.4 \cdot 50} \cdot 0.4 \cdot 150 = 2.86\,^\circ\text{C},$$
$$t_{i,k} = t_i + \Delta t_i = 20 + 2.86 = 22.86\,^\circ\text{C}.$$

(2) for the thermoregulator with a thermostatic head with the design proportional range of 1 K:

(a)
$$\Delta t_i = \frac{1}{1 + k_{s,\%} \cdot k_{m,\%}} \cdot k_{s,\%} \cdot \Delta\dot{Q}_i = \frac{1}{1 + 0.4 \cdot 100} \cdot 0.4 \cdot 50 = 0.49\,^\circ\text{C},$$
$$t_{i,k} = t_i + \Delta t_i = 20 + 0.49 = 20.49\,^\circ\text{C},$$

(b)
$$\Delta t_i = \frac{1}{1 + k_{s,\%} \cdot k_{m,\%}} \cdot k_{s,\%} \cdot \Delta\dot{Q}_i = \frac{1}{1 + 0.4 \cdot 50} \cdot 0.4 \cdot 100 = 0.975\,^\circ\text{C},$$
$$t_{i,k} = t_i + \Delta t_i = 20 + 0.975 = 20.975\,^\circ\text{C},$$

(c)
$$\Delta t_i = \frac{1}{1 + k_{s,\%} \cdot k_{m,\%}} \cdot k_{s,\%} \cdot \Delta\dot{Q}_i = \frac{1}{1 + 0.4 \cdot 50} \cdot 0.4 \cdot 150 = 1.46\,^\circ\text{C},$$
$$t_{i,k} = t_i + \Delta t_i = 20 + 1.46 = 21.46\,^\circ\text{C}.$$

The characteristics are gathered in Fig. 6.5.

Analysing the calculation results it can be seen that the variations in the room temperature t_i are approximately linearly dependent on changes in design proportional range X_p. The smaller the range, the smaller the temperature variations.

A comment is also necessary on the calculation results for cases (b) and (c), where gains in the heat flux are equal to or higher than the radiator design heat output and the room thermal load. If the two quantities are equal, neither of the thermoregulators closes completely and due to that the temperature rises above the design value. As a result, the radiator proportional thermoregulator is unable to make a 100% use of additional gains in the heat flux. However, the share of the gains can be maximized by reducing the proportional range value. For (c), where heat gains are higher than the design value, the values of the temperature increment Δi_i calculated taking account of the thermoregulator operation, do not give correct results. It is first necessary

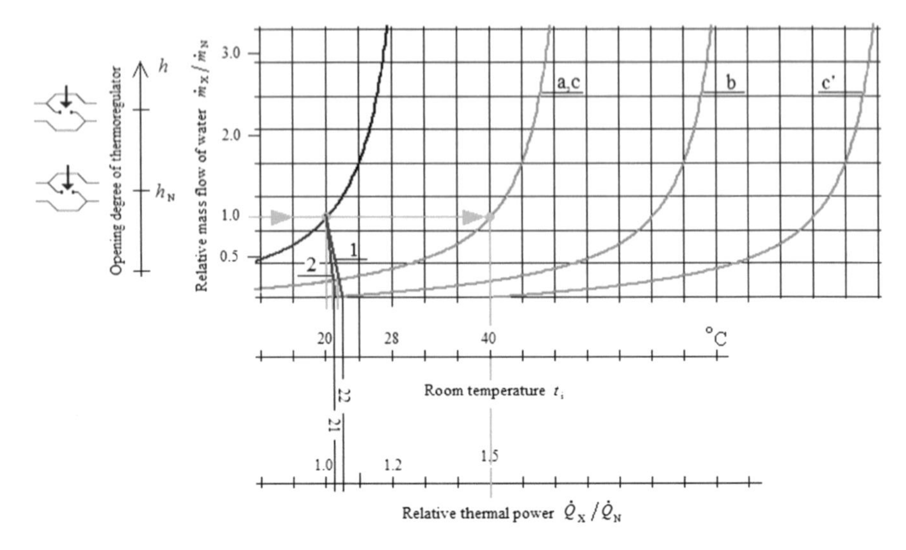

Fig. 6.5 Changes in the room temperature at the operation of a radiator thermoregulator with a proportional effect: a, b, c, d—curves illustrating the resultant thermal characteristic of heat sources in the room; (1) static characteristic of the thermoregulator with proportional range $X_p = 2$ K; (2) static characteristic of the thermoregulator with proportional range $X_p = 1$

to define the potential increment in the room temperature Δt_i for the case with no thermoregulator and a turned-off radiator. According to formula (2.10) describing the room static characteristic and formula (2.8) for the final temperature value, the result would be:

$$\Delta t_i = k_s \cdot \Delta \dot{Q}_i = 0.04 \cdot 1500 = 60\,°\text{C},$$

$$t_{i,k} = t_i + \Delta t_i = -20 + 60 = 40\,°\text{C}.$$

The resultant temperature $t_{i,k}$ will be higher compared to the value calculated from the formula taking account of the thermoregulator operation. This is due to the fact that the relation takes no account of the device operation beyond the working range, i.e. at a temperature higher than in the closing point, where temperature is determined by the initial value t_i and by the defined proportional range X_p, i.e. $t_i = 20 + 2 = 22$ °C and $t_i = 20 + 1 = 21$ °C, respectively. The fact that the calculated result is a value from beyond the device operating range means that the formula is invalid, the thermoregulator will be fully closed and resultant temperature $t_{i,k}$ will be a function of the room response only.

In order to make the analysis easier, the results of the calculations performed for variants (a) and (c) are marked on a single curve because they are very similar when it comes to the surplus resultant heat output compared to the initial value. For (a), there is a heat output surplus of 500 W, so changes in the room temperature t_i

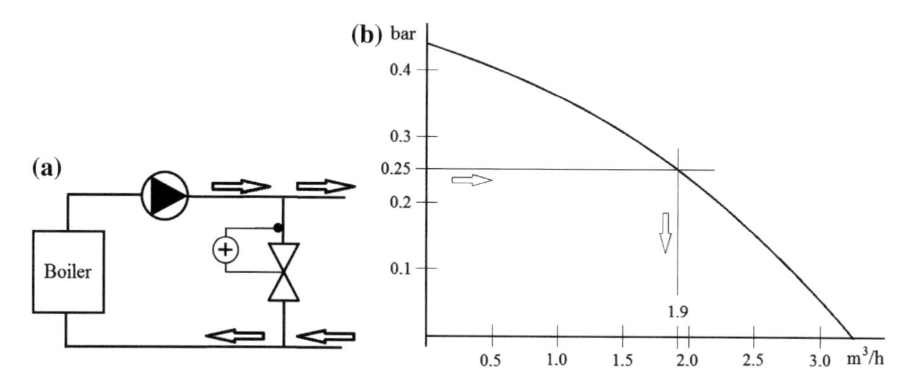

Fig. 6.6 **a** the relief valve location in the analysed system; **b** throttling characteristic of the selected pump

occur according to curve {a}, illustrating aggregated values of the heat flux from the radiator and from additional heat sources (1500 W). But the same curve describes case (c), where the heat flux surplus totals 1500 W. However, the totalized thermal power will not be 1000 + 1500 W = 2500 W. It will not run along curve {c′}, but along the curve plotted for case (a). This is due to the fact that the heat gains are higher than the room thermal load and the heat flux supplied to the room by the radiator. As a consequence, the thermoregulator closes completely and turns off the radiator. The room is thus supplied with an additional heat flux of 1500 W, which in this case is by 500 W higher than the required heat output value. These 500 W cause a surplus rise in the room temperature to $t_\mathrm{i} = 40$ °C.

Example 6.4

The heating system equipped with radiator thermoregulators includes a pump with a known throttling characteristic. A relief valve should be selected that, installed right downstream the pump, will prevent a rise in the active pressure value in the system above the permissible level of 25 kPa. The system diagram and the characteristic of the selected pump are presented in Fig. 6.6. Characteristics of the relief valves under consideration are presented in Fig. 6.7.

Solution

At such location of the valve, the pressure drop on the boiler-pump-valve-boiler section may be omitted as it is relatively small. The throttling curve indicates that for the maximum permissible value of the pressure difference at the level of 0.25 bar, the pump output totals about $\dot{V} = 1.9\,\mathrm{m}^3/\mathrm{h}$. The selected valve must therefore ensure appropriate flow capacity for a defined difference in pressure. Due to the proportional character of the valve operation, this means that it must be set to a lower pressure value of the zero opening (zero volume flow), as marked in Fig. 6.8.

The value of the pressure difference setting should be selected at the level of 0.05 and 0.1 kPa, respectively. The parameters in this example are selected to enable both of the valves to meet the problem conditions. The smaller valve (*AVDO* 15) operates

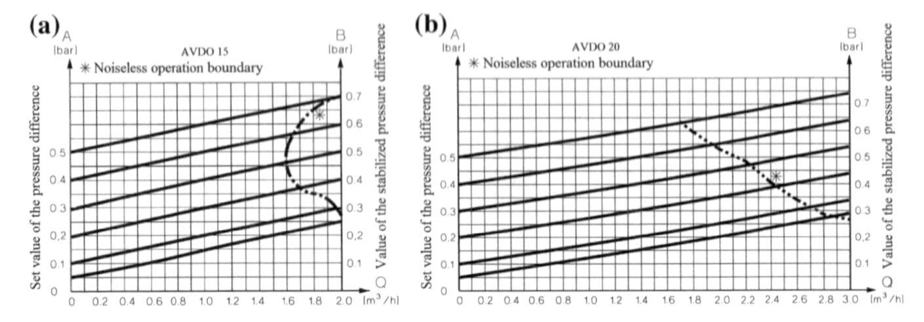

Fig. 6.7 Hydraulic characteristics of the analysed valves: **a** *Danfoss AVDO* 15; **b** *Danfoss AVDO* 20 [1]

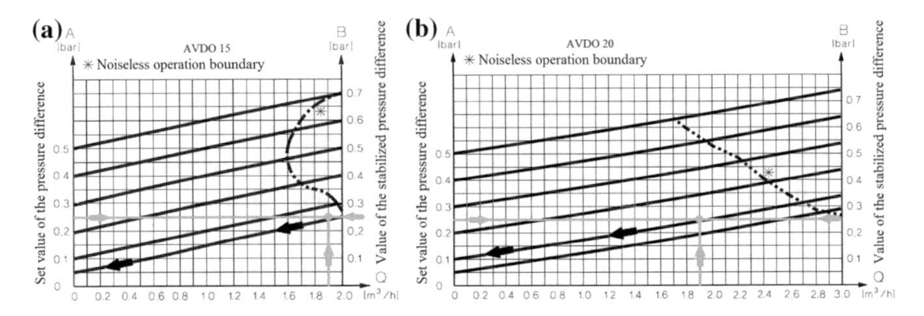

Fig. 6.8 Readout of the relief valve pressure difference pre-setting: **a** *Danfoss* AVDO 15; **b** *Danfoss* AVDO 20

already on the limit of its flow capacity, at almost full opening. If it turned out that the required volume flow of the medium for a given pressure value in the system was higher, either a bigger valve (with a higher flow capacity) would have to be selected or two valves would have to be connected in parallel.

Example 6.5

The heating system equipped with radiator thermoregulators includes a pump with a known throttling characteristic. A relief valve installed downstream the pump should be selected so that the medium minimal circulation should be ensured in the primary circuit (boiler and pump) at the level of $\dot{V} = 0.5\,\mathrm{m^3/h}$ if the thermoregulators in the pipework secondary part close. The system diagram and characteristics of the selected pump and of the analysed valve are the same as in the previous example.

Solution

Putting the required minimal value of the medium volume flow on the pump throttling characteristic, a pressure value is obtained at the level of about 0.4 bar, as presented in Fig. 6.9.

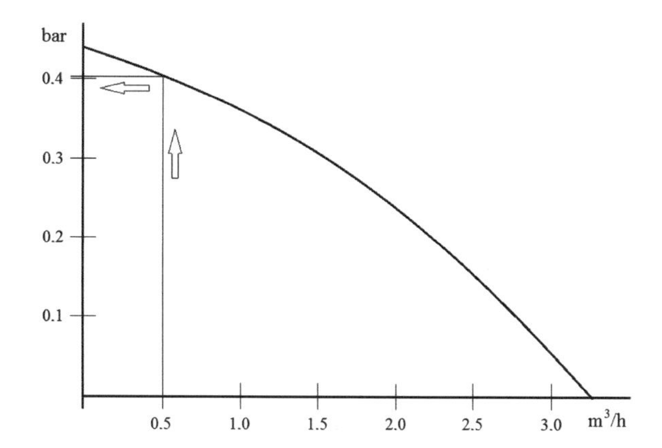

Fig. 6.9 Selected pump throttling characteristic

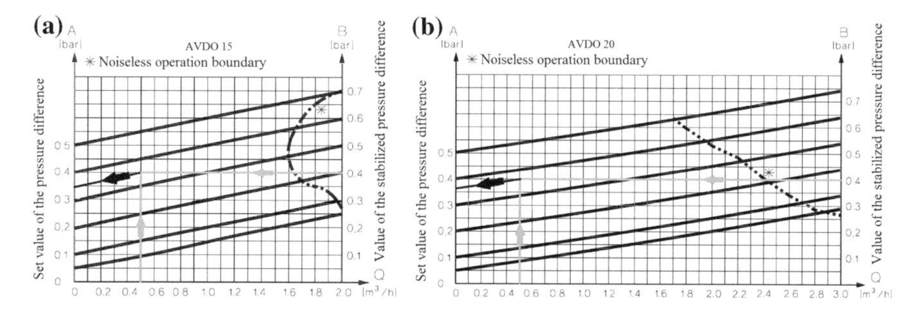

Fig. 6.10 Readout of the relief valve pressure difference pre-setting: **a** *Danfoss* AVDO 15; **b** *Danfoss* AVDO 20

Therefore a valve and its pre-setting should be selected to ensure the required flow at this value of the difference in pressure (the pump active pressure). According to Fig. 6.10, these are the values of 0.35 and 0.37 kPa.

If both the conditions—of maintaining at least a defined minimal value of the medium volume flow and of not exceeding a specific permissible pressure value—are to be satisfied, the valve should be set according to the condition for which the pressure level required for the beginning of opening is lower.

Example 6.6

Select the pressure difference stabilizer and its pre-setting for the following data:

- within the range of its usable proportional range X_p, the valve should be able to operate both as a closing and an opening element,
- required value of stabilized pressure: $\Delta p = 250$ mbar $= 0.25$ bar,
- design value of the medium volume flow: $\dot{V} = 650 \mathrm{l/h} = 0.65\,\mathrm{m^3/h}$,
- consider application of *Herz 4007* valves with the DN15 and DN25 nominal diameters.

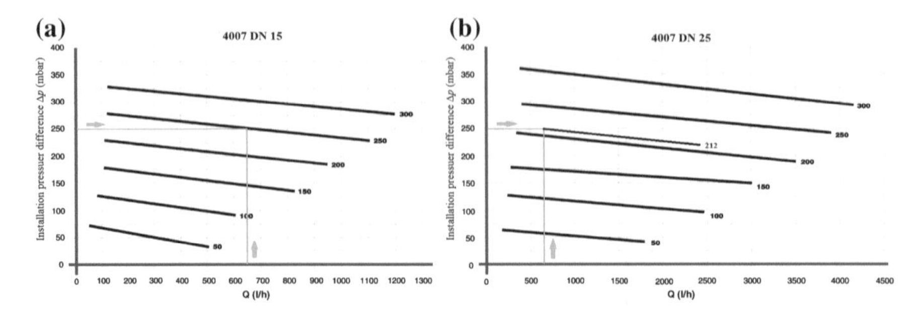

Fig. 6.11 Readout of the pressure difference stabilizer pre-setting: 4007 DN 15 valve (left) and 4007 DN 25 valve (right) Installation pressure difference

Solution

The nomograms of the valves hydraulic characteristics presented in Fig. 6.11 indicate that the selected working point is in the operating range of both of the valves.

The pressure difference pre-setting for the *Herz 4007* DN 15 and for the *Herz 4007* DN 25 valve should be $\Delta p = 250$ mbar and $\Delta p = 212$ mbar, respectively. However, for the *Herz 4007* DN25 valve this means that the working point is always located almost at the very beginning of the static characteristic, i.e. at a small usable proportional range X_{p} and a small opening degree—the regulating plug lift h_{x} over the valve seat. The valve will then close in a rather small range of travel, entering regions of unstable operation and poor quality of regulation. For the *Herz 4007* DN15 valve, the working point is located more or less halfway through proportional range X_{p}. As a result, the range of travel for closing and a reduction in pressure in the secondary part of the pipework and for opening and an increase in pressure in this part is similar. For this reason, this is the valve that should be selected. However, specific values do not result only from the division of the plug available travel into the closing and the opening range, but they are also dependent on the shape of the valve final regulation characteristic determined by the shape of the initial regulation characteristic and the value of the valve total authority a_{c}. The division of the plug travel into the closing and the opening range may therefore not correspond—in fact it never does—to the division into the ratios between the medium volume flows and the stabilized pressure values in the system. The issue is discussed in detail in Sect. 4.1.2.

Example 6.7

Define the hydraulic characteristic exponent m and the pipe hydraulic resistance r_{l} related to the medium volume flow and pressure loss expressed in m^3/h and bar, respectively, using the following data:

– pipe length: $l = 25$ m,

(a) *Kisan* multi-layered 25 × 2.5 mm pipe,
(b) *Herz* multi-layered 26 × 20 mm pipe,
(c) 22 × 1 mm copper pipe with relative roughness $k = 0.01$ mm,
(d) 26.9 × 2.65 mm steel pipe with relative roughness $k = 0.2$ mm,

Solution

Nomograms of pressure losses for the pipes under consideration are presented in Sect. 3.2. If the hydraulic resistance of a pipe section with a selected diameter and length is to be determined using such a nomogram, the following should be done:

- read out the nomogram values of unit linear pressure losses R for two different points on the line of flow-related losses in pressure and the values of the medium volume flow corresponding to them,
- determine the value of exponent m,
- calculate pressure loss $\Delta p_{l,i}$, according to relation (3.6) as the product of the pipe assumed length l_i and unit linear pressure loss R_i read out from the nomogram at a given value of the medium flow mass, volume or velocity,
- calculate hydraulic resistance transforming formula (3.32) and dividing the pressure loss calculated value by the medium volume or mass flow value raised to the m-th power, according to the following formula:

$$r_{l,i} = \frac{R_i \cdot l_i}{\dot{V}_i^m} \tag{6.1}$$

(a) For the *Kisan* 25 × 2.5 mm multi-layered pipe, the points with the mass flow values of $\dot{m}_{\mathrm{i},1} = 0.07\,\mathrm{kg/s}$ and $\dot{m}_{\mathrm{i},2} = 0.3\,\mathrm{kg/s}$ are selected for the analysis. The unit linear pressure losses corresponding to them total $R_{1,\mathrm{i}} = 40$ Pa/m and $R_{2,\mathrm{i}} = 510$ Pa/m, respectively. According to formula (3.33), the characteristic exponent m totals:

$$m = \frac{\log \frac{R_{1,i}}{R_{2,i}}}{\log \frac{\dot{m}_{1,i}}{\dot{m}_{2,i}}} = \frac{\log \frac{40}{510}}{\log \frac{0.07}{0.3}} = 1.75.$$

In order to determine hydraulic resistance in relation to the volume flow, the set mass flow has to be converted to the medium volume flow values. This can be done by means of the following formula:

$$\dot{V}_\mathrm{i} = \frac{\dot{m}_\mathrm{i}}{\rho}. \tag{6.2}$$

In this case, it is enough to perform calculations for one selected value of the medium mass flow because the other is related to the linear pressure loss through the already known, calculated, exponent m. Substituting the first point data, the following is obtained:

$$\dot{V}_\mathrm{i} = \frac{0.07}{977} = 7.16 \times 10^{-5}\,\mathrm{m^3/s} = 0.258\,\mathrm{m^3/h}.$$

According to formula (6.1), hydraulic resistance totals:

$$r_{l,i} = \frac{R_i \cdot l_i}{\dot{V}_i^m} = \frac{40 \cdot 25}{0.258^{1.75}} = 10,706\,(\text{h}^2\,\text{Pa})/\text{m}^6 = 0.107\,(\text{h}^2\,\text{bar})/\text{m}^6.$$

Checking the obtained result, the pressure loss can be calculated for a different value of the medium flow and compared with the nomogram. For example, for the medium mass flow of 0.03 kg/s = 0.1105 m^3/h, formula (3.32) gives:

$$\Delta p_{l,i} = r_{l,i} \cdot V_i^{1.75} = 0.107 \cdot 0.1105^{1.75} = 0.00227\,\text{bar} = 227\,\text{Pa},$$

$$R_i = \Delta p_{l,i}/l_i = 227/25 = 9.1\,\text{Pa/m}.$$

This is in agreement with the nomogram readout for the turbulent flow range. For the laminar flow range, the following values of the previously calculated parameters are obtained:

$$m = \frac{\log \frac{R_{1,i}}{R_{2,i}}}{\log \frac{\dot{m}_{1,i}}{\dot{m}_{2,i}}} = \frac{\log \frac{0.22}{1.1}}{\log \frac{0.002}{0.01}} = 1.0.$$

$$\dot{V}_\text{i} = \frac{0.002}{977} = 2.05 \times 10^{-6}\,\text{m}^3/\text{s} = 0.0074\,\text{m}^3/\text{h}.$$

$$r_{l,i} = \frac{R_i \cdot l_i}{\dot{V}_i^m} = \frac{0.22 \cdot 25}{0.0074^{1.03}} = 861\,(\text{h}^2\,\text{Pa})/\text{m}^6 = 0.00861\,(\text{h}^2\,\text{bar})/\text{m}^6.$$

(b) For the *Herz* 26 × 20 mm multi-layered pipe, the analysis is conducted for points with the selected mass flow values of $\dot{m}_{\text{i},1} = 100\,\text{kg/h}$ and $\dot{m}_{\text{i},2} = 1000\,\text{kg/h}$. The unit linear pressure losses corresponding to them total $R_{1,\text{i}} = 10$ Pa/m and $R_{2,\text{i}} = 500$ Pa/m, respectively. According to formula (3.33), the characteristic exponent m totals:

$$m = \frac{\log \frac{R_{1,i}}{R_{2,i}}}{\log \frac{\dot{m}_{1,i}}{\dot{m}_{2,i}}} = \frac{\log \frac{10}{500}}{\log \frac{100}{1000}} = 1.7.$$

According to formula (6.1), after mass flow values are converted to the medium volume flow values, hydraulic resistance totals:

$$r_{l,i} = \frac{R_i \cdot l_i}{\dot{V}_i^m} = \frac{10 \cdot 25}{\left(\frac{100}{977}\right)^{1.7}} = 12,043\,(\text{h}^2\,\text{Pa})/\text{m}^6 = 0.1204\,(\text{h}^2\,\text{bar})/\text{m}^6.$$

As indicated by the manufacturer's data put onto the nomogram of pressure losses, the value is constant in the entire range of flows.

(c) For the 22 × 1.0 mm copper pipe with relative roughness at the level of $k = 0.01$ mm, the analysis is conducted for points with selected mass flow values of $\dot{m}_{i,1} = 100\,\text{kg/h}$ and $\dot{m}_{i,2} = 1000\,\text{kg/h}$. The unit linear pressure losses corresponding to them total $R_{1,i} = 8.0$ Pa/m and $R_{2,i} = 450$ Pa/m, respectively. According to formula (3.33), the characteristic exponent m totals:

$$m = \frac{\log \frac{R_{1,i}}{R_{2,i}}}{\log \frac{\dot{m}_{1,i}}{\dot{m}_{2,i}}} = \frac{\log \frac{8}{450}}{\log \frac{100}{1000}} = 1.75.$$

According to formula (6.1), after mass flow values are converted to the medium volume flow values, hydraulic resistance totals:

$$r_{l,i} = \frac{R_i \cdot l_i}{\dot{V}_i^m} = \frac{8 \cdot 25}{\left(\frac{100}{977}\right)^{1.75}} = 10{,}800\,(\text{h}^2\,\text{Pa})/\text{m}^6 = 0.108\,(\text{h}^2\,\text{bar})/\text{m}^6.$$

(d) For the 26.9 × 2.65 mm steel pipe with relative roughness at the level of $k = 0.2$ mm, the analysis is conducted for points with selected mass flow values of $\dot{V}_{i,1} = 100\,\text{m}^3/\text{h}$ and $\dot{V}_{i,2} = 1000\,\text{m}^3/\text{h}$. The unit linear pressure losses corresponding to them total $R_{1,i} = 6.1$ Pa/m and $R_{2,i} = 500$ Pa/m, respectively. According to formula (3.33), the characteristic exponent m totals:

$$m = \frac{\log \frac{R_{1,i}}{R_{2,i}}}{\log \frac{\dot{m}_{1,i}}{\dot{m}_{2,i}}} = \frac{\log \frac{6{,}1}{500}}{\log \frac{100}{1000}} = 1.91.$$

According to formula (6.1), after mass flow values are converted to the medium volume flow values, hydraulic resistance totals:

$$r_{l,i} = \frac{R_i \cdot l_i}{\dot{V}_i^m} = \frac{6.1 \cdot 25}{\left(\frac{100}{977}\right)^{1.91}} = 11857\,(\text{h}^2 \cdot \text{Pa})/\text{m}^6 = 0.118\,(\text{h}^2 \cdot \text{bar})/\text{m}^6.$$

The value of the pipe hydraulic characteristic exponent m depends directly on the pipe inner surface roughness k (and, to some extent, on inner diameter d). The pipe material and the quality of the pipe inner surface finish are therefore insignificant. Due to the technological process particularities and properties of materials subjected to treatment/machining, steel is characterized by the greatest roughness. For copper, the surface is more uniform and the roughness of the inner layer of a multi-layered pipe is usually even smaller. Comparing the values of the hydraulic characteristic exponent m for different types of pipes, it turns out that it is the highest for steel pipes and lower for copper and multi-layered pipes. But this also means that for a smoother pipe, the medium turbulent flow is fully developed at higher values of the *Reynolds* number, i.e. at bigger values of the medium volume flow. Sometimes, it does not occur at all. This can be proved analysing the Moody chart presented in Fig. 3.3. It can be seen that the smaller the value of relative roughness e, i.e. of the ratio between absolute roughness k and inner diameter d, the smaller the value of the coefficient of linear pressure losses λ (the friction factor) and, thereby, of hydraulic resistance r_l on the one hand. But on the other, the values of the *Reynolds* number at which the flow becomes fully turbulent get higher, and the value of λ stabilizes and pressure losses become dependent only on the medium volume flow value raised to the second power. Due to the fact that the exponent (as previously described in a simplified manner) is related to the full range of flows, until the flow becomes fully turbulent, for smoother pipes the mean value from the entire range is lower.

It often happens that for plastic (multi-layered) or copper pipes with diameters typically applied in heating installations, the value of relative roughness e is so small that turbulent flows do not arise and the pipe operates as hydraulically smooth. In such a situation, according to formula (3.31), the hydraulic characteristic exponent value is $m = 1.75$. This is the case in the presented example with copper and plastic pipes with a very smooth surface. If the values of relative roughness e are calculated for them and then compared to the value of boundary roughness e_{gr} calculated by means of formulae (3.27–3.29), it is found that in fact $e < e_{gr}$, i.e. the pipe is hydraulically smooth. It follows from the *Moody* chart (cf. Fig. 3.3), the *Blasius* formula (3.18) and formula (3.31) that for the turbulent flow the minimum value of the pipe hydraulic characteristic exponent is $m = 1.75$. But for the *Herz* multi-layered pipes the value is lower and totals $m = 1.7$. This is due to the previously discussed linear interpolation of the line of pressure losses. As indicated e.g. by the characteristics and calculations made for the *Kisan* pipes from the example, the slope of the line of pressure losses in the chart, and consequently the value of exponent m, differs between the laminar and the turbulent flow. For the former, the exponent is lower and totals, according to formula (3.30), $m = 1.0$. For the *Herz* pipes, equivalent lines are plotted jointly for both the ranges and therefore their resultant slope, and consequently the value of exponent m, averaged for the two ranges, is smaller than the exponent value for the medium turbulent flow range only.

Both the hydraulic characteristic exponent m and the pipe hydraulic resistance r_l can be found knowing the value of absolute roughness k and using the computational relations presented in Sect. 3.1. However, practical engineering calculations are often performed using the manufacturers' nomograms prepared based on those relations or based on the results of experimental testing. Helpful as they are, they still have a number of limitations. They are prepared for specific values of the medium temperature, usually for one or two values of the parameter. Considering the wide variety of water temperatures adopted for calculations of heating installations, this can be a significant drawback. Sometimes, manufacturers provide additional formulae or tables of correction factors that make it possible to carry out calculations for other temperatures of the medium. This, however, is rare in practice and does not solve the problem completely because the water temperature values in pipes are the effect of the heat transfer process taking place in the installation. They are known only after full thermal and hydraulic calculations are performed, i.e. after a specific pipe or radiator is selected. Moreover, specialist computer programs are now commonly used to balance the installation thermally and hydraulically. They may utilize full mathematical models that make it possible to determine pressure losses in pipes. The nomograms are therefore useful to carry out a preliminary estimation of pressure losses and determine the hydraulic characteristic exponent.

Example 6.8
The water volume flow in a heating installation circuit with constant active pressure $\Delta p_{ob} = 20$ [kPa] is $\dot{V} = 0.8\,\mathrm{m^3/h}$. Calculate the flow factor and the hydraulic resistance of an additional valve that will reduce the volume flow value to $\dot{V}' = 0.4\,\mathrm{m^3/h}$.

Solution
There are a few ways of solving this problem. The different approaches are presented below.

In order to find the additional valve flow factor, it is first necessary to calculate the initial flow factor of the pipework. According to formula (3.57), it totals:

$$k_{v,ob} = \frac{\dot{V}}{\sqrt{\Delta p_{ob}}} = \frac{0.8}{\sqrt{0.2}} = 1.789\,\mathrm{m^3/h}.$$

This corresponds to the following hydraulic resistance value calculated according to formula (3.61):

$$r_{ob} = \frac{1}{k^2_{v,ob}} = \frac{1}{1.789^2} = 0.3125\,(\mathrm{h^2\,bar})/\mathrm{m^6}.$$

(hydraulic resistance is written here taking account of the flow factor full dimension). The circuit hydraulic resistance can also be found from the relation between the pressure loss and the medium volume flow. Using relation (3.42), the following equation is obtained:

$$\Delta p_{ob} = r_{ob} \cdot \dot{V}^2 \Rightarrow r_{ob} = \frac{\Delta p_{ob}}{\dot{V}^2}.$$

Substituting the data, the result is:

$$r_{ob} = \frac{0.2}{0.8^2} = 0.3125\,(\mathrm{h^2\,bar})/\mathrm{m^6}.$$

As posed by the problem, the medium volume flow is to be halved. According to the flow factor definition, considering the set constancy of pressure, this means a twice smaller flow factor of the circuit with the additional valve. The condition can be written as follows:

$$k'_{v,ob} = \frac{1}{2}k_{v,ob} = 0.8945\,\mathrm{m^3/h}.$$

This corresponds to the following hydraulic resistance value calculated according to formula (3.61):

$$r'_{ob} = \frac{1}{k^2_{v,ob}} = \frac{1}{0.8945^2} = 1.25\,(\mathrm{h^2\,bar})/\mathrm{m^6}.$$

The value is four times smaller than the initial one, which is fairly natural as both the medium volume flow and the flow factor are halved. The new hydraulic resistance is a resultant value for the pipework including the additional valve. Due to the fact that in this case hydraulic resistances are connected in series, the value is a sum of individual resistances according to formula (VII):

$$r'_{\mathrm{ob}} = r_{\mathrm{ob}} + r_{\mathrm{z}}.$$

The valve hydraulic resistance totals:

$$r_{\mathrm{z}} = r'_{\mathrm{ob}} - r_{\mathrm{ob}} = 1.25 - 0.3125 = 0.9372\,(\mathrm{h^2\,bar})/\mathrm{m^6}.$$

Transforming it to express the flow factor, according to formula (3.62), the result is:

$$k_{\mathrm{v,z}} = \frac{1}{\sqrt{r_{\mathrm{z}}}} = \frac{1}{\sqrt{0.9372}} = 1.033\,\mathrm{m^3/h}.$$

And this is the required value of the valve flow factor. It may be an additional valve, or a valve already operating in the pipework but with a changed pre-setting corresponding to this flow factor value.

The valve flow factor can also be determined writing an equation expressing this parameter directly, using the factor dependence on hydraulic resistance for an in-series configuration. Using formulae (VII) and (3.61), the following is obtained:

$$\frac{1}{k'^2_{\mathrm{v,ob}}} = \frac{1}{k^2_{\mathrm{v,ob}}} + \frac{1}{k^2_{\mathrm{v,z}}} \Rightarrow k_{\mathrm{v,z}} = \frac{1}{\sqrt{\frac{1}{k'^2_{\mathrm{v,ob}}} - \frac{1}{k^2_{\mathrm{v,ob}}}}}.$$

Substituting the data, the result is:

$$k_{\mathrm{v,z}} = \frac{1}{\sqrt{\frac{1}{0.8945^2} - \frac{1}{1.789^2}}} = 1.033\,\mathrm{m^3/h}.$$

Alternatively, the solution starting point can be the relation between the pressure loss and the medium volume flow. According to relation (3.44), the following equation is obtained for the case under discussion:

$$\Delta p_{\mathrm{ob}} = r'_{\mathrm{ob}} \cdot \dot{V}'^2. = (r_{\mathrm{ob}} + r_{\mathrm{z}}) \cdot \dot{V}'^2,$$

Transforming the equation to express the sought value of the valve hydraulic resistance, the following is obtained:

$$r_{\mathrm{z}} = \frac{\Delta p_{\mathrm{ob}}}{\dot{V}'^2} - r_{\mathrm{ob}} = \frac{\Delta p_{\mathrm{ob}}}{\dot{V}'^2} - \frac{\Delta p_{\mathrm{ob}}}{\dot{V}^2}.$$

Substituting the data, the result is:

$$r_z = \frac{0.2}{0.4^2} - \frac{0.2}{0.8^2} = 0.9375\,(\mathrm{h^2\,bar})/\mathrm{m^6},$$

which corresponds to the flow factor at the level of:

$$k_{v,z} = \frac{1}{\sqrt{r_z}} = \frac{1}{\sqrt{0.9375}} = 1.033\,\mathrm{m^3/h}.$$

The last approach, which uses hydraulic resistances directly, seems to be the most intuitive one.

Example 6.9
The water volume flow in a heating installation circuit with active pressure $\Delta p_{ob} = 20$ [kPa] totals $\dot{V} = 0.8\mathrm{m^3/h}$. Calculate the flow factor value of a relief valve installed close to the pressure source needed to double the medium volume flow value to $\dot{V}' = 1.6\mathrm{m^3/h}$. Next, calculate the medium volume flow value for the case with no relief valve but an in-series element characterized by the same flow factor installed in the circuit.

Solution
As posed by the problem, the valve in the system is installed in parallel. Consequently, regardless of the setting, it will each and every time increase the total volume flow of the medium. Due to the fact that the valve is installed close to the pressure source, i.e. at the beginning of the pipework, such a configuration may actually be treated as parallel, taking no account of possible resistances connected in series between the pressure source and the system and a valve.

If the volume flow value is to be doubled, the flow factor has to be twice bigger. For the initial conditions and according to formula (3.57), the factor totals:

$$k_{v,ob} = \frac{\dot{V}}{\sqrt{\Delta p_{ob}}} = \frac{0.8}{\sqrt{0.2}} = 1.789\,\mathrm{m^3/h}.$$

The new volume flow value will then total:

$$k'_{v,ob} = 2k_{v,ob} = 3.578\,\mathrm{m^3/h}.$$

In a system connected in parallel, the flow factors of individual elements are summed up. It may therefore be written that:

$$k'_{v,ob} = k_{v,ob} + k_{v,z}.$$

This means that the sought value of the valve flow factor is:

$$k_{v,z} = k'_{v,ob} - k_{v,ob} = 3.578 - 1.789 = 1.789\,\mathrm{m^3/h}.$$

The fact that flow factor values in the parallel configuration are simply summed up can be proved starting with the relation describing the equivalent resistance of a parallel connection of hydraulic resistances. For the system under consideration, according to formula (IX′), it may be written that:

$$r'_{ob} = \frac{1}{\left(\sqrt{\frac{1}{r_{ob}}} + \sqrt{\frac{1}{r_z}}\right)^2}.$$

Replacing hydraulic resistances with the flow factor, according to formula (3.61), the following equation is obtained:

$$\frac{1}{k'^2_{v,ob}} = \frac{1}{\left(k_{v,ob} + k_{v,z}\right)^2} \Rightarrow k'_{v,ob} = k_{v,ob} + k_{v,z}.$$

Connecting the elements with the flow factor values mentioned above in series, and using Eq. (3.43), the following value of the medium volume flow is obtained:

$$\dot{V}' = \sqrt{\frac{\Delta p_{ob}}{r_{ob} + r_z}} = \sqrt{\frac{\Delta p_{ob}}{\frac{1}{k^2_{v,ob}} + \frac{1}{k^2_{v,z}}}} = \sqrt{\frac{0.2}{\frac{1}{1.789^2} + \frac{1}{1.789^2}}} = 0.565\,\mathrm{m^3/h}.$$

Comparing the calculation results obtained for the parallel and the in-series configuration, it can be seen that identical changes in the medium volume flow require different changes in the flow factor values. Connecting the same hydraulic resistances in series, the drop in the volume flow will differ from the rise in it occurring in a parallel configuration. A parallel connection of two identical hydraulic resistances, i.e. elements characterized by identical flow factor values, doubles the initial medium volume flow value (in this case: a rise from 0.8 to 1.6 $\mathrm{m^3/h}$, at constant pressure). If the elements are connected in series, a drop occurs by the root of two times—here: from 0.8 to 0.565 $\mathrm{m^3/h}$.

Solving this type of problems, some general useful guidelines can be formulated:

- if the hydraulic system elements are connected in parallel, their flow factors have to be summed up,
- connecting the hydraulic system elements in series, their hydraulic resistances have to be summed up,
- in a parallel system with a constant available pressure value, in order to obtain an i-time rise in the medium volume flow (a reduction is impossible by principle) by connecting an additional element, the equivalent hydraulic resistance should be i^2 times smaller compared to the initial system. The hydraulic resistance of an additional element connected in parallel should therefore be the $1/(i - 1)^2$ multiplicity of the initial system hydraulic resistance, whereas the equivalent flow factor should be i times higher. The flow factor of an additional element (e.g. a valve) connected in parallel should therefore be the $(i - 1)$ multiplicity of the initial system flow factor. For example, if the system initial hydraulic resistance

is $r_{ob} = 2$, and a 3-time rise is required in the medium volume flow value, the equivalent hydraulic resistance will have to total $r'_{ob} = r_{ob}/i^2 = 2/3^2 = 2/9$, i.e. it will need to get nine times smaller, and the resistance of the additional element (e.g. a valve) will then have to total $r_z = r_{ob} \cdot (1/(i-1)^2) = 2 \cdot (1/(3-1)^2) = 1$. The initial system flow factor is $k_{v,ob} = 1/r_{ob}^{0.5} = 1/2^{0.5} = 0.707$; for the equivalent system, it totals $k'_{v,ob} = k_{v.ob} \cdot 3 = 0.707 \cdot 3 = 2.121$, i.e. it is three times higher, and for an element added in parallel it will have to be equal to $k_{v.z} = (i-1) \cdot k_{v,ob} = (3-1) \cdot 0.707 = 1.414$.

- in a series system with a constant available pressure value, in order to obtain an i-time reduction in the medium volume flow (an increase is impossible by principle) by means of an additional element, the equivalent hydraulic resistance should be i^2 times higher compared to the initial system. The hydraulic resistance of an additional element connected in series should therefore be the $(i^2 - 1)$ multiplicity of the initial system hydraulic resistance, whereas the equivalent flow factor should be i times smaller. The flow factor of the additional element (e.g. a valve) should therefore be the $1/(i^2 - 1)^{0.5}$ multiplicity of the initial system flow factor. For example, if the system initial hydraulic resistance is $r_{ob} = 2$, and a 3-time decrease is required in the medium volume flow value, the equivalent hydraulic resistance will have to total $r'_{ob} = r_{ob} \cdot i^2 = 2 \cdot 3^2 = 18$, i.e. it will need to get nine times bigger, and the hydraulic resistance of the additional element (e.g. a valve) will then have to total $r_z = r_{ob} \cdot (i^2 - 1) = r_z = 2 \cdot (3^2 - 1) = 16$. The initial system flow factor is $k_{v,ob} = 1/r_{ob}^{0.5} = 1/2^{0.5} = 0.707$; for the equivalent system, it totals $k'_{v,ob} = k_{v,ob}/3 = 0.707/3 = 0.236$, i.e. it is three times smaller, and for an element added in parallel it will have to be equal to $k_{v.z} = 1/(i^2 - 1)^{0.5} \cdot k_{v,ob} = 1/(3^2 - 1)^{0.5} \cdot 0.707 = 0.25$.

Example 6.10

The water volume flow in the heating installation circuit is $\dot{V} = 0.8\,\text{m}^3/\text{h}$. Calculate the required flow factor and hydraulic resistance of the following elements:

(a) an additional in-series throttling valve that will reduce the medium volume flow to $\dot{V}' = 0.4\,\text{m}^3/\text{h}$,

(b) a relief valve installed close to the pressure source that will cause a rise in the volume flow to $\dot{V}' = 1.6\,\text{m}^3/\text{h}$.

The calculations are to be performed taking account of the fact that the pump pressure varies and the changes are a function of the medium volume flow, which is expressed as: $\Delta p_x = 23 - a \cdot \dot{V}_x$, $a = 3.75$ kPa/(m^3/h).

Solution

In this case, the variation in the pump pressure depending on the medium volume flow, i.e. the pump throttling characteristic, is described by a linear relation. This assumption is explained and a more detailed discussion of throttling characteristics is presented in the following, more comprehensive, problems.

In order to determine the valve flow factor required in this case, the calculations should be carried out in the same manner as in the previous computational example.

The only difference is that the first thing to do is to define the new pressure level in the system after the valve installation because the quantity varies, as stated in the problem. After the pressure value is determined, calculations can be performed of the new distribution of pressures between the pipework and the additional valve, and the valve flow factor can be found.

(a) The medium volume flow halved using a throttling valve installed in series.

For the medium initial volume flow value, the pump pressure is:

$$\Delta p_{\mathrm{x}} = 23 - 3.75 \cdot \dot{V}_{\mathrm{x}} = 23 - 3.75 \cdot 0.8 = 20\,\mathrm{kPa}.$$

The value complies with the problem data. The new value of the system pressure for the medium required volume flow will total:

$$\Delta p_{\mathrm{x}} = 23 - 3.75 \cdot \dot{V}_{\mathrm{x}} = 23 - 3.75 \cdot 0.4 = 21.5\,\mathrm{kPa}.$$

Using formula (3.44), which relates the medium volume flow to the drop in pressure (or, in this case, to available pressure), it may be written that:

$$\Delta p'_{\mathrm{ob}} = r'_{\mathrm{ob}} \cdot \dot{V}'^2. = (r_{\mathrm{ob}} + r_{\mathrm{z}}) \cdot \dot{V}'^2.$$

The relation is the same as for the case of the active pressure constancy. The only difference is that the circuit active pressure changes. For this reason, its initial value $\Delta p_{\mathrm{ob.}}$ cannot be written on the left side. Instead, the final value $\Delta p'_{\mathrm{ob}}$ is used. Transforming the equation to express the sought value of the valve hydraulic resistance, the following is obtained:

$$r_{\mathrm{z}} = \frac{\Delta p'_{\mathrm{ob}}}{\dot{V}'^2} - r_{\mathrm{ob}} = \frac{\Delta p'_{\mathrm{ob}}}{\dot{V}'^2} - \frac{\Delta p_{\mathrm{ob}}}{\dot{V}^2}$$

Substituting the data, the result is:

$$r_{\mathrm{z}} = \frac{0.215}{0.4^2} - \frac{0.2}{0.8^2} = 1.03\,(\mathrm{h}^2\,\mathrm{bar})/\mathrm{m}^6,$$

which corresponds to the flow factor at the level of:

$$k_{\mathrm{v,z}} = \frac{1}{\sqrt{r_{\mathrm{z}}}} = \frac{1}{\sqrt{1.03}} = 0.985\,\mathrm{m}^3/\mathrm{h}.$$

(b) The medium volume flow doubled using a relief valve installed in parallel.

The new value of the system pressure for the medium required volume flow is:

$$\Delta p_{\mathrm{x}} = 23 - 3.75 \cdot \dot{V}_{\mathrm{x}} = 23 - 3.75 \cdot 1.6 = 17\,\mathrm{kPa}.$$

Knowing the pressure final value and the required volume flow of the medium, it is possible to calculate the required equivalent flow factor of the system with the additional valve. According to formula (3.57), it will total:

$$k'_{\mathrm{v,ob}} = \frac{\dot{V}'}{\sqrt{\Delta' p_{\mathrm{ob}}}} = \frac{1.6}{\sqrt{0.17}} = 3.88\,\mathrm{m}^3/\mathrm{h}.$$

The initial value of the system flow factor is:

$$k_{\mathrm{v,ob}} = \frac{\dot{V}}{\sqrt{\Delta p_{\mathrm{ob}}}} = \frac{0.8}{\sqrt{0.2}} = 1.789\,\mathrm{m}^3/\mathrm{h}.$$

As mentioned earlier, in a system connected in parallel, the flow factors of individual elements are summed up. It may therefore be written that:

$$k'_{\mathrm{v,ob}} = k_{\mathrm{v,ob}} + k_{\mathrm{v,z}}.$$

This means that the sought value of the valve flow factor is:

$$k_{\mathrm{v,z}} = k'_{\mathrm{v,ob}} - k_{\mathrm{v,ob}} = 3.88 - 1.789 = 2.091\,\mathrm{m}^3/\mathrm{h}.$$

Analysing the calculation results and comparing them to the results obtained under the assumption of the pump pressure constancy, the following conclusions can be formulated:

- in a series configuration, if active pressure rises with a drop in the medium volume flow, which is the case in practice, in order to obtain the assumed volume flow of the medium, the additional valve flow factor value has to be lower compared to the situation where the active pressure value is constant. The higher the increment in the system pressure at a drop in the medium volume flow, i.e. the steeper the pump characteristic, the bigger the difference between the calculated flow factor of the valve at constant pressure and the required value.
- in a parallel configuration, if active pressure drops with a rise in the medium volume flow, which is the case in practice, in order to obtain the assumed volume flow of the medium, the additional valve flow factor value has to be higher compared to the situation where the active pressure value is constant. The higher the drop in the system pressure at a rise in the medium volume flow, i.e. the steeper the pump characteristic, the bigger the difference between the calculated flow factor of the valve at constant pressure and the required value.

Example 6.11
A circulating pump with a known hydraulic characteristic operates in a heating installation. The results of its throttling characteristic analysis indicate that for the required volume flow of the medium it supplies pressure higher than required by the pipework. The pressure has to be reduced to the required value. Estimate in which of

the three variants of the applied regulation method the pump drive needs the smallest power:

(a) excess pressure reduced by means of a regulation valve installed in series,
(b) excess pressure reduced by means of a relief valve installed in parallel (a bypass),
(c) required pressure achieved by changing the pump rotational speed.

Determine the required value of the valve flow factor for cases {a} and {b} and the power loss arising on the devices. Then, as item {d}, calculate the medium volume flow value and the power needed to drive the pump if none of the above-mentioned regulation methods is applied. Assume that the exponent of the pipework hydraulic characteristic is $n = 2$.

Installation data:

- pressure losses of the circuit for design conditions: $\Delta p_{\mathrm{ob}} = 36$ kPa,
- thermal load of the serviced facility: $\dot{Q}_{\mathrm{ob}} = 24\,\mathrm{kW}$,
- heating water parameters: $t_z/t_p = 80/60$ °C,
- water computational density: $\rho = 972$ kg/m^3,
- assume a constant value of the pump total efficiency at the level of: $\eta_c = 80\%$,
- omit gravitational pressure.

In order to simplify the calculations, the pump throttling characteristic is approximated by a broken line spanned by the following four points:

(1) $\dot{V}_{\mathrm{pl}} = 0.25\ \mathrm{dm}^3/\mathrm{s}$ and $\Delta p_{\mathrm{pl}} = 40\,\mathrm{kPa}$,
(2) $\dot{V}_{\mathrm{pl}} = 0.333\,\mathrm{dm}^3/\mathrm{s}$ and $\Delta p_{\mathrm{pl}} = 38\,\mathrm{kPa}$,
(3) $\dot{V}_{\mathrm{pl}} = 0.417\ \mathrm{dm}^3/\mathrm{s}$ and $\Delta p_{\mathrm{pl}} = 35\,\mathrm{kPa}$,
(4) $\dot{V}_{\mathrm{pl}} = 0.5\,\mathrm{dm}^3/\mathrm{s}$ and $\Delta p_{\mathrm{pl}} = 32\,\mathrm{kPa}$.

Solution

The medium pumping power is equal to the product of the volume flow and pressure generated in the i-th point of the throttling characteristic. Taking account of the device total efficiency, the power needed to drive the pump can be expressed as:

$$P_{\mathrm{el}} = \frac{\Delta p_{\mathrm{i}} \cdot \dot{V}_{\mathrm{i}}}{\eta_{\mathrm{c,i}}}. \tag{6.3}$$

The values of the parameters have to be found. In order to determine the location of the required point on the pipework characteristic, the medium volume flow should be calculated; pressure losses are given. The medium volume flow can be calculated from the following relation:

$$\dot{Q}_{ob} = \dot{m}_{\mathrm{ob}} \cdot c_{\mathrm{w}} \cdot \left(t_z - t_p\right) = \dot{V}_{\mathrm{ob}} \cdot \rho \cdot c_{\mathrm{w}} \cdot (t_z - t_p) \Rightarrow \dot{V}_{\mathrm{ob}} = \frac{\dot{Q}_{\mathrm{ob}}}{\rho \cdot c_{\mathrm{w}} \cdot (t_z - t_p)}. \tag{6.4}$$

Substituting relevant data, the following is obtained:

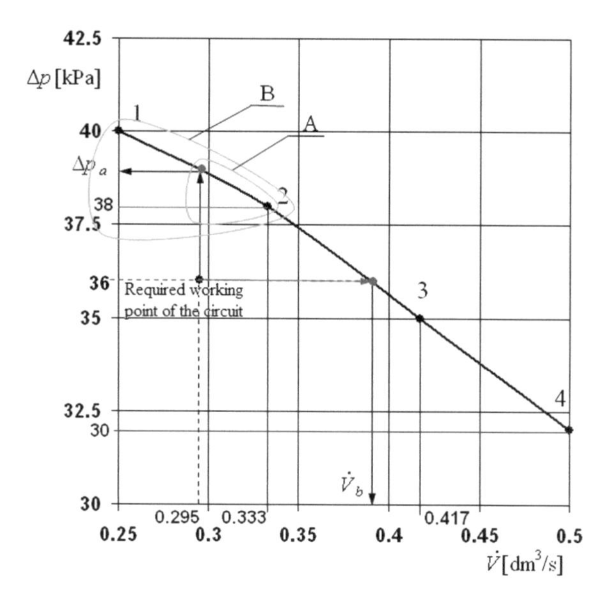

Fig. 6.12 Pump throttling characteristic and location of the pipework required working point

$$\dot{V}_{\mathrm{ob}} = \frac{\dot{Q}_{\mathrm{ob}}}{\rho \cdot c_{\mathrm{w}} \cdot (t_{\mathrm{z}} - t_{\mathrm{p}})} = \frac{24{,}000}{972 \cdot 4190 \cdot 20} = 0.295\,\mathrm{dm}^3/\mathrm{s}$$
$$= 0.000295\,\mathrm{m}^3/\mathrm{s} = 1.062\,\mathrm{m}^3/\mathrm{h}.$$

Figure 6.12 presents the pump characteristic and the location of the pipework required working point.

The chart also presents the lines of the excess of pressure and of the medium volume flow (vertical and horizontal lines, respectively). The lines intersect the pump throttling characteristic in points whose coordinates need to be established. The first line corresponds to variant {a} of the problem, i.e. the application of an in-series regulation valve to reduce the excess of pressure; the second represents variant {b}, i.e. the application of a bypass reducing the pressure generated by the pump by increasing the volume flow of the pumped medium.

(a) Excess pressure reduced by means of a regulation valve installed in series.

The vertical segment between the required point and the pipework characteristic reflects the pressure value that the valve will need to throttle. The value is unknown and has to be determined, which can be done in several ways. Knowing the coordinates of individual points of the pump characteristic, it is possible to create a mathematical description of a linear function for each segment, also for the first, in the area of which the sought point is located. Knowing the medium volume flow, i.e. the function independent variable, it is possible to establish the function value, i.e. the sought pressure Δp_{a}. An appropriate proportion can also be set up using the principle of the similarity of triangles. As it can be seen in the figure, triangle {A} created based on the added lines of the medium required volume flow and the pressure value is inscribed in triangle {B}, for which the coordinates of all points are known. According to the principle of the similarity of triangles, the ratios between corresponding sides are equal. Considering that, it is possible to set up a few identical proportions where the sought value of pressure Δp_{a} will appear as an unknown. The following will be obtained for example:

$$\frac{0.295 - 0.25}{0.333 - 0.25} = \frac{40 - \Delta p_{\mathrm{a}}}{40 - 38}.$$

After transformations the result is:

$$\Delta p_a = 40 - \frac{0.295 - 0.25}{0.333 - 0.25}(40 - 38) = 38.91\,\mathrm{kPa} = 38910\,\mathrm{Pa} = 0.391\,\mathrm{bar}.$$

Consequently, the pressure to be throttled totals:

$$\Delta p_z = \Delta p_{\mathrm{a}} - \Delta p_{\mathrm{ob}} = 38.91 - 36 = 2.91\,\mathrm{kPa} = 0.0291\,\mathrm{bar}.$$

Because the elements are connected in series, the medium volume flow through the valve $\dot{V}_z$ equals the volume flow in the entire circuit $\dot{V}_{\mathrm{ob}}$. The valve flow factor thus totals:

$$k_{\mathrm{v,z}} = \frac{\dot{V}_z}{\sqrt{\Delta p_z}} = \frac{1.062}{\sqrt{0.0291}} = 6.22\,\mathrm{m}^3/\mathrm{h}.$$

The power loss arising on the valve is:

$$P_{\mathrm{tr,a}} = \Delta p_z \cdot \dot{V}_z = 2910 \cdot 0.000295 = 0.86\,\mathrm{W}.$$

The power taken by the pump totals:

$$P_{\mathrm{el,a}} = \frac{\Delta p_a \cdot \dot{V}_a}{\eta_{\mathrm{c,a}}} = \frac{39,100 \cdot 0.000295}{0.8} = 14.42\,\mathrm{W}.$$

The power loss arising on the valve in this case makes up about 6% of the total pumping power.

(b) Excess pressure reduced by means of a relief valve installed in parallel (a bypass).

The point on the pump characteristic is found in the same manner, setting up an appropriate proportion:

$$\frac{0.417 - \dot{V}_{\mathrm{b}}}{0.417 - 0.333} = \frac{36 - 35}{38 - 35}.$$

After relevant transformations, the result is:

$$\dot{V}_b = 0.417 - \frac{0.417 - 0.333}{38 - 35}(36 - 35) = 0.389\,\mathrm{dm^3/s}$$
$$= 0.000389\,\mathrm{m^3/s} = 1.4\,\mathrm{m^3/h}.$$

The valve required flow factor can be calculated from the same relation as in the first case. However, this time the elements are connected in parallel so the medium volume flow through the valve is not equal to the volume flow in the entire circuit. It equals the difference between the total volume flow and the volume flow through the main circuit, according to the following relation:

$$\dot{V}_z = \dot{V}_b - \dot{V}_{ob} = 0.389 - 0.295 = 0.094\,\mathrm{dm^3/s} = 0.000094\,\mathrm{m^3/s} = 0.3384\,\mathrm{m^3/h}.$$

In an in-parallel connection the pressure on individual elements is identical. The pressure loss on the valve Δp_z is equal to the pressure loss in the circuit Δp_{ob}. Using the calculated value of the medium flow through the valve, the valve flow factor totals:

$$k_{v,z} = \frac{\dot{V}_z}{\sqrt{\Delta p_z}} = \frac{0.3384}{\sqrt{0.036}} = 1.78\,\mathrm{m^3/h}.$$

The power loss arising on the valve is:

$$P_{tr,b} = \Delta p_z \cdot \dot{V}_z = 36{,}000 \cdot 0.000094 = 3.38\,\mathrm{W}.$$

The power taken by the pump totals:

$$P_{el,b} = \frac{\Delta p_b \cdot \dot{V}_b}{\eta_{c,b}} = \frac{36{,}000 \cdot 0.000389}{0.8} = 17.5\,\mathrm{W}.$$

The power loss arising on the valve in this case makes up about 19.3% of the total pumping power.

(c) Required pressure achieved by changing the pump rotational speed.

In this case, there are no elements diminishing the excess of energy and the pump characteristic is lowered to the intersection with the required working point. This occurs due to a reduction in the impeller rotational speed, as described previously. Here, the power totals:

$$P_{el,c} = \frac{\Delta p_b \cdot \dot{V}_b}{\eta_{c,b}} = \frac{36{,}000 \cdot 0.000295}{0.8} = 13.27\,\mathrm{W}.$$

(d) No elements reducing pressure and the medium volume flow in the system.

If none of the above-described regulation methods are applied, the medium volume flow can be determined by finding the point of intersection of the circuit characteristic and the pump throttling characteristic. The pump characteristic mathematical description is known. According to formula (3.44) and assuming that $m = 2$, the description of the circuit characteristic can be written as:

$$\Delta p_{\mathrm{ob}} = r_{\mathrm{ob}} \cdot \dot{V}_{\mathrm{ob}}^{2}.$$

It is an equation defining a quadratic function with the origin of coordinates in point (0,0) of the system of coordinates. The function cannot be plotted unless hydraulic resistance r_{ob} is known. It can be determined based on the data on the coordinates of the second point. After relevant substitutions, the above formula gives:

$$r_{\mathrm{ob}} = \frac{\Delta p_{\mathrm{ob}}}{\dot{V}_{\mathrm{ob}}^{2}} = \frac{36}{0.295^{2}} = 413.7\,\mathrm{kPa}/(\mathrm{dm}^{3}/\mathrm{s})^{2}.$$

The circuit characteristic will have the form of the function expressed as:

$$\Delta p_{\mathrm{ob}} = 413.7 \cdot \dot{V}_{\mathrm{ob}}^{2}.$$

In order to find the point of intersection with the pump characteristic, it is first necessary to determine the segment of the broken curve describing the characteristic where the intersection will take place. For this purpose, the value of pressure losses arising in the circuit has to be calculated for given values of the medium volume flow and then it has to be compared to the pressure produced in those points by the pump. If the calculated pressure drop is higher than the pressure produced by the pump for the point with a given value of the medium volume flow, it will mean that the curve lies on the point left side. Otherwise, it will lie on the point right side (cf. Fig. 6.12). Checking the first two points {1} and {2}, between which the circuit characteristic may pass, the following is obtained:

$$\Delta p_{\mathrm{ob},1} = 413.7 \cdot 0.25^{2} = 25.85\,\mathrm{kPa}.$$

This is a value lower than the pressure generated by the pump for this particular value of the medium volume flow. The circuit characteristic is thus on the right side of point {1}. It should now be checked if it lies on the left or on the right side of point {2}. In order to do so, the volume flow value from point {2} is substituted in the equation defining the circuit characteristic:

$$\Delta p_{\mathrm{ob},2} = 413.7 \cdot 0.333^{2} = 45.87\,\mathrm{kPa}.$$

This is a value higher than the pressure generated by the pump for this particular value of the medium volume flow. The circuit characteristic is thus on the left side of the point, which also means that it lies on the {1–2} segment of the pump throttling characteristic.

In order to find the intersection point for the characteristics, a proportion is set up in the same manner as in the previous cases, which gives the following expression:

$$\frac{\dot{V}_d - 0.25}{0.333 - 0.25} = \frac{40 - \Delta p_d}{40 - 38} \Rightarrow \Delta p_d = 40 - \frac{\left(\dot{V}_d - 0.25\right) \cdot (40 - 38)}{0.333 - 0.25}$$

It is an equation with two unknowns—$\dot{V}_d$ i Δp_d (identical with $\dot{V}_{ob}$ i Δp_{ob}, subscript "d" denotes the problem variant—{d}). Another equation should therefore be added to it that will not depend on it linearly from the mathematical point of view and that will relate the two unknowns to each other. The equation is the equation of the circuit characteristic $\Delta p_{ob} = 413.7 \cdot \dot{V}_{ob}^2$. Substituting it, the following formula is obtained:

$$40 - \frac{\left(\dot{V}_{ob} - 0.25\right) \cdot (40 - 38)}{0.333 - 0.25} = 413.7 \cdot \dot{V}_{ob}^2 \Rightarrow 34.34 \cdot \dot{V}_{ob}^2 + 2 \cdot \dot{V}_{ob} - 3.82 = 0.$$

It is a quadratic equation, the solution of which are the following values of the medium volume flow:

$$\dot{V}_{ob} = (-0.364;\ 0.3057)\ \mathrm{dm}^3/\mathrm{s}.$$

The negative solution is rejected by definition, which means that the result is $\dot{V}_{ob} \equiv \dot{V}_d = 0.3057\ \mathrm{dm}^3/\mathrm{s}$. The pressure generated by the pump for this value of the medium volume flow can be calculated by substituting the calculated volume flow either in the pump characteristic equation or in the equation defining the circuit characteristic. Using the latter, simpler equation, the following is obtained:

$$\Delta p_d = 413.7 \cdot 0.3057^2 = 38.66\ \mathrm{kPa}.$$

Figure 6.13 presents the characteristics of the pump throttling and of the circuit. In this case, the power taken by the pump totals:

$$P_{el,d} = \frac{\Delta p_d \cdot \dot{V}_d}{\eta_{c,d}} = \frac{38{,}660 \cdot 0.0003057}{0.8} = 14.77\ \mathrm{W}.$$

Analysing the calculation results, it can be seen that the lowest power needed to drive the pump occurs in the case of adjusting the pump impeller rotational speed, and, thereby, the generated pressure value, to the required conditions. The use of a throttling valve involves a bigger rise in the power demand, and the highest power is needed if a relief valve (a bypass) is applied. This is natural, since as a rule the two elements operate to diminish excess pressure. Which of the two is less advantageous depends on a specific case, but most often it is the bypass application.

Despite the fact that in reality efficiency values vary, the problem assumes a constant value of the pump total efficiency, regardless of the coordinates of the point on the pump throttling characteristic. The assumption is made because actual changes in efficiency and the efficiency specific values can be determined only by means of an analysis of a specific pump. Typical curves illustrating changes in the

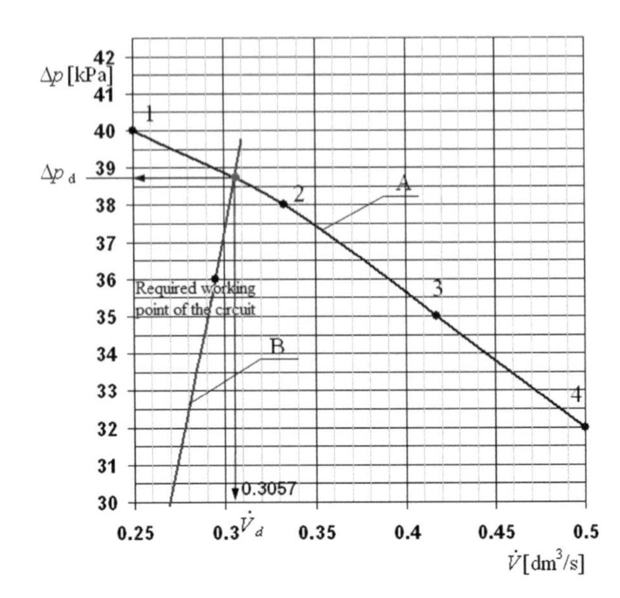

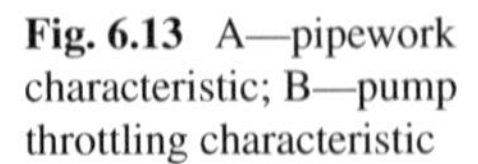
Fig. 6.13 A—pipework characteristic; B—pump throttling characteristic

pump efficiency are illustrated in Fig. 2.50. The pump reaches its highest efficiency values in the middle areas of its throttling characteristic.

The presented discretization of the pump throttling characteristic in the form of a broken line is a simplification. In this case, the real curve is also known only after a pump-specific analysis is performed. However, manufacturers usually do not specify mathematical relations that determine the characteristic shape. They provide relevant charts instead. In such a situation, it is possible to create a mathematical model using the chart readouts, but in practice the chart readout itself is sufficient. The characteristic of the pipework is then put thereon and an approximate working point is read out (the intersection point of the two curves). The same method can be used to estimate efficiency in a given working point.

The pump throttling curve can also be approximated in a way different from the one presented in the example. The following linear relation is sometimes used:

$$\Delta p_{\mathrm{x}} = \Delta p_0 - a \cdot (\dot{V}_{\mathrm{x}} - \dot{V}_0). \tag{6.5}$$

Parameter Δp_{x} is the sought pressure value for the set value of the medium volume flow $\dot{V}_{\mathrm{x}}$, a is the pump throttling characteristic slope factor, expressed as the ratio between changes in pressure and the medium volume flow values. Transforming formula (6.5), the parameter is described as follows:

$$a = -\frac{\Delta p_{\mathrm{x}} - \Delta p_0}{(\dot{V}_{\mathrm{x}} - \dot{V}_0)}. \tag{6.6}$$

Quantities Δp_0 and $\dot{V}_0$ are, respectively, the generated pressure and the medium volume flow for the pump initial operating conditions. This may be the origin of the pump throttling characteristic. If so, $\dot{V}_0 = \dot{V}_{\text{min}} = 0$ and $\Delta p_0 = \Delta p_{\text{max}}$. Relation (6.5) then gives the following equation:

$$\Delta p_{\text{x}} = \Delta p_{\text{max}} - a \cdot \dot{V}_{\text{x}} = -a \cdot \dot{V}_{\text{x}} + \Delta p_{\text{max}}. \tag{6.7}$$

It is thus a notation of a linear function with a negative slope—a decreasing function with its beginning in point $(\dot{V}_0, \Delta p_{\text{max}}) = (0, \Delta p_{\text{max}})$. If the end of the pump throttling characteristic, i.e. point $(\dot{V}_{\text{max}}, \Delta p_{\text{min}}) = (\dot{V}_{\text{max}}, 0)$, is taken into account in the notation, the line spanned on points $(0, \Delta p_{\text{max}})$ (initial point) and $(\dot{V}_{\text{max}}, 0)$ (final point) enables determination of the characteristic slope factor a. The following relation is then obtained:

$$a = \frac{\Delta p_{\text{max}} - \Delta p_{\text{min}}}{\dot{V}_{\text{max}} - \dot{V}_0} = \frac{\Delta p_{\text{max}} - 0}{\dot{V}_{\text{max}} - 0} = \frac{\Delta p_{\text{max}}}{\dot{V}_{\text{max}}}. \tag{6.8}$$

Substitution of the above in equation $\left(\Delta p_{\text{x}} = \Delta p_{\text{max}} - a \cdot \dot{V}_{\text{x}} = -a \cdot \dot{V}_{\text{x}} + \Delta p_{\text{max}}.\right)$ results in:

$$\Delta p_{\text{x}} = \Delta p_{\text{max}} - a \cdot \dot{V}_{\text{x}} = -\frac{\Delta p_{\text{max}}}{\dot{V}_{\text{max}}} \cdot \dot{V}_{\text{x}} + \Delta p_{\text{max}} = \Delta p_{\text{max}} \cdot \left(1 - \frac{\dot{V}_{\text{x}}}{\dot{V}_{\text{max}}}\right). \tag{6.9}$$

A more accurate approximation of the pump characteristic curve is also frequently used. It is assumed that the impeller pump is a combination of an ideal pressure source, keeping a constant pressure value depending on the flow, with an internal hydraulic resistance connected in series. This is an analogy with a real source of voltage in an electric circuit. If such a mathematical model is assumed, outlet pressure will be zero-flow initial pressure Δp_{max} reduced by the pressure drop arising on the internal hydraulic resistance. The bigger the medium volume flow, the higher the pressure loss on the resistance and the smaller the pressure output value. Equation (6.9) assumes a linear pressure drop on the pump internal hydraulic resistance as a function of the medium volume flow. Considering the fact that, as indicated by formula (3.43), in a hydraulic system a change in the volume flow is proportional to the square root of the pressure change (on the internal hydraulic resistance), the following can be written:

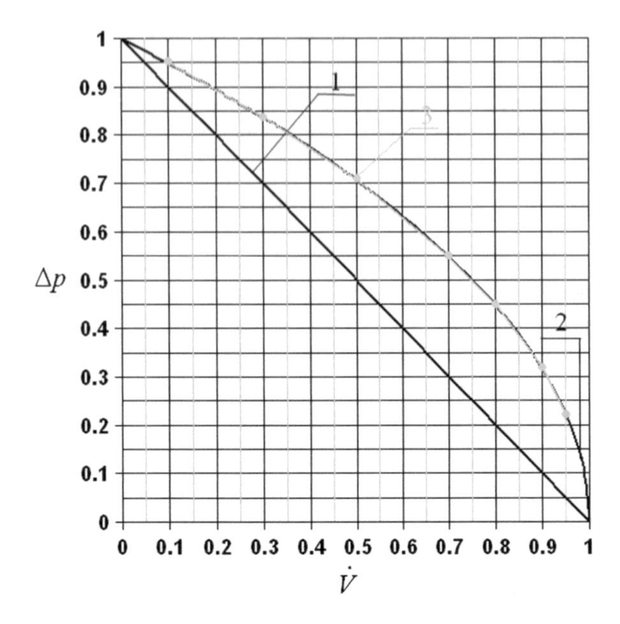

Fig. 6.14 Pump throttling characteristic: (1) linear, according to relation 6.9; (2) root-dependent, according to relation 6.10; (3) discretization of the root-dependent curve by means of a broken line

$$\Delta p_{\mathrm{x}} = \Delta p_{\max} \cdot \sqrt{1 - \frac{\dot{V}_{\mathrm{x}}}{\dot{V}_{\max}}}. \tag{6.10}$$

In a system of conventional units of pressure and of the volume flow, Fig. 6.14 presents throttling characteristics plotted according to the two relations.

It can be seen in the figure that if sufficiently small intervals are assumed in the form of segments of the broken line discretizing the root-dependent curve, the readout from the broken line is accurate enough. Consequently, if it is known that changes in the working point location in the installation occur in a small range of the pump throttling characteristic, the curve approximation may be used in the form of a fragment of a linear function, and not of the broken line. A situation like this is presented in the computational example under consideration. In practice, the pump throttling curve is also often approximated using a second-degree polynomial.

Example 6.12
In a pump central heating installation with three risers (marked as P1, P2 and P3) the flow through riser P3 is stopped. Calculate the relative percentage change in the medium flow rate in the other two risers compared to nominal flow values using the data given below. Assume that the exponent of the pipework hydraulic characteristic is $n = 2$.

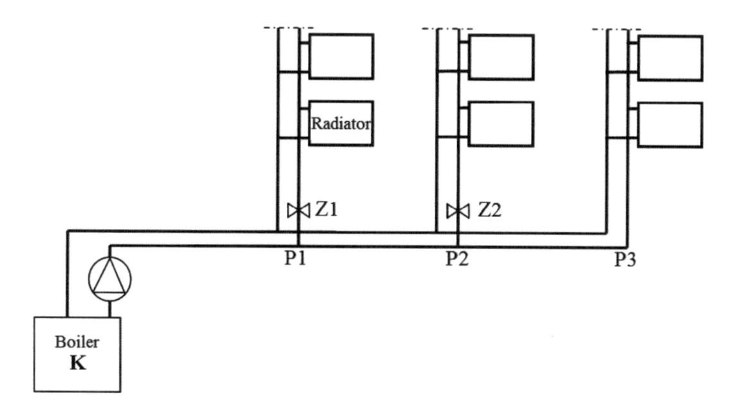

Fig. 6.15 Diagram of the installation under analysis

Installation data:

- thermal load of each riser: $\dot{Q}_p = 20\,\text{kW}$,
- heating water parameters: $t_z/t_p = 80/60\,°\text{C}$,
- required available pressure difference on every riser: $\Delta p_P = 10\,\text{kPa}$,
- pressure drops of the sections distributing the medium into individual riser (jointly for the supply and the return pipe), for initial conditions: $\Delta p_{K-P1} = 4.8\,\text{kPa}$; $\Delta p_{P1-P2} = 3\,\text{kPa}$; $\Delta p_{P2-P3} = 1.8\,\text{kPa}$,
- the circulation pump throttling characteristic is described by the following relation:

(a) $\Delta p_x = \Delta p_0 - a \cdot (\dot{m}_x - \dot{m}_0)$, $a = 2\,\text{kPa}/(\text{kg/s})$,
(b) $\Delta p_x = \text{const} = \Delta p_0$,

- Δp_0, $\dot{m}_0$—parameters of the installation initial working point,
- the installation diagram is presented in Fig. 6.15.

Solution
In the diagram the installation risers P1 and P2 are equipped with valves. There is no such element under riser P3, which is the effect of the general practice of the installation hydraulic balancing and of the requirement of selecting the pump active pressure for what is referred to as the most unfavourable, critical circuit. This circuit requires the highest value of the pump pressure. If therefore each of the circuits requires for example an identical value of the medium volume flow, resulting from the same thermal power value, the most unfavourable of them is the one located the farthest away because flowing through the longest pipe sections, water loses the most pressure. However, the pump pressure values determined under this condition are too high for the receivers located closer, which in this case are risers P1 and P2. Valves Z1 and Z2 are to throttle this surplus of pressure.

Before the task is solved, it should be considered whether after riser P3 is excluded, a change in the medium volume flow will actually occur and, if so, what causes the change. First, consider case "a".

(a) Calculations performed for the pump throttling characteristic described by a linear decreasing function.

Cutting off the flow through one riser means a decrease in the total flow rate of the medium pumped by the pump. According to the pump throttling characteristic, the pump pressure will then increase due to the occurrence of internal hydraulic resistance. This will naturally cause a change, a rise to be exact, in the medium volume/mass flow through the installation other risers. The increment can be determined by first defining the initial values of the medium mass flow and the pump initial working point, and then—the new working point of the pump-pipework system and the mass flows in individual risers. For this purpose, it is necessary to determine individual hydraulic resistances of the system and the system equivalent resistance.

The medium total initial mass flow in the installation can be found from the following relation:

$$\dot{Q}_{\mathrm{c}} = \dot{m}_0 \cdot c_{\mathrm{w}} \cdot \left(t_{\mathrm{z}} - t_{\mathrm{p}}\right) \Rightarrow \dot{m}_0 = \frac{\dot{Q}_{\mathrm{c}}}{c_{\mathrm{w}} \cdot \left(t_{\mathrm{z}} - t_{\mathrm{p}}\right)}. \tag{6.11}$$

Substituting relevant data, the following is obtained:

$$\dot{m}_0 = \frac{\dot{Q}_{\mathrm{c}}}{c_{\mathrm{w}} \cdot \left(t_{\mathrm{z}} - t_{\mathrm{p}}\right)} = \frac{3 \cdot 20,000}{4186 \cdot 20} = 0.717\,\mathrm{kg/s}.$$

The medium mass flow in individual risers totals:

$$\dot{m}_{\mathrm{P1}} = \dot{m}_{\mathrm{P2}} = \dot{m}_{\mathrm{P3}} = \frac{\dot{m}_0}{3} = 0.239\,\mathrm{kg/s}.$$

The required pressure of the pump for the initial conditions (for the reasons presented above, it must at least equal the pressure losses arising in the most unfavourable circuit) is:

$$\Delta p_0 = \Delta p_{\mathrm{P3}} + \Delta p_{\mathrm{K-P1}} + \Delta p_{\mathrm{P1-P2}} + \Delta p_{\mathrm{P2-P3}} = 10 + 4.8 + 3 + 1.8 = 19.6\,\mathrm{kPa}.$$

Substituting relevant data, the pump throttling characteristic is therefore described by the following equation:

$$\Delta p_{\mathrm{x}} = 19.6 - 2 \cdot (\dot{m}_{\mathrm{x}} - 0.717).$$

In order to maintain the required pressure value of 10 kPa in risers P1 and P2, valves Z1 and Z2 must respectively diminish the pressure by:

$$\Delta p_{\mathrm{Z1}} = \Delta p_0 - \Delta p_{\mathrm{K-P1}} - \Delta p_{\mathrm{P1}} = 19.6 - 4.8 - 10 = 4.8\,\mathrm{kPa},$$

$$\Delta p_{\mathrm{Z2}} = \Delta p_0 - \Delta p_{\mathrm{K-P1}} - \Delta p_{\mathrm{P1-P2}} - \Delta p_{\mathrm{P2}} = 19.6 - 4.8 - 3 - 10 = 1.8\,\mathrm{kPa}.$$

The hydraulic resistances of the risers, the distributing sections and the valves located under the risers total:

$$r_P = \frac{\Delta p_{\mathrm{P}}}{\dot{m}_{\mathrm{P}}^2} = \frac{10}{0.239^2} = 175\,\mathrm{kPa/(kg/s)^2},$$

$$r_{\mathrm{P2-P3}} = \frac{\Delta p_{\mathrm{P2-P3}}}{\dot{m}_{\mathrm{P2-P3}}^2} = \frac{1.8}{(0.717/3)^2} = 31.5\,\mathrm{kPa/(kg/s)^2},$$

$$r_{\mathrm{P1-P2}} = \frac{\Delta p_{\mathrm{P1-P2}}}{\dot{m}_{\mathrm{P1-P2}}^2} = \frac{3}{(2 \cdot 0.717/3)^2} = 13.13\,\mathrm{kPa/(kg/s)^2},$$

$$r_{\mathrm{K-P1}} = \frac{\Delta p_{\mathrm{K-P1}}}{\dot{m}_{\mathrm{K-P1}}^2} = \frac{4.8}{(0.717)^2} = 9.34\,\mathrm{kPa/(kg/s)^2},$$

$$r_{\mathrm{Z,1}} = \frac{\Delta p_{\mathrm{Z1}}}{\dot{m}_{\mathrm{P1}}^2} = \frac{4.8}{(0.717/3)^2} = 84\,\mathrm{kPa/(kg/s)^2},$$

$$r_{\mathrm{Z,2}} = \frac{\Delta p_{\mathrm{Z2}}}{\dot{m}_{P2}^2} = \frac{1.8}{(0.717/3)^2} = 31.5\,\mathrm{kPa/(kg/s)^2}.$$

If all the risers are active, the installation hydraulic resistance for the initial conditions is:

$$r_0 = \frac{\Delta p_0}{\dot{m}_0^2} = \frac{19,6}{(0.717)^2} = 38.1\,\mathrm{kPa/(kg/s)^2}.$$

The installation equivalent resistance after riser P3 is cut off cannot be calculated in the same manner as for the entire installation for the initial conditions because the pressure value and the flow rate are sought quantities and not ones which are set. The equivalent resistance value should be calculated "curling up" the system to a single element with an equivalent hydraulic resistance. In order to facilitate the analysis of the connections between individual elements, the system diagram can be presented in an equivalent form, such as the one shown in Fig. 6.16.

Analysing the diagram, it can be seen that riser P2 is connected with valve Z1 and section P1–P2 in series. The resultant resistance of this part is connected in parallel with in-series connected valve Z1 and riser P1. The resultant resistance of this part is connected in series with section K-P1. The installation equivalent hydraulic resistance r' after riser P3 is cut off can thus be expressed using the following relation:

$$r' = r_{\mathrm{K-P1}} + \frac{1}{\left(\frac{1}{\sqrt{r_{\mathrm{P2}}+r_{\mathrm{Z2}}+r_{\mathrm{P1-P2}}}} + \frac{1}{\sqrt{r_{\mathrm{P1}}+r_{\mathrm{Z1}}}}\right)^2}.$$

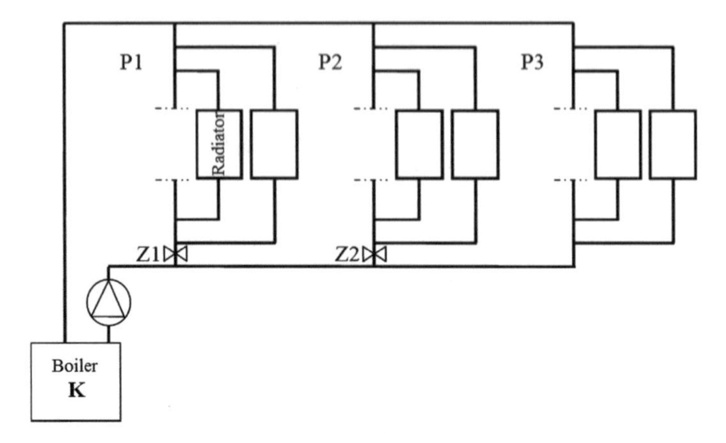

Fig. 6.16 Equivalent diagram of the installation under analysis

Substituting the data, the result is:

$$r' = 9.34 + \frac{1}{\left(\frac{1}{\sqrt{175+31.5+13.13}} + \frac{1}{\sqrt{175+84}}\right)^2} = 68.9\,\text{kPa}/(\text{kg/s})^2.$$

The circuit characteristic after riser P3 is cut off can be expressed as:

$$\Delta p_{\text{ob}} = 68.9 \cdot \dot{m}_{\text{ob}}^2.$$

The system new working point, as the point of the pipework and the pump characteristic intersection, should be calculated comparing the mathematical relations of the two functions. The following equation is then obtained:

$$\Delta p_{\text{x}} = \Delta p_{\text{ob}} \Rightarrow 19.6 - 2 \cdot (\dot{m}_{\text{x}} - 0.717) = 68.9 \cdot \dot{m}_{\text{x}}^2.$$

Rearranging the terms, the following is obtained:

$$68.9 \cdot \dot{m}_{\text{x}}^2 + 2 \cdot \dot{m}_{\text{x}} - 21.03 = 0.$$

It is a quadratic equation whose positive solution is:

$$\dot{m}_{\text{x}} = 0.538\,\text{kg/s}.$$

The pump pressure for this value of the medium mass flow is:

$$\Delta p_{\text{x}} = 68.9 \cdot 0.538^2 = 19.95\,\text{kPa}.$$

Knowing the pressure value at the pipework beginning, calculations can be started to find pressure values in other locations and the medium flow rate values resulting therefrom. The medium flow rate in riser P1 results from the pressure value at the beginning of the riser. This is the pressure value at the pipework beginning reduced by a new pressure drop occurring on section K-P1 (different from the initial one because the value of the medium mass flow in this section is different), according to the following relation:

$$\Delta p_{\text{P1}} = \Delta p_{\text{x}} - \Delta p'_{\text{K-P1}} = \Delta p_{\text{x}} - r_{\text{K-P1}} \cdot \dot{m}_x^2.$$

Substituting relevant data, the following is obtained:

$$\Delta p_{\text{P1}} = 19.95 - 9.34 \cdot 0.538^2 = 17.25\,\text{kPa}.$$

According to formula (3.46), the relation defining the medium mass flow through riser P1 takes the following form:

$$\dot{m}'_{\text{P1}} = \sqrt{\frac{\Delta p_{\text{P1}}}{r_{\text{P1}} + r_{\text{Z1}}}}.$$

Substituting relevant data, the following is obtained:

$$\dot{m}'_{\text{P1}} = \sqrt{\frac{17.25}{175 + 84}} = 0.258\,\text{kg/s}.$$

Compared to the initial value, the percentage rise in the medium mass flow through riser P1 totals:

$$\Delta \dot{m}_{\text{P1}} = \frac{\dot{m}'_{\text{P1}} - \dot{m}_{\text{P1}}}{\dot{m}_{\text{P1}}} \cdot 100\% = \frac{0.258 - 0.239}{0.239} \cdot 100\% = 7.95\,\%.$$

For riser P2 the same pressure value can be used as for riser P1. It can be noticed that the inflow to riser P2 occurs also from the point of connection of riser P1, but through section P1-P2, which is incorporated in series. The following equation is thus obtained:

$$\dot{m}'_{\text{P2}} = \sqrt{\frac{\Delta p_{\text{P1}}}{r_{\text{P1-P2}} + r_{\text{P2}} + r_{\text{Z2}}}} = \sqrt{\frac{17.25}{13.13 + 175 + 31.5}} = 0.28\,\text{kg/s}.$$

However, the value of the mass flow can be calculated in a simpler form. Since there are two risers operating in the system and the value of the total flow rate of the medium and of the medium mass flow through the first riser are already known, the mass flow through the second riser will be the difference between the two, according to the following formula:

$$\dot{m}'_{\mathrm{P2}} = \dot{m}_{\mathrm{x}} - \dot{m}'_{\mathrm{P1}} = 0.538 - 0.258 = 0.28\,\mathrm{kg/s}.$$

Compared to the initial value, the percentage rise in the flow rate through riser P1 totals:

$$\Delta\dot{m}_{\mathrm{P2}} = \frac{\dot{m}'_{\mathrm{P2}} - \dot{m}_{\mathrm{P2}}}{\dot{m}_{\mathrm{P2}}} \cdot 100\% = \frac{0.28 - 0.239}{0.239} \cdot 100\% = 17.1\%.$$

(b) Calculations performed for the pump throttling characteristic described by a constant function.

Here, the calculations start from a later step compared to the previous case because the pressure value for the system new operating conditions, after riser P3 is cut off, is known and, as stated in the contents of the problem, equal to initial pressure. The new working point of the system can therefore be determined using the following relation:

$$\Delta p_{\mathrm{x}} = \Delta p_{\mathrm{ob}} \Rightarrow 19.6 = 68.9 \cdot \dot{m}_{\mathrm{x}}^2.$$

The positive solution of the equation is:

$$\dot{m}_{\mathrm{x}} = 0.533\,\mathrm{kg/s}.$$

Naturally, this value is lower than before. Nonetheless, it is still not the value that would result from reducing the total mass flow of the medium to the three risers by the mass flow to the cut-off one, i.e. 0.717 − 0.239 = 0.478 kg/s, and the value is higher.

Next, the calculations are performed in the same way as previously, and the following is obtained one by one:

$$\Delta p_{\mathrm{P1}} = \Delta p_{\mathrm{x}} - \Delta p'_{\mathrm{K-P1}} = \Delta p_{\mathrm{x}} - r_{\mathrm{K-P1}} \cdot \dot{m}_x^2 = 19.6 - 9.34 \cdot 0.533^2 = 16.95\,\mathrm{kPa},$$

$$\dot{m}'_{\mathrm{P1}} = \sqrt{\frac{\Delta p_{\mathrm{P1}}}{r_{\mathrm{P1}} + r_{\mathrm{Z1}}}} = \sqrt{\frac{16.95}{175 + 84}} = 0.256\,\mathrm{kg/s},$$

$$\Delta\dot{m}_{\mathrm{P1}} = \frac{\dot{m}'_{\mathrm{P1}} - \dot{m}_{\mathrm{P1}}}{\dot{m}_{\mathrm{P1}}} \cdot 100\% = \frac{0.256 - 0.239}{0.239} \cdot 100\% = 7.1\%,$$

$$\dot{m}'_{\mathrm{P2}} = \dot{m}_{\mathrm{x}} - \dot{m}'_{\mathrm{P1}} = 0.533 - 0.256 = 0.277\,\mathrm{kg/s},$$

$$\Delta\dot{m}_{\mathrm{P2}} = \frac{\dot{m}'_{\mathrm{P2}} - \dot{m}_{\mathrm{P2}}}{\dot{m}_{\mathrm{P2}}} \cdot 100\% = \frac{0.277 - 0.239}{0.239} \cdot 100\% = 15.9\%.$$

Analysing the calculations for {a}, it can be noticed that after one of the risers is cut off, which involves a drop in the total mass flow of the medium, the pressure generated by the pump increases. This is due to the occurrence of the pump non-zero internal hydraulic resistance on which the pressure drop arises. The bigger the mass flow, the bigger the pressure drop, which is subtracted from the pump maximum pressure at a zero flow and which results in an increasingly lower output pressure. This means that the smaller the medium mass flow, the higher the pressure. In turn, a rise in pressure causes an increase in the flow rate in the other active risers.

However, analysing item {b}, it can be seen that a similar phenomenon occurs here, despite the fact that the pump pressure is constant, i.e. it has a zero value of internal hydraulic resistance and constitutes an ideal source of pressure. However, there is no mistake here. The medium flowing from the pump to the first and subsequent risers suffers a pressure loss on section K-P1, as the section is characterized by a certain non-zero hydraulic resistance. Consequently, a pressure drop occurs on it and the phenomenon is the same as for a real pump with a non-zero value of internal hydraulic resistance on which a pressure drop was assumed earlier. As a matter of fact, for case {a} a pressure drop occurs both on the pump internal resistance and on section K-P1, and the two resistances sum up with each other. In the other case, it is only the resistance of section K-P1 that matters. For this reason, for this case the pressure drop from the pump to the first riser is smaller and so are the increments in the medium mass flow values in the risers. Consequently, the bigger the hydraulic resistance of the section from the pump to the first riser, the bigger the increments in the medium mass flow values if more risers are cut off.

The example can be related to a typical phenomenon occurring in installations with radiator thermoregulators where, as some of the thermoregulators close, the pressure difference and, thereby, the medium mass flow increase on the others. In this case, the analysis concerns whole risers but it can be directly transferred to apply to single radiators.

Figure 6.17 presents characteristics of the pump and of the pipework under consideration, together with marked working points.

Example 6.13

For the installation from the previous case, for variant {a}, calculate the relative percentage change in the medium flow rate in risers P2 and P3 compared to nominal flow values after the flow is cut off in riser P1. Compare the results to the results obtained in the previous example.

Solution

Like in the previous example, the new pump pressure value should be calculated first. Then, the pressure drops and the medium volume/mass flow values have to be calculated in the same manner. In order to establish the new pump pressure value, the mathematical form of the new hydraulic characteristic of the pipework has to be determined. In order to do so, the pipework hydraulic resistance has to be found. "Curling up" the diagram, with riser P1 cut off, the following relation is obtained:

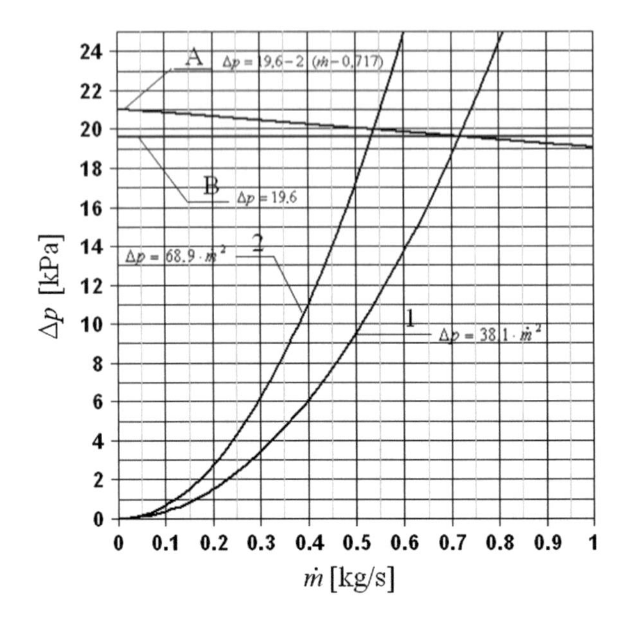

Fig. 6.17 Pump throttling characteristics and characteristics of the pipework: (1) pipework original characteristic; (2) pipework final characteristic; (A) pump throttling characteristic assuming a linear decreasing function of the dependence of pressure on the flow; (B) pump throttling characteristic assuming a constant function

$$r' = r_{\text{K}-\text{P1}} + r_{\text{P1}-\text{P2}} + \frac{1}{\left(\frac{1}{\sqrt{r_{\text{P2}}+r_{\text{Z2}}}} + \frac{1}{\sqrt{r_{\text{P2 - P3}}+r_{\text{P3}}}}\right)^2}.$$

Substituting the data, the result is:

$$r' = 9.34 + 13.13 + \frac{1}{\left(\frac{1}{\sqrt{175+31.5}} + \frac{1}{\sqrt{31.5+175}}\right)^2} = 74.1\,\text{kPa}/(\text{kg/s})^2.$$

Carrying out the same calculations as in the previous example, the obtained result is:

$$74.1 \cdot \dot{m}_{\text{x}}^2 + 2 \cdot \dot{m}_{\text{x}} - 21.03 = 0.$$

It is a quadratic equation whose positive solution is:

$$\dot{m}_{\text{x}} = 0.519\,\text{kg/s}.$$

The pump pressure for this value of the medium mass flow is:

$$\Delta p_{\text{x}} = 74.1 \cdot 0.519^2 = 19.99\,\text{kPa}.$$

The pressure in riser P2 totals:

$$\Delta p_{\text{P2}} = \Delta p_{\text{x}} - \Delta p'_{\text{K}-\text{P1}} - \Delta p'_{\text{P1 - P2}} = \Delta p_{\text{x}} - (r_{\text{K}-\text{P1}} + r_{\text{P1}-\text{P2}}) \cdot \dot{m}_x^2.$$

Substituting relevant data, the following is obtained:

$$\Delta p_{P2} = 19.99 - (9.34 + 13.13) \cdot 0.519^2 = 13.94\,\text{kPa}.$$

The medium mass flow in riser P2 totals:

$$\dot{m}'_{P2} = \sqrt{\frac{\Delta p_{P2}}{r_{P2} + r_{Z2}}} = \sqrt{\frac{13.94}{175 + 31.5}} = 0.2595\,\text{kg/s}.$$

Compared to the initial value, the percentage rise in the flow rate through riser P2 totals:

$$\Delta \dot{m}_{P2} = \frac{\dot{m}'_{P2} - \dot{m}_{P2}}{\dot{m}_{P2}} \cdot 100\% = \frac{0.2595 - 0.239}{0.239} \cdot 100\% = 8.58\%.$$

The medium mass flow in riser P3 totals:

$$\dot{m}'_{P3} = \dot{m}_x - \dot{m}'_{P3} = 0.519 - 0.2595 = 0.2595\,\text{kg/s}.$$

Compared to the initial value, the percentage rise in the flow rate through riser P3 totals:

$$\Delta \dot{m}_{P3} = \frac{\dot{m}'_{P3} - \dot{m}_{P3}}{\dot{m}_{P3}} \cdot 100\% = \frac{0.2595 - 0.239}{0.239} \cdot 100\% = 8.58\%.$$

Analysing the calculation results, it can be seen that in this case the increments in the medium mass flow are identical in both risers, whereas in the previous example they were different. This results directly from the location of the cut off receiver in relation to the other risers. If a given receiver (e.g. a riser) is cut off, the increments in the mass flow values in the receivers located downstream (looking from the pressure source onwards) and characterized by the same pressure loss in initial conditions are equal to each other. If the receivers are located upstream, the values are different and become bigger as the distance between the analysed receiver gets closer to the receiver where the flow is stopped. This may be an indication if the location of the relief valve reducing the system pressure is to be selected. Due to the phenomenon described above, if the differences in the medium mass flow values in individual receivers are to be reduced, it is better to locate such a valve at the end of the system, in parallel to the last receiver, and not at the beginning. The effect will be an increased mass flow in the sections distributing the medium to all the receivers. Consequently, the farther the receiver, the bigger the additional pressure loss. The relief valve location that increases the pressure loss for the receivers located farther away makes it possible to compensate for the pressure increment and, at the same time, for the rise in the medium mass flow in the receivers, compared to the cut off receiver. Thereby, it also makes it possible to reduce the differences between the values of the quantities in individual receivers.

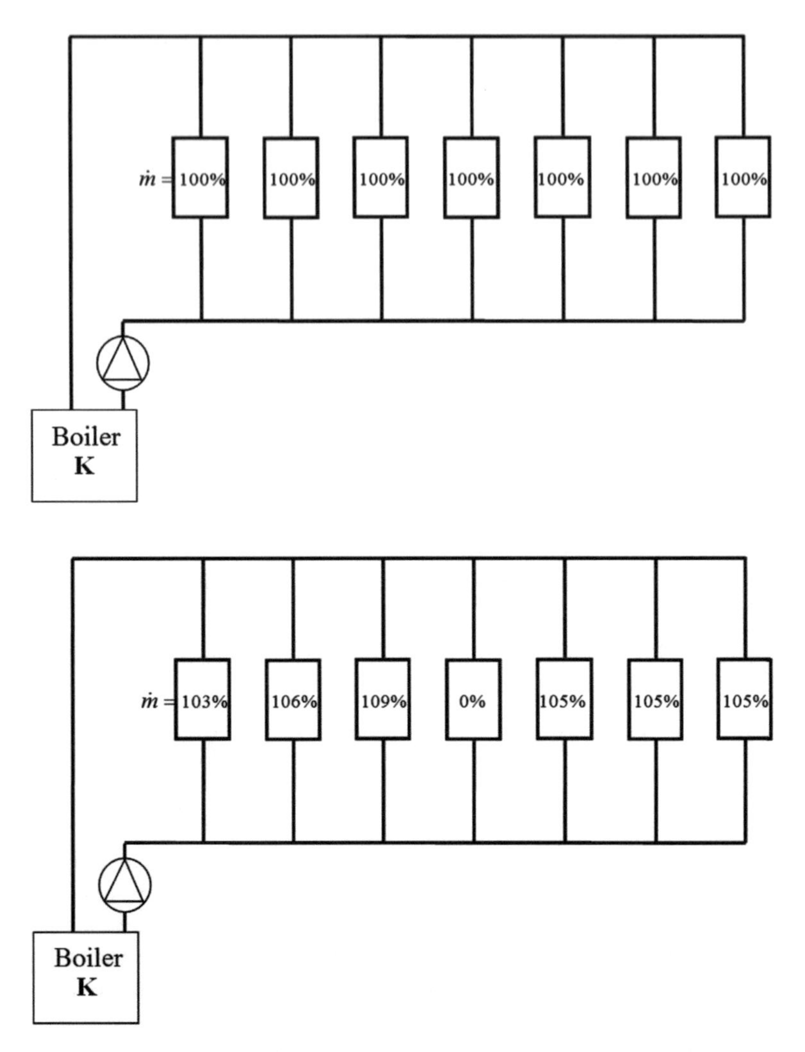

Fig. 6.18 Example illustration of changes in the medium mass flow in individual receivers for a system connected in parallel when a selected, central, receiver is cut off

Figure 6.18 presents example changes in the medium mass flow values in individual receivers for a system connected in parallel when a selected, central, receiver is cut off.

Example 6.14
For the installation from the previous example, with all risers operating, check how the medium mass flow values vary in individual risers at a change in the inlet pressure to the following values:

(a) $p_0 = 20.6$ kPa,
(b) $p_0 = 18.6$ kPa.

Solution

(a) For pump pressure $p_0 = 20.6$ kPa.

If all the risers are active, the pipework hydraulic resistance was already calculated in the penultimate example. It totals $r_0 = 38.1\,[\text{kPa}/(\text{kg/s})^2]$. One coordinate of the system working point—inlet pressure—is set. The other can be found using the following relation:

$$\Delta p_0 = \Delta p_{\text{ob}} \Rightarrow 20.6 = 38.1 \cdot \dot{m}_0^2.$$

Therefore:

$$\dot{m}_0 = 0.735\,[\text{kg/s}].$$

Next, the calculations are performed in the same way as before, but it is taken into account that all the risers are active.

$$\Delta p_{\text{P1}} = \Delta p_0 - \Delta p'_{\text{K-P1}} = \Delta p_0 - r_{\text{K-P1}} \cdot \dot{m}_0^2 = 20.6 - 9.34 \cdot 0.735^2 = 15.55\,\text{kPa},$$

$$\dot{m}'_{\text{P1}} = \sqrt{\frac{\Delta p_{\text{P1}}}{r_{\text{P1}} + r_{\text{Z1}}}} = \sqrt{\frac{15.55}{175 + 84}} = 0.245\,\text{kg/s},$$

$$\Delta \dot{m}_{\text{P1}} = \frac{\dot{m}'_{\text{P1}} - \dot{m}_{\text{P1}}}{\dot{m}_{\text{P1}}} \cdot 100\% = \frac{0.245 - 0.239}{0.239} \cdot 100\% = 2.536\%.$$

The pressure value at the beginning of the second riser is equal to the pressure at the beginning of the first, reduced by the new pressure drop arising on section P1-P2 in between the risers. The pressure drop on this section is caused by the medium flow in it. It is the total mass flow value reduced by the mass flow through riser P1. It may therefore be written that:

$$\Delta p_{\text{P2}} = \Delta p_{\text{P1}} - \Delta p'_{\text{P1-P2}} = \Delta p_{\text{P1}} - r_{\text{P1-P2}} \cdot \dot{m}^2_{\text{P1-P2}} = \Delta p_{\text{P1}} - r_{\text{P1-P2}} \cdot (\dot{m}_0 - \dot{m}_{P1})^2.$$

Substituting the data, the result is:

$$\Delta p_{\text{P2}} = 15.55 - 13.13 \cdot (0.735 - 0.245)^2 = 12.4\,\text{kPa}.$$

Consequently, the medium mass flow in riser P2 and its change compared to the initial value totals:

$$\dot{m}'_{\text{P2}} = \sqrt{\frac{\Delta p_{\text{P2}}}{r_{\text{P2}} + r_{\text{Z2}}}} = \sqrt{\frac{12.4}{175 + 31.5}} = 0.245\,\text{kg/s},$$

$$\Delta \dot{m}_{\text{P2}} = \frac{\dot{m}'_{\text{P2}} - \dot{m}_{\text{P2}}}{\dot{m}_{\text{P2}}} \cdot 100\% = \frac{0.245 - 0.239}{0.239} \cdot 100\% = 2.536\%.$$

For riser P3, the following values are obtained one by one:

$$\begin{aligned}\Delta p_{\text{P3}} &= \Delta p_{\text{P2}} - \Delta p'_{\text{P2-P3}} \\ &= \Delta p_{\text{P2}} - r_{\text{P2-P3}} \cdot \dot{m}^2_{\text{P2-P3}} \\ &= \Delta p_{\text{P2}} - r_{\text{P2-P3}} \cdot (\dot{m}_0 - \dot{m}_{P1} - \dot{m}_{P2})^2,\end{aligned}$$

$$\Delta p_{\text{P3}} = 12.4 - 31.5 \cdot (0.735 - 0.245 - 0.245)^2 = 10.51\,\text{kPa},$$

$$\dot{m}'_{\text{P3}} = \sqrt{\frac{\Delta p_{\text{P3}}}{r_{\text{P3}}}} = \sqrt{\frac{10.51}{175}} = 0.245\,\text{kg/s},$$

$$\Delta \dot{m}_{\text{P3}} = \frac{\dot{m}'_{\text{P3}} - \dot{m}_{\text{P3}}}{\dot{m}_{\text{P3}}} \cdot 100\% = \frac{0.245 - 0.239}{0.239} \cdot 100\% = 2.536\%.$$

(a) For pump pressure $p_0 = 18.6$ kPa.

Carrying out the same calculations, the result is:

$$\Delta p_0 = \Delta p_{\text{ob}} \Rightarrow 18.6 = 38.1 \cdot \dot{m}^2_{\text{x}}.$$

Therefore:

$$\dot{m}_0 = 0.699\,\text{kg/s}.$$

$$\Delta p_{\text{P1}} = \Delta p_0 - \Delta p'_{\text{K-P1}} = \Delta p_0 - r_{\text{K-P1}} \cdot \dot{m}^2_0 = 18.6 - 9.34 \cdot 0.699^2 = 14.04\,\text{kPa},$$

$$\dot{m}'_{\text{P1}} = \sqrt{\frac{\Delta p_{\text{P1}}}{r_{\text{P1}} + r_{\text{Z1}}}} = \sqrt{\frac{14.04}{175 + 84}} = 0.233\,\text{kg/s},$$

$$\Delta \dot{m}_{\text{P1}} = \frac{\dot{m}'_{\text{P1}} - \dot{m}_{\text{P1}}}{\dot{m}_{\text{P1}}} \cdot 100\% = \frac{0.233 - 0.239}{0.239} \cdot 100\% = -2.51\%.$$

$$\Delta p_{\text{P2}} = \Delta p_{\text{P1}} - \Delta p'_{\text{P1-P2}} = \Delta p_{\text{P1}} - r_{\text{P1-P2}} \cdot \dot{m}^2_{\text{P1-P2}} = \Delta p_{\text{P1}} - r_{\text{P1-P2}} \cdot (\dot{m}_0 - \dot{m}_{P1})^2.$$

Substituting the data, the following is obtained:

$$\Delta p_{\text{P2}} = 14.04 - 13.13 \cdot (0.699 - 0.233)^2 = 11.9\,\text{kPa},$$

$$\dot{m}'_{P2} = \sqrt{\frac{\Delta p_{P2}}{r_{P2} + r_{Z2}}} = \sqrt{\frac{11.9}{175 + 31.5}} = 0.233\,\text{kg/s},$$

$$\Delta \dot{m}_{P2} = \frac{\dot{m}'_{P2} - \dot{m}_{P2}}{\dot{m}_{P2}} \cdot 100\% = \frac{0.233 - 0.239}{0.239} \cdot 100\% = -2.51\%,$$

$$\dot{m}'_{P3} = \dot{m}_0 - \dot{m}'_{P1} - \dot{m}'_{P2} = 0.699 - 0.233 - 0.233 = 0.233\,\text{kg/s},$$

$$\Delta \dot{m}_{P3} = \frac{\dot{m}'_{P3} - \dot{m}_{P3}}{\dot{m}_{P3}} \cdot 100\% = \frac{0.233 - 0.239}{0.239} \cdot 100\% = -2.51\%.$$

The solution can also be obtained in a much simpler way if it is realized that the changes in the medium mass flow values in individual receivers at a change in the inlet pressure are the same. It is then enough to use relation (3.46), indicating that changes in the medium flow rate are proportional to the square root of changes in the pressure difference. For two different pressure values it may then be written that:

$$\dot{m}_1 = \sqrt{\frac{\Delta p_1}{r}},$$

$$\dot{m}_2 = \sqrt{\frac{\Delta p_2}{r}}.$$

Dividing the two equations by each other, the following notation is obtained:

$$\frac{\dot{m}_2}{\dot{m}_1} = \sqrt{\frac{\Delta p_2}{\Delta p_1}}. \tag{6.12}$$

If point 1 is assumed as initial conditions and the medium initial mass flow, and point 2 as the conditions after the change in pressure, the sought mass flow can be determined by transforming the above relation to the following form:

$$\dot{m}_2 = \dot{m}_1 \cdot \sqrt{\frac{\Delta p_2}{\Delta p_1}}. \tag{6.13}$$

Substituting the data, the result is:

(a) For pump pressure $p_0 = 20.6$ kPa

$$\dot{m}'_{P1} = \dot{m}'_{P2} = \dot{m}'_{P3} = \dot{m}_{P1} \cdot \sqrt{\frac{\Delta p_2}{\Delta p_1}} = 0.239 \cdot \sqrt{\frac{20.6}{19.6}} = 0.245\,\text{kg/s},$$

$$\Delta \dot{m}_{\mathrm{P1}} = \Delta \dot{m}_{\mathrm{P2}} = \Delta \dot{m}_{\mathrm{P3}} = \frac{\dot{m}'_{\mathrm{P3}} - \dot{m}_{\mathrm{P3}}}{\dot{m}_{\mathrm{P3}}} \cdot 100\% = \frac{0.245 - 0.239}{0.239} \cdot 100\% = 2.536\%.$$

(a) For pump pressure $p_0 = 18.6$ kPa

$$\dot{m}'_{P1} = \dot{m}'_{P2} = \dot{m}'_{P3} = \dot{m}_{P1} \cdot \sqrt{\frac{\Delta p_2}{\Delta p_1}} = 0.239 \cdot \sqrt{\frac{18.6}{19.6}} = 0.233\,\mathrm{kg/s},$$

$$\Delta \dot{m}_{\mathrm{P1}} = \Delta \dot{m}_{\mathrm{P2}} = \Delta \dot{m}_{\mathrm{P3}} = \frac{\dot{m}'_{\mathrm{P3}} - \dot{m}_{\mathrm{P3}}}{\dot{m}_{\mathrm{P3}}} \cdot 100\% = \frac{0.233 - 0.239}{0.239} \cdot 100\% = -2.51\%.$$

It can be seen that at a change in the system inlet pressure, the changes in pressure on individual receives, and, consequently, in the medium mass flow in them, are identical. It can also be seen that external changes, i.e. those arising due to global changes in inlet pressure, have a smaller impact on the system operation than internal changes, i.e. those arising due to cut off receivers.

Example 6.15
Variant 1

Find the required values of the *Danfoss MSV-I DN20* valve pre-setting and total authority using the following data:

- active (differential) pressure in a given part of the circuit: $\Delta p_{\mathrm{cz}} = 0.8$ bar,
- pipework pressure loss for the required volume flow of the medium: $\Delta p_{\mathrm{str,x}} = 0.55$ bar,
- required volume flow of the medium:

(a) $\dot{V}_x = 0.15\,\mathrm{m^3/h}$,
(b) $\dot{V}_x = 0.65\,\mathrm{m^3/h}$,
(c) $\dot{V}_x = 1\,\mathrm{m^3/h}$,
(d) $\dot{V}_x = 0.5\,\mathrm{m^3/h}$.

The *Danfoss MSV-I DN20* valve throttling/closing characteristic data (selected values from Table 5.1) are listed in Table 6.1.

Table 6.1 Throttling/closing characteristic data of the *Danfoss MSV-I DN20* valve [1]

Number of pre-setting n_i	The valve flow factor k_{vx} [m^3/(h bar$^{0.5}$)]	Number of pre-setting n_i	The valve flow factor k_{vx} [m^3/(h bar$^{0.5}$)]
0.2	0.3	2.0	2.0
0.5	0.7	2.5	2.3
1.0	1.3	3.0	2.5
1.5	1.7	3.2	2.5

Solution

Calculations performed by means of the common method

In order to carry out the calculations in this variant, it is necessary to find the pressure that will have to be throttled on the valve for the sought x-th opening degree of the valve plug corresponding to the i-th pre-setting. In the case of a manual valve, this always corresponds to the maximum relative range of the closing element travel. The sought pressure value can be determined using relation (4.108) transformed to the following expression:

$$\Delta p_{z,x} = \Delta p_{cz} - \Delta p_{str,x}. \tag{6.14}$$

Substituting relevant data, the following is obtained:

$$\Delta p_{z,x} = \Delta p_{cz} - \Delta p_{str,x} = 0.25 \text{ bar}.$$

Knowing this pressure, the required value of flow factor k_v can be calculated:

(a)	$k_{v,x} = \frac{\dot{V}_x}{\sqrt{\Delta p_{z,x}}} = \frac{0.15}{\sqrt{0.25}} = 0.3$
(b)	$k_{v,x} = \frac{\dot{V}_x}{\sqrt{\Delta p_{z,x}}} = \frac{0.65}{\sqrt{0.25}} = 1.3$
(c)	$k_{v,x} = \frac{\dot{V}_x}{\sqrt{\Delta p_{z,x}}} = \frac{1}{\sqrt{0.25}} = 2.0$
(d)	$k_{v,x} = \frac{\dot{V}_x}{\sqrt{\Delta p_{z,x}}} = \frac{0.5}{\sqrt{0.25}} = 1.0$

According to the valve measuring data, the required flow factor is ensured if the following pre-settings are selected, respectively:

(a)	$n_i = 0.2$
(b)	$n_i = 1.0$
(c)	$n_i = 2.0$
(d)	$n_i = (0.5; 1.0)$

The valve authority a' will take the same value for every case because the pressure loss on the valve is identical. According to formula (4.1), the following is obtained:

$$a' = \frac{\Delta p_{z,x}}{\Delta p_{ob}} = \frac{\Delta p_{z,x}}{\Delta p_{z,x} + \Delta p_{str,x}} = \frac{0.25}{0.25 + 0.55} = 0.3125.$$

The parameter value calculated in this manner is sufficiently high and satisfies the set condition of $a' > 0.3$. Several criteria for the elements selection are used to balance heating installations hydraulically, but obtaining the value of $a' > 0.3$ is a basic one. Therefore, the input data in the example are established assuming a distribution of pressures that meet the condition. In this case, since a constant pressure distribution is adopted but the medium volume/mass flows are different (e.g. different thermal power values in circuits), different pipe diameters are selected, which results in different hydraulic resistances of the elements. And this is what happens in practice.

Calculations performed according to the proposed methods

The required pre-setting value can be calculated using either of the two proposed computational algorithms.

Method #1:

Using relation (4.125c), transformed to express the sought value of $k_{v,x}$, and comparing the result with the values for the defined values of pre-setting n_i, the correct value of the pre-setting can be found. However, this method is not direct and in fact it is identical with the previous one from the moment that $k_{v,x}$ is calculated onwards. But if Method #1 and the notation using function $k_v(n_i)$ are used, the sought value of pre-setting n_i can be determined directly. Based on the data from the valve throttling/closing characteristic measurements presented in Table 6.1, function $k_v(n_i)$ can be written as:

$$k_v(n_i) = 0.007938n_i^6 - 0.13668n_i^5 + 0.7473n_i^4 - 1.729n_i^3 + 1.506n_i^2 + 0.8157n_i + 0.0893.$$

The relation was developed for pre-setting values from the range of $n_i = 0.2$–3.0 because the pre-setting value of $n_i = 3.2$ is characterized by the same value of k_v as a lower pre-setting. Naturally, solutions should be looked for in the set of real numbers and in the interval of pre-settings which are available for a given valve and which are used to create function $k_v(n_i)$. According to formula (4.127b), the required pre-setting value calculated using Method #1 totals:

$$\dot{V}_x - \sqrt{\frac{\Delta p_{cz}}{\frac{1}{(k_v(n_i))^2} + \frac{\Delta p_{str,x}}{\dot{V}_x^2}}} = 0.$$

(a)	$0.15 - \sqrt{\frac{0.8}{\frac{1}{\left(0.007938n_i^6 - 0.13668n_i^5 + 0.7473n_i^4 - 1.729n_i^3 + 1.506n_i^2 + 0.8157n_i + 0.0893\right)^2} + \frac{0.55}{0.15^2}}} = 0$ $\underline{n_i = 0.2}$

(continued)

(continued)

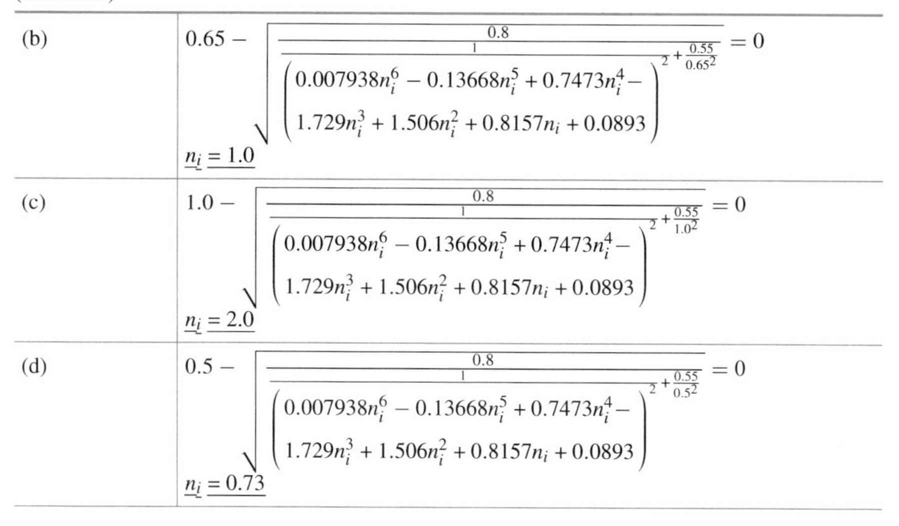

(b)	$0.65 - \sqrt{\frac{0.8}{\frac{1}{\left(0.007938n_i^6 - 0.13668n_i^5 + 0.7473n_i^4 - 1.729n_i^3 + 1.506n_i^2 + 0.8157n_i + 0.0893\right)^2} + \frac{0.55}{0.65^2}}} = 0$ $\underline{n_i = 1.0}$
(c)	$1.0 - \sqrt{\frac{0.8}{\frac{1}{\left(0.007938n_i^6 - 0.13668n_i^5 + 0.7473n_i^4 - 1.729n_i^3 + 1.506n_i^2 + 0.8157n_i + 0.0893\right)^2} + \frac{0.55}{1.0^2}}} = 0$ $\underline{n_i = 2.0}$
(d)	$0.5 - \sqrt{\frac{0.8}{\frac{1}{\left(0.007938n_i^6 - 0.13668n_i^5 + 0.7473n_i^4 - 1.729n_i^3 + 1.506n_i^2 + 0.8157n_i + 0.0893\right)^2} + \frac{0.55}{0.5^2}}} = 0$ $\underline{n_i = 0.73}$

Method #2

If the pre-setting value is determined by means of Method #2, using the relation that defines the quantity directly, it is additionally necessary to determine the value of a few parameters. The method of setting initial throttling should also be known. The valve under consideration is a device where initial pre-setting is realized by reducing the lift of a single plug moving above the valve seat. It is thus an element with one adjustable section of the medium flow. The valve structure is presented in Fig. 2.26. Due to the structure and the mutual geometrical relation between the plug and the valve seat, its initial characteristic is linear. The data of flow factor k_v depending on pre-setting n_i, together with the values of the valve inner authority a_w calculated from relation (4.48), are listed in Table 6.2.

Table 6.2 Throttling/closing characteristic data of the *Danfoss MSV-I DN20* valve and calculated inner authority

Number of pre-setting n_i	The valve flow factor k_{vx} [m^3/(h bar$^{0.5}$)]	Inner authority a_w
0.2	0.3	0.268
0.5	0.7	0.294
1.0	1.3	0.292
1.5	1.7	0.327
2.0	2.0	0.361
2.5	2.3	0.284
3.0	2.5	–
3.2	2.5	–

Calculating the value of pre-setting n_{i} by means of Method #2 proposed herein and relation (4.131) written to define the quantity directly, one of the things that have to be done is to determine the value of inner authority $a_{\mathrm{w,min}}$ if there is no reduction in the closing element maximum range of travel h_{max}, and the minimum value of the valve outer authority a_{z100} resulting therefrom. The inner authority mean value for the set initial linear throttling/closing characteristic totals $a_{\mathrm{w,min}} \approx 0.3$. If there is no reduction in the opening degree, the quantity variability as a function of the pre-setting value written in Table 6.1 can be expressed as follows:

$$a(n_i)_{w,\min} = 0.02812n_i^5 - 0.25592n_i^4 + 0.76968n_i^3 - 0.95078n_i^2 + 0.5003n_i + 0.20062.$$

The values produced by the function agree with the values of $a_{\mathrm{w,min}}$ calculated for the defined values of pre-setting n_i to the fourth significant digit. In order to calculate the valve outer authority at full opening $a_{z,100}$, the value of the pressure drop on the valve should be determined first.

The valve hydraulic resistance at full opening is:

$$r_{z,100} = \frac{1}{k_{vs}^2} = \frac{1}{2.5^2} = 0.16\,\frac{\mathrm{h^2\,bar}}{\mathrm{m^6}}.$$

(a)	$r_{str} = \frac{\Delta p_{str,x}}{\dot{V}_x^2} = \frac{0.55}{0.15^2} = 24.444\,\frac{\mathrm{h^2\,bar}}{\mathrm{m^6}}$
(b)	$r_{str} = \frac{\Delta p_{str,x}}{\dot{V}_x^2} = \frac{0.55}{0.65^2} = 1.302\,\frac{\mathrm{h^2\,bar}}{\mathrm{m^6}}$
(c)	$r_{str} = \frac{\Delta p_{str,x}}{\dot{V}_x^2} = \frac{0.55}{1^2} = 0.55\,\frac{\mathrm{h^2\,bar}}{\mathrm{m^6}}$
(d)	$r_{str} = \frac{\Delta p_{str,x}}{\dot{V}_x^2} = \frac{0.55}{0.5^2} = 2.2\,\frac{\mathrm{h^2\,bar}}{\mathrm{m^6}}$

The maximum volume flow of the medium is:

(a)	$\dot{V}_{100} = \sqrt{\frac{\Delta p_{cz}}{r_{z,100}+r_{str}}} = \sqrt{\frac{0.8}{0.16+24.444}} = 0.1803\,\frac{\mathrm{m^3}}{\mathrm{h}}$
(b)	$\dot{V}_{100} = \sqrt{\frac{\Delta p_{cz}}{r_{z,100}+r_{str}}} = \sqrt{\frac{0.8}{0.16+1.302}} = 0.74\,\frac{\mathrm{m^3}}{\mathrm{h}}$
(c)	$\dot{V}_{100} = \sqrt{\frac{\Delta p_{cz}}{r_{z,100}+r_{str}}} = \sqrt{\frac{0.8}{0.16+0.55}} = 1.0615\,\frac{\mathrm{m^3}}{\mathrm{h}}$
(d)	$\dot{V}_{100} = \sqrt{\frac{\Delta p_{cz}}{r_{z,100}+r_{str}}} = \sqrt{\frac{0.8}{0.16+2.2}} = 0.5822\,\frac{\mathrm{m^3}}{\mathrm{h}}$

The pressure losses on the elements for the valve full opening are as follows:

(a)	$\Delta p_{z,100} = r_{z,100} \cdot \dot{V}_{100}^2 = 0.16 \cdot 0.1803^2 = 0.0052\,\text{bar}$
(b)	$\Delta p_{z,100} = r_{z,100} \cdot \dot{V}_{100}^2 = 0.16 \cdot 0.74^2 = 0.0876\,\text{bar}$
(c)	$\Delta p_{z,100} = r_{z,100} \cdot \dot{V}_{100}^2 = 0.16 \cdot 1.0615^2 = 0.1803\,\text{bar}$
(d)	$\Delta p_{z,100} = r_{z,100} \cdot \dot{V}_{100}^2 = 0.16 \cdot 0.5822^2 = 0.05423\,\text{bar}$

The valve outer authority at full opening totals:

(a)	$a_{z,100} \equiv a_{z,\min} = \frac{r_{z,100}}{r_{z,100}+r_{str}} = \frac{0.16}{0.16+24.444} = 0.0065$
(b)	$a_{z,100} \equiv a_{z,\min} = \frac{r_{z,100}}{r_{z,100}+r_{str}} = \frac{0.16}{0.16+1.302} = 0.1094$
(c)	$a_{z,100} \equiv a_{z,\min} = \frac{r_{z,100}}{r_{z,100}+r_{str}} = \frac{0.16}{0.16+0.55} = 0.2253$
(d)	$a_{z,100} \equiv a_{z,\min} = \frac{r_{z,100}}{r_{z,100}+r_{str}} = \frac{0.16}{0.16+2.2} = 0.0678$

According to formula (4.131c), the required pre-setting values are as follows:

$$n_i = \frac{n_{\max}}{\sqrt{\frac{\frac{\Delta p_{cz}}{\left(\frac{\dot{V}_x}{k_{vs}}\right)^2 + \Delta p_{str,x}} - 1}{a_{z,\min} \cdot a(n_i)_{w,\min}} + 1}}.$$

(a)	$n_i = \frac{3.2}{\sqrt{\frac{\frac{0.8}{\left(\frac{0.15}{2.5}\right)^2+0.55}-1}{0.0065\cdot\left(0.02812n_i^5-0.25592n_i^4+0.76968n_i^3-0.95078n_i^2+0.5003n_i+0.20()62\right)}+1}}$ $\underline{n_i = 0.2}$
(b)	$n_i = \frac{3.2}{\sqrt{\frac{\frac{0.8}{\left(\frac{0.65}{2.5}\right)^2+0.55}-1}{0{,}1094\cdot\left(0.02812n_i^5-0.25592n_i^4+0.76968n_i^3-0.95078n_i^2+0.5003n_i+0.20062\right)}+1}}$ $\underline{n_i = 1.0}$
(c)	$n_i = \frac{3.2}{\sqrt{\frac{\frac{0.8}{\left(\frac{1.0}{2.5}\right)^2+0.55}-1}{0{,}2253\cdot\left(0.02812n_i^5-0.25592n_i^4+0.76968n_i^3-0.95078n_i^2+0.5003n_i+0.20062\right)}+1}}$ $\underline{n_i = 2.0}$
(d)	$n_i = \frac{3.2}{\sqrt{\frac{\frac{0.8}{\left(\frac{0.5}{2.5}\right)^2+0.55}-1}{0{,}0678\cdot\left(0.02812n_i^5-0.25592n_i^4+0.76968n_i^3-0.95078n_i^2+0.5003n_i+0.20062\right)}+1}}$ $\underline{n_i = 0.734}$

Calculating the required value of pre-setting n_i, it is necessary to make use of the relation describing variability $a(n_i)_w$ for the set initial closing characteristic so that the curve actual shape should be obtained. This is required only if the variability occurs. In practice, however, this is always the case, even though inner authority a_w, being a reflection of the distribution of hydraulic resistances in the valve for the set position of the throttling element and/or the set reduction in the closing element maximum range of travel h_{max}, is constant. The variability is related to any deviations/irregularities of the elements making, fitting tolerance, the mathematical approximation of the characteristic used for the calculations, measuring accuracy, etc., as explained earlier herein. If the actual value of the valve inner authority a_w has to be determined, the above scheme should not be used. The difference results from the fact that the value of pre-setting n_i can be calculated without having to know the actual value of the valve inner authority a_w and the actual mathematical description of the valve original closing and/or throttling characteristic. It is only necessary to find the combination of $k_{v,x}/k_{v100} = f(h_x/h_{100})$ and $a(n_i)_w$ giving a resultant curve which agrees with the curve measured for the valve. Therefore, other than the real shape of the original characteristic can be assumed for these calculations and a function of the variability in the valve inner authority a_w can be created such that, in combination with the other, will ensure the curve shape agreeing with the curve illustrating measured values for the valve. It is not sufficient, though, to know the mathematical description of the final curve only because in this case it is impossible to establish its deformation if the valve is incorporated into the pipework, where total authority a_c and, thereby, inner authority a_w have to be known. Obviously, the original curve, the description of which will be assumed for the calculations, must in its entire domain return values lower than the target one, considering the curve upward deformation due to the impact of inner authority a_w. This is illustrated by Fig. 4.1 for example, where the valve initial equal-percentage curve, at an appropriate (even constant) value of a_w, becomes almost linear in a wide range, and at this point it must be remembered that the descriptions of the two functions differ considerably. It is also possible, for the assumed value of inner authority a_w, to look for an appropriate description of the closing function, but this is rather troublesome.

The bigger the difference between the hypothetical and the actual mathematical description of the valve original characteristic used for the calculations, the bigger the oscillations in the determined value of $a_{w,min}$, and, thereby, the bigger the deviation from the quantity real value. Therefore, if the valve inner authority a_w is to be found to determine the curve according to which the valve installed in the pipework will perform the regulation task, this scheme, which gives correct results in the procedure for selecting the value of pre-setting n_i, should not be used. If the mathematical description of the valve original regulation characteristic is not provided by the manufacturer, the real value of inner authority a_w can be defined using the methods described herein, either directly—by measuring hydraulic parameters of the valve with and without the closing element installed in it (cf. Sect. 4.1.2.1) or indirectly—by measuring geometrical quantities and using this way to define the description of the original closing characteristic and inner authority a_w, after the curve is confronted with the measured characteristic (cf. Sect. 4.1.2.2). If results of such measurements are unavailable, the best-grounded method is to use the mean value of inner authority a_w, calculated for the entire range of available values of pre-setting n_i.

Summing up, the value of total authority a_c can be determined after real values of the inner and outer authority (a_w and a_z, respectively) are found. The value of the valve outer authority a_z will differ from the value used to calculate pre-setting n_i because the latter relates to the valve full opening and no reduction in the regulating element maximum range of travel h_{max}, and not to the case of selecting the required value of pre-setting n_i that the valve will actually operate with. So that total authority a_c can be determined correctly, the value of the valve outer authority a_z has to be found for the selected value of pre-setting n_i.

The valve inner authority a_w can be established using the proposed computational relation (4.54), considering the variability in the calculated value of inner authority $a(n_i)_{w,min}$. The following relation is then obtained:

$$a_{w,x} = a_{w,\min} + \left(1 - a_{w,\min}\right) \cdot \left(1 - \frac{1}{1 + a(n_i)_{w,\min}\left[\left(\frac{n_{\max}}{n_i}\right)^2 - 1\right]}\right)$$

(a)	$a_{w,x} = 0.3 + (1 - 0.3) \cdot \left(1 - \frac{1}{1+0.268\cdot\left[\left(\frac{3.2}{0.2}\right)^2-1\right]}\right) = 0.99$
(b)	$a_{w,x} = 0.3 + (1 - 0.3) \cdot \left(1 - \frac{1}{1+0.292\cdot\left[\left(\frac{3.2}{1.0}\right)^2-1\right]}\right) = 0.811$
(c)	$a_{w,x} = 0.3 + (1 - 0.3) \cdot \left(1 - \frac{1}{1+0.361\cdot\left[\left(\frac{3.2}{2.0}\right)^2-1\right]}\right) = 0.552$
(d)	$a_{w,x} = 0.3 + (1 - 0.3) \cdot \left(1 - \frac{1}{1+0.292\cdot\left[\left(\frac{3.2}{0.734}\right)^2-1\right]}\right) = 0.888$

The value of inner authority a_w varies with changes in the value of pre-setting n_i, which is understandable because a change in the throttling/closing element position involves changes in its hydraulic resistance at a constant hydraulic resistance of the valve body. Therefore, the higher the resistance of the regulated section of the medium flow, i.e. the bigger the reduction in the plug maximum range of travel h_{max} and the lower the value of pre-setting n_i, the higher the value of inner authority a_w of a valve with one adjustable section of the medium flow, as indicated by the performed experimental analysis.

The value of inner authority a_w for case a is very high and totals almost one. This is the effect of the fact that the closing element position is very close to full closing ($h_x/h_{100} = 0.2/3.2 = 6.25\%$), which naturally results in a very high relative value of this part hydraulic resistance in the valve total resistance. Moreover, another consequence is that almost the entire pressure loss falls on the closing element. The effect of selecting pre-setting n_i that makes it possible to achieve the required flow factor k_v means that the required pressure loss arises on the valve. In this case, the loss totals 0.25 bar.

The valve outer authority for the selected pre-setting totals:

(a)	$a_{z,x} = \frac{r_{z,x}}{r_{z,x}+r_{str}} = \frac{1/k_{v,x}^2}{1/k_{v,x}^2+r_{str}} = \frac{1/0.3^2}{1/0.3^2+24.444} = 0.3125$
(b)	$a_{z,x} = \frac{r_{z,x}}{r_{z,x}+r_{str}} = \frac{1/k_{v,x}^2}{1/k_{v,x}^2+r_{str}} = \frac{1/1.3^2}{1/1.3^2+1.302} = 0.3125$
(c)	$a_{z,x} = \frac{r_{z,x}}{r_{z,x}+r_{str}} = \frac{1/k_{v,x}^2}{1/k_{v,x}^2+r_{str}} = \frac{1/2.0^2}{1/2.0^2+0.55} = 0.3125$
(d)	$a_{z,x} = \frac{r_{z,x}}{r_{z,x}+r_{str}} = \frac{1/k_{v,x}^2}{1/k_{v,x}^2+r_{str}} = \frac{1/1.0^2}{1/1.0^2+2.2} = 0.3125$

The valve total authority for the selected pre-setting totals:

(a)	$a_{c,x} = a_{w,x} \cdot a_{z,x} = 0.99 \cdot 0.3125 = 0.31$
(b)	$a_{c,x} = a_{w,x} \cdot a_{z,x} = 0.811 \cdot 0.3125 = 0.253$
(c)	$a_{c,x} = a_{w,x} \cdot a_{z,x} = 0.552 \cdot 0.3125 = 0.1725$
(d)	$a_{c,x} = a_{w,x} \cdot a_{z,x} = 0.888 \cdot 0.3125 = 0.278$

The value of total authority a_c calculated in this manner and resulting from determination of the values of $a_{w,100}$, $a_{w,x}$ and $a_{z,min}$ can be checked against the result that would be obtained if the regulated section hydraulic resistance was compared to the total resistance in the circuit, using relations (4.56–4.60) and (4.101). In such a situation, it is necessary to find the regulated section hydraulic resistance for every selected value of pre-setting n_i, starting with determination of the regulated section hydraulic resistance at the maximum travel range and the hydraulic resistance of the valve body. The following values are then obtained:

$$r_{reg,100} = a_{w,\min} \cdot \frac{1}{k_{vs}^2} = 0.3 \cdot \frac{1}{2.5^2} = 0.048 \frac{\mathrm{h^2\,bar}}{\mathrm{m^6}},$$

$$r_k = r_{z,100} - r_{reg,100} = \frac{1}{k_{vs}^2} - r_{reg,100} = \frac{1}{2.5^2} - 0.048 = 0.112 \frac{\text{h}^2\,\text{bar}}{\text{m}^6}.$$

The regulated section hydraulic resistance for the calculated pre-setting values is as follows:

(a)	$r_{reg,x} = \frac{1}{k_{v,x}^2} - r_k = \frac{1}{0.3^2} - 0.112 = 11 \frac{\text{h}^2\,\text{bar}}{\text{m}^6}$
(b)	$r_{reg,x} = \frac{1}{k_{v,x}^2} - r_k = \frac{1}{1.3^2} - 0.112 = 0.48 \frac{\text{h}^2\,\text{bar}}{\text{m}^6}$
(c)	$r_{reg,x} = \frac{1}{k_{v,x}^2} - r_k = \frac{1}{2.0^2} - 0.112 = 0.138 \frac{\text{h}^2\,\text{bar}}{\text{m}^6}$
(d)	$r_{reg,x} = \frac{1}{k_{v,x}^2} - r_k = \frac{1}{1.0^2} - 0.112 = 0.888 \frac{\text{h}^2\,\text{bar}}{\text{m}^6}$

According to formula (4.101), the valve total authority for the selected pre-setting totals:

$$a_{c,x} = \frac{r_{reg,x}}{r_{z,x} + r_{str}} = \frac{r_{reg,x}}{r_{reg,x} + r_k + r_{str}}.$$

(a)	$a_{c,x} = \frac{r_{reg,x}}{r_{reg,x}+r_k+r_{str}} = \frac{11}{11+0.112+24.444} = 0.31$
(b)	$a_{c,x} = \frac{r_{reg,x}}{r_{reg,x}+r_k+r_{str}} = \frac{0.48}{0.48+0.112+1.302} = 0.253$
(c)	$a_{c,x} = \frac{r_{reg,x}}{r_{reg,x}+r_k+r_{str}} = \frac{0.138}{0.138+0.112+0.55} = 0.1725$
(d)	$a_{c,x} = \frac{r_{reg,x}}{r_{reg,x}+r_k+r_{str}} = \frac{0.888}{0.888+0.112+2.2} = 0.278$

The analysis of the obtained results points to their full agreement.

Calculations performed by means of the method proposed by *Pyrkov* and *Szaflik*

This method is described in detail in [12, 13]. Only the indispensable computational relations will be presented here, together with the results they produce.

The pressure losses on the elements for the valve full opening are as follows:

(a)	$\Delta p_{z,100} = \left(\frac{\dot{V}_x}{k_{vs}}\right)^2 = \left(\frac{0.15}{2.5}\right)^2 = 0.0036\,\text{bar}$ $\Delta p_{str,100} = 0.55\,\text{bar}$
(b)	$\Delta p_{z,100} = \left(\frac{\dot{V}_x}{k_{vs}}\right)^2 = \left(\frac{0.65}{2.5}\right)^2 = 0.0676\,\text{bar}$ $\Delta p_{str,100} = 0.55\,\text{bar}$
(c)	$\Delta p_{z,100} = \left(\frac{\dot{V}_x}{k_{vs}}\right)^2 = \left(\frac{1.0}{2.5}\right)^2 = 0.16\,\text{bar}$ $\Delta p_{str,100} = 0.55\,\text{bar}$
(d)	$\Delta p_{z,100} = \left(\frac{\dot{V}_x}{k_{vs}}\right)^2 = \left(\frac{0.5}{2.5}\right)^2 = 0.04\,\text{bar}$ $\Delta p_{str,100} = 0.55\,\text{bar}$

The valve outer authority at full opening totals:

(a)	$a_{z,100} \equiv a_{z,\min} = \frac{\Delta p_{z,100}}{\Delta p_{z,100}+\Delta p_{str,100}} = \frac{0.0036}{0.0036+0.55} = 0.0065$
(b)	$a_{z,100} \equiv a_{z,\min} = \frac{\Delta p_{z,100}}{\Delta p_{z,100}+\Delta p_{str,100}} = \frac{0.0676}{0.0676+0.55} = 0.1094$
(c)	$a_{z,100} \equiv a_{z,\min} = \frac{\Delta p_{z,100}}{\Delta p_{z,100}+\Delta p_{str,100}} = \frac{0.16}{0.16+0.55} = 0.2253$
(d)	$a_{z,100} \equiv a_{z,\min} = \frac{\Delta p_{z,100}}{\Delta p_{z,100}+\Delta p_{str,100}} = \frac{0.04}{0.04+0.55} = 0.0678$

The valve inner authority for the selected pre-setting (mean value) totals:

(a)	$a_{w,x} = a_{w,\min} = 0.3$
(b)	$a_{w,x} = a_{w,\min} = 0.3$
(c)	$a_{w,x} = a_{w,\min} = 0.3$
(d)	$a_{w,x} = a_{w,\min} = 0.3$

The valve outer authority for the selected pre-setting totals:

(a)	$a_{z,x} = a_{z,100} = 0.0065$
(b)	$a_{z,x} = a_{z,100} = 0.1094$
(c)	$a_{z,x} = a_{z,100} = 0.2253$
(d)	$a_{z,x} = a_{z,100} = 0.0678$

The valve total authority for the selected pre-setting totals:

(a)	$a_{c,x} = a_{w,x} \cdot a_{z,x} = 0.3 \cdot 0.0065 = 0.00195$
(b)	$a_{c,x} = a_{w,x} \cdot a_{z,x} = 0.3 \cdot 0.1094 = 0.03282$
(c)	$a_{c,x} = a_{w,x} \cdot a_{z,x} = 0.3 \cdot 0.2253 = 0.06759$
(d)	$a_{c,x} = a_{w,x} \cdot a_{z,x} = 0.3 \cdot 0.0678 = 0.02034$

The required pre-setting values are as follows:

$$n_i = \frac{n_{\max}}{\sqrt{1 - \frac{1}{a_{z,\min} \cdot a_{w,\min}} + \frac{\Delta p_{cz}}{a_{w,\min} \cdot \left(\frac{\dot{V}_x}{k_{vs}}\right)^2}}}.$$

(a)	$n_i = \frac{3.2}{\sqrt{1 - \frac{1}{0.0065 \cdot 0.3} + \frac{0.8}{0.3 \cdot \left(\frac{0.15}{2.5}\right)^2}}} = 0.212$
(b)	$n_i = \frac{3.2}{\sqrt{1 - \frac{1}{0.1094 \cdot 0.3} + \frac{0.8}{0.3 \cdot \left(\frac{0.65}{2.5}\right)^2}}} = 1.013$
(c)	$n_i = \frac{3.2}{\sqrt{1 - \frac{1}{0.2253 \cdot 0.3} + \frac{0.8}{0.3 \cdot \left(\frac{1.0}{2.5}\right)^2}}} = 1.888$
(d)	$n_i = \frac{3.2}{\sqrt{1 - \frac{1}{0.0678 \cdot 0.3} + \frac{0.8}{0.3 \cdot \left(\frac{0.5}{2.5}\right)^2}}} = 0.744$

The results of the parameters calculation are listed in Table 6.3.

All the calculation results, for all the analysed examples of the pre-setting selection, will be discussed collectively at the end of this chapter.

Table 6.3 Calculation results for Example 6.15, variant 1

Method	Sought quantity			
	Pre-setting n_i	Inner authority a_W	Outer authority a'/a_Z	Total authority a_C
The common method	Required (a) $n_i = 0.2$ (b) $n_i = 1.0$ (c) $n_i = 2.0$ (d) $n_i = (0.5; 1.0)$	No computational algorithm No requirements	Required: $a' \geq 0.3$ (a) $a' = 0.3125$ (b) $a' = 0.3125$ (c) $a' = 0.3125$ (d) $a' = 0.3125$	No computational algorithm No requirements
Method proposed by *Pyrkov* and *Szaflik* in [12, 13]	(a) $n_i = 0.212$ (b) $n_i = 1.013$ (c) $n_i = 1.888$ (d) $n_i = 0.744$	No requirements (a) $a_W = 0.3$ (b) $a_W = 0.3$ (c) $a_W = 0.3$ (d) $a_W = 0.3$	No requirements (a) $a_Z = 0.0065$ (b) $a_Z = 0.1094$ (c) $a_Z = 0.2253$ (d) $a_Z = 0.0678$	Required: $a_C \geq 0.3$ (a) $a_C = 0.00195$ (b) $a_C = 0.03282$ (c) $a_C = 0.06759$ (d) $a_C = 0.02034$
Proposed method	(a) $n_i = 0.2$ (b) $n_i = 1.0$ (c) $n_i = 2.0$ (d) $n_i = 0.734$	No requirements (a) $a_W = 0.99$ (b) $a_W = 0.814$ (c) $a_W = 0.523$ (d) $a_W = 0.89$	No requirements (a) $a_Z = 0.3125$ (b) $a_Z = 0.3125$ (c) $a_Z = 0.3125$ (d) $a_Z = 0.3125$	Required: $a_C \geq 0.3$ (a) $a_C = 0.31$ (b) $a_C = 0.254$ (c) $a_C = 0.163$ (d) $a_C = 0.278$

Variant 2

Find the required values of the *Danfoss MSV-I DN20* valve pre-setting and total authority using the following data:

- required volume flow of the medium: $\dot{V}_x = 0.65\,\mathrm{m}^3/\mathrm{h}$,
- hydraulic resistance of the pipework: $r_{str} = 1.302$ (h^2 bar)/m^6,
- active (differential) pressure in the circuit (Δp_{cz}):

(a) 0.63 bar,
(b) 0.696 bar,
(c) 0.8 bar.

Solution
This time the set value is not the pipework pressure loss Δp_{str} but the pipework hydraulic resistance r_{str}, i.e. a characteristic parameter of the pipework. Moreover, different pressure values are set for which the same value of the medium volume flow has to be maintained. In practice for example, this corresponds to a change in the pressure supplied by the pump to the installation or the pump replacement with a different one, or the use of local pressure-stabilizing systems. The calculations are performed using the same relations as in the previous variant. The final relations and results are presented further below.

Calculations performed by means of the common method

In order to carry out the calculations in this variant, it is necessary to find the pressure that will have to be throttled on the valve for the sought i-th pre-setting. Because there is no direct information on the pressure value or the pressure distribution in the circuit, it is impossible to determine the required value of pre-setting n_i by means of the commonly applied calculation method.

Calculations performed by means of the method proposed by *Pyrkov* and *Szaflik*

The situation is similar to the described and commonly applied method. The two researchers put forward computational relations for which the pressure loss in the pipework should be defined for the required volume flow of the medium. But the set problem data do not define the values. For this reason, the required computational relations cannot directly be used in this case, either.

For both methods, the pressure loss arising on the valve and in the pipework, as the required quantities, can be found using the hydraulic resistance concept described in Chap. 3. However, such a procedure is not a part of the methods being compared. Moreover, while calculating required pressures, for natural reasons the results of the selection of the value of pre-setting n_i and of the valve authority are the same as presented in variant 1 of the example, which is discussed in the summary of all the computational examples.

Calculations performed by means of the proposed method

Looking for the required value of pre-setting n_i, both computational algorithms proposed herein and applied earlier can be used. The use of the first—Method #1—is presented below.

Method #1

According to formula (4.127a), the required pre-setting value is:

$$\dot{V}_x - \sqrt{\frac{\Delta p_{cz}}{\frac{1}{(k_v(n_i))^2} + r_{str}}} = 0$$

(a)	$0.65 - \sqrt{\frac{0.63}{\frac{1}{\left(\begin{array}{l}0.007938n_i^6 - 0.13668n_i^5 + 0.7473n_i^4 - \\ 1.729n_i^3 + 1.506n_i^2 + 0.8157n_i + 0.0893\end{array}\right)^2} + 1.302}} = 0$ $\underline{n_i = 2.5}$
(b)	$0.65 - \sqrt{\frac{0.696}{\frac{1}{\left(\begin{array}{l}0.007938n_i^6 - 0.13668n_i^5 + 0.7473n_i^4 - \\ 1.729n_i^3 + 1.506n_i^2 + 0.8157n_i + 0.089\end{array}\right)^2} + 1.302}} = 0$ $\underline{n_i = 1.5}$
(c)	$0.65 - \sqrt{\frac{0.8}{\frac{1}{\left(\begin{array}{l}0.007938n_i^6 - 0.13668n_i^5 + 0.7473n_i^4 - \\ 1.729n_i^3 + 1.506n_i^2 + 0.8157n_i + 0.089\end{array}\right)^2} + 1.302}} = 0$ $\underline{n_i = 1.0}$

The valve inner authority for the selected pre-setting totals:

$$a_{w,x} = a_{w,\min} + \left(1 - a_{w,\min}\right) \cdot \left(1 - \frac{1}{1 + a(n_i)_{w,\min}\left[\left(\frac{n_{\max}}{n_i}\right)^2 - 1\right]}\right).$$

(a)	$a_{w,x} = 0.3 + (1 - 0.3) \cdot \left(1 - \frac{1}{1+0.284\cdot\left[\left(\frac{3.2}{2.5}\right)^2 - 1\right]}\right) = 0.407$
(b)	$a_{w,x} = 0.3 + (1 - 0.3) \cdot \left(1 - \frac{1}{1+0.327\cdot\left[\left(\frac{3.2}{1.5}\right)^2 - 1\right]}\right) = 0.676$
(c)	$a_{w,x} = 0.3 + (1 - 0.3) \cdot \left(1 - \frac{1}{1+0.292\cdot\left[\left(\frac{3.2}{1.0}\right)^2 - 1\right]}\right) = 0.811$

The valve outer authority for the selected pre-setting totals:

(a)	$a_{z,x} = \frac{r_{z,x}}{r_{z,x}+r_{str}} = \frac{1/k_{v,x}^2}{1/k_{v,x}^2+r_{str}} = \frac{1/2.3^2}{1/2.3^2+1.302} = 0.127$
(b)	$a_{z,x} = \frac{r_{z,x}}{r_{z,x}+r_{str}} = \frac{1/k_{v,x}^2}{1/k_{v,x}^2+r_{str}} = \frac{1/1.7^2}{1/1.7^2+1.302} = 0.21$
(c)	$a_{z,x} = \frac{r_{z,x}}{r_{z,x}+r_{str}} = \frac{1/k_{v,x}^2}{1/k_{v,x}^2+r_{str}} = \frac{1/1.3^2}{1/1.3^2+1.302} = 0.3125$

The valve total authority for the selected pre-setting totals:

(a)	$a_{c,x} = a_{w,x} \cdot a_{z,x} = 0.407 \cdot 0.127 = 0.0517$
(b)	$a_{c,x} = a_{w,x} \cdot a_{z,x} = 0.676 \cdot 0.21 = 0.142$
(c)	$a_{c,x} = a_{w,x} \cdot a_{z,x} = 0.811 \cdot 0.3125 = 0.253$

The calculation results are listed in Table 6.4.

Table 6.4 Calculation results for Example 6.15, variant 2

Method	Sought quantity			
	Pre-setting n_i	Inner authority a_W	Outer authority a'/a_Z	Total authority a_C
The common method	(a) n_i = unknown (b) n_i = unknown (c) n_i = unknown	No computational algorithm No requirements	Required: $a' \geq 0.3$ (a) a' = unknown (b) a' = unknown (c) a' = unknown	No computational algorithm No requirements
Method proposed by *Pyrkov* and *Szaflik* in [12, 13]	(a) n_i = unknown (b) n_i = unknown (c) n_i = unknown	No requirements (a) a_W = unknown (b) a_W = unknown (c) a_W = unknown	No requirements (a) a_Z = unknown (b) a_Z = unknown (c) a_Z = unknown	Required: $a_C \geq 0.3$ (a) a_C = unknown (b) a_C = unknown (c) a_C = unknown
Proposed method	Required (a) $n_i = 2.5$ (b) $n_i = 1.5$ (c) $n_i = 1.0$	No requirements (a) $a_W = 0.407$ (b) $a_W = 0.676$ (c) $a_W = 0.811$	No requirements (a) $a_Z = 0.127$ (b) $a_Z = 0.21$ (c) $a_Z = 0.3125$	Required: $a_C \geq 0.3$ (a) $a_C = 0.0517$ (b) $a_C = 0.142$ (c) $a_C = 0.253$

Example 6.16
Variant 1

Find the required values of pre-setting and total authority of the *Danfoss RTD-N15* valve co-operating with a thermostatic head using the following data:

- active (differential) pressure in the circuit: $\Delta p_{cz} = 0.3$ bar,
- pipework pressure loss for the required volume flow of the medium: $\Delta p_{str,x} = 0.2$ bar,
- required volume flow of the medium:

(a) $\dot{V}_x = 0.01265\,\text{m}^3/\text{h}$,
(b) $\dot{V}_x = 0.03795\,\text{m}^3/\text{h}$,
(c) $\dot{V}_x = 0.08538\,\text{m}^3/\text{h}$,
(d) $\dot{V}_x = 0.0506\,\text{m}^3/\text{h}$.

Solution
The case concerns a thermostatic, double-regulation valve with two adjustable sections of the medium flow, as presented in Fig. 2.29b. The values of flow factor k_v as a function of pre-setting n_i for the analysed valve co-operating with a thermostatic head are listed in Table 6.5.

Calculations performed by means of the common method

In order to carry out the calculations in this variant, it is necessary to find the pressure that will have to be throttled on the valve for the sought i-th pre-setting and the nominal opening degree of the closing plug corresponding to proportional range $X_p = 2$ K. According to formula (6.14), it totals:

$$\Delta p_{z,x} = \Delta p_{cz} - \Delta p_{str,x} = 0.1\ \text{bar}.$$

Table 6.5 Throttling characteristic data of the *Danfoss RTD-N15* valve [1]

Number of pre-setting n_i		The valve flow factor $k_{v,x}$ [$m^3/(h\ bar^{0.5})$]
$X_p = 2$ K	1	0.04
	2	0.08
	3	0.12
	4	0.20
	5	0.27
	6	0.36
	7	0.45
	N	0.60
$X_p = \max$	N	0.90

Knowing this pressure, the required value of flow factor k_v can be calculated:

(a)	$k_{v,x} = \frac{\dot{V}_x}{\sqrt{\Delta p_{z,x}}} = \frac{0.01265}{\sqrt{0.1}} = 0.04$
(b)	$k_{v,x} = \frac{\dot{V}_x}{\sqrt{\Delta p_{z,x}}} = \frac{0.03795}{\sqrt{0.1}} = 0.12$
(c)	$k_{v,x} = \frac{\dot{V}_x}{\sqrt{\Delta p_{z,x}}} = \frac{0.08538}{\sqrt{0.1}} = 0.27$
(d)	$k_{v,x} = \frac{\dot{V}_x}{\sqrt{\Delta p_{z,x}}} = \frac{0.0506}{\sqrt{0.1}} = 0.16$

According to the data listed in Table 6.5, the required value of flow factor k_v is ensured if the following pre-settings are selected, respectively:

(a)	$n_i = 1.0$
(b)	$n_i = 3.0$
(c)	$n_i = 5.0$
(d)	$n_i = (3.0; 4.0)$

The valve authority a' will take the same value for every case because the pressure loss on the valve is identical. According to formula (4.1), it totals:

$$a' = \frac{\Delta p_{z,x}}{\Delta p_{ob}} = \frac{\Delta p_{z,x}}{\Delta p_{z,x} + \Delta p_{str,x}} = \frac{0.1}{0.1 + 0.2} = 0.333-$$

The parameter value calculated in this manner is sufficiently high and satisfies the set condition of $a' > 0.3$. In this case, since a constant pressure distribution is adopted but the medium volume flows are different (e.g. different thermal power values in

Table 6.6 Throttling/closing characteristic data of the *Danfoss RTD-N15* valve [1] and calculated inner authority

Number of pre-setting n_i		The valve flow factor k_{vx} [m^3/(h bar$^{0.5}$)]	Closing element inner authority a_{wx}
$X_p = 2$ K	1	0.04	0.000074
	2	0.08	0.00029
	3	0.12	0.00066
	4	0.20	0.0018
	5	0.27	0.0034
	6	0.36	0.0060
	7	0.45	0.0093
	$8 \equiv$ N	0.60	0.0166
$X_p = \max$	$8 \equiv$ N	0.90	–

circuits), different pipe diameters are selected, which results in different hydraulic resistances of the elements. And this is what happens in practice.

Calculations performed by means of the method proposed by *Pyrkov* and *Szaflik*

The two researchers do not offer a method of determining the pre-setting value for double-regulation valves with two adjustable sections of the medium flow, which makes it impossible to determine the valve total authority a_c. However, they present a method of calculating inner authority a_w for such valves. The relation is the same as formula (4.77), except that the values of flow factor k_v are related to a partial opening degree of the closing element, corresponding to the assumed proportional range X_p:

$$a_{w,x} = a_{w,\max} \cdot \left(\frac{k_{v,x}}{k_{vs}}\right)^2_{Xp=2K}.$$

The value of inner authority $a_{w,\max}$ is calculated by means of a relation similar to the one proposed herein for such valves, which is expressed as:

$$a_{w,\max} = \frac{\left(\frac{k_{vN,Xp=\max}}{k_{vN,Xp=2K}}\right)^2 - 1}{\exp[2c(1 - h_x/h_{\max})] - 1},$$

which gives $a_{w,\max} = 0.0166$. The parameter calculation results obtained by means of the method proposed by *Pyrkov* and *Szaflik* in [12, 13] are gathered in Table 6.6.

Calculations performed by means of the proposed method

The value of pre-setting n_i can be calculated in the same manner as described in the introduction to the calculations performed by means of the proposed algorithm, in Example 6.15.

Method #1:

Based on the data from the valve throttling/closing characteristic measurements presented in Table 6.5, function $k_v(n_i)$ for the pre-setting value of $n_i = 1.0–7.0$ can be written as:

$$[k_v\,(n_i)]_{Xp=2} = -0.0004167\,n_i^6 + 0.010167\,n_i^5 - 0.09792\,n_i^4 \\ + 0.47083\,n_i^3 - 1.16667\,n_i^2 + 1.424\,n_i - 0.6.$$

According to formula (4.141b), the required pre-setting values are as follows:

$$\dot{V}_x - \sqrt{\frac{\Delta p_{cz}}{\frac{1}{\left([k_v(n_i)]_{Xp=2}\right)^2} + \frac{\Delta p_{str,x}}{\dot{V}_x^2}}} = 0.$$

(a)	$0.01265 - \sqrt{\frac{0.3}{\frac{1}{\left(\begin{array}{l}-0.0004167n_i^6 + 0.010167n_i^5 - 0.09792n_i^4 + \\ 0.47083n_i^3 - 1.16667n_i^2 + 1.424n_i - 0.6\end{array}\right)^2} + \frac{0.2}{0.01265^2}}} = 0$ $\underline{\underline{n_i = 1.0}}$
(b)	$0.03795 - \sqrt{\frac{0.3}{\frac{1}{\left(\begin{array}{l}-0.0004167n_i^6 + 0.010167n_i^5 - 0.09792n_i^4 + \\ 0.47083n_i^3 - 1.16667n_i^2 + 1.424n_i - 0.6\end{array}\right)^2} + \frac{0.2}{0.03795^2}}} = 0$ $\underline{\underline{n_i = 3.0}}$
(c)	$0.08538 - \sqrt{\frac{0.3}{\frac{1}{\left(\begin{array}{l}-0.0004167n_i^6 + 0.010167n_i^5 - 0.09792n_i^4 + \\ 0.47083n_i^3 - 1.16667n_i^2 + 1.424n_i - 0.6\end{array}\right)^2} + \frac{0.2}{0.08538^2}}} = 0$ $\underline{\underline{n_i = 5.0}}$
(d)	$0.0506 - \sqrt{\frac{0.3}{\frac{1}{\left(\begin{array}{l}-0.0004167n_i^6 + 0.010167n_i^5 - 0.09792n_i^4 + \\ 0.47083n_i^3 - 1.16667n_i^2 + 1.424n_i - 0.6\end{array}\right)^2} + \frac{0.2}{0,\backslash.0506^2}}} = 0$ $\underline{\underline{n_i = 3.51}}$

Method #2:

Like in Example 6.15, the value of pre-setting n_i can be calculated by means of the other method, after relevant quantities are determined. It is necessary to find the values of inner and total authority ($a_{w,I}$ and $a_{c,I}$, respectively) of the valve throttling element responsible for the pre-setting in this case, as well as establish the variability in the element inner authority as a function of pre-setting ($a_w(n_i)$), for the closing element set opening degree for which the circuit is to be designed. Moreover, in order to define the current regulation characteristic, it is also necessary to determine inner and total authority (a_w and a_c, respectively) of the closing element responsible for the regulation. The manufacturer provides the values of flow factor k_v depending on pre-setting n_i, but each and every time it is for the closing element opening degree $h_x/h_{100} \equiv h_x/h_{max}$ corresponding to proportional range $X_p = 2$ K, the value for which, as described earlier, it is recommended that thermostatic heating units should be selected. The exception is pre-setting $n_i = \mathrm{N}$, for which the flow factor value concerns two cases. One where $X_p = 2$ K and the other—for the closing element full lift h_{max}, i.e. for $X_p = \max$. As long as the amplification factor k_m of the thermostatic head co-operating with the analysed valve is unknown, it is impossible to determine the closing element opening degree h_x/h_{max} and, consequently, the element authority. Moreover, the knowledge of the closing plug maximum range of travel h_{max} is required because this is the value to which each intermediate position of the plug is related in the authority calculations. Alternatively, for a known nominal proportional range X_p, it is possible to use the information about the maximum proportional range $X_{p,max}$ of the analysed thermostatic unit (because the ratio between the two quantities is equivalent to the ratio between opening degrees).

The throttling element inner authority a_w can be calculated using relation (4.49) and replacing the relative lift with the relative angle of rotation because a change in throttling is effected by rotation a cylinder with a shutter. First and foremost, it is necessary to define the mathematical description of the original throttling characteristic in relation to which inner authority $a_{w,I}$ and the authority variability will be calculated depending on pre-setting n_i. Like in the previous case, the required value of pre-setting n_i can be determined without knowing the actual original characteristic (of throttling in this case); it is enough to know the appropriate combination of this curve with the curve illustrating changes in inner authority $a_{w,I}$, which will result in a curve agreeing with the one measured for the valve. It is assumed that, in the range of the defined values of pre-setting n_i, the original throttling characteristic is an equal-percentage curve with factor $c = 3.5$. Using this assumption, the throttling element inner authority is calculated by means of formula (4.49), for a given proportional range X_p:

$$\left[a_{w,I,\min}\right]_{Xp=2} = \frac{\left(\frac{k_{vs}}{k_{v,x}}\right)^2_{Xp=2} - 1}{\exp[2c(1 - n_i/n_{\max})] - 1}.$$

It should be remembered that the value calculated by means of this formula is the minimum one related to the throttling element full available range of travel, but the minimum one for the closing element given, partial, position corresponding to a given proportional range X_p. It is therefore not the minimum value in the absolute scale for a given valve—it would be such only in relation to the closing element minimal lift. Compared to manual valves with two adjustable sections of the medium flow, there is an essential difference here. The value calculated for them in the same manner would be the minimum one because there it is related only to one (maximum) travel range h_{max} of the closing element.

The closing element inner authority a_w can be calculated in the same way as for the throttling part, replacing the n_i/n_{max} ratio with relative lift h_x/h_{max} and the ratios between flow factors with values corresponding to the replaced opening degree ratio. In this case, the maximum value is obtained for full opening of the throttling element, i.e. for the second pre-setting N, and the minimum one—for the element minimum opening degree, according to the following formula:

$$a_{w,\max} = \frac{\left(\frac{k_{vN,Xp=\max}}{k_{vN,Xp=2K}}\right)^2 - 1}{\exp[2c \cdot (1 - h_x/h_{\max})] - 1}.$$

For the defined partial pre-settings ($n_i = 1$–7), the quantity can be calculated by means of formula (4.77), where the ratio between the flow factors is related to the throttling element at the closing element full lift, which can be expressed as:

$$a_{w,x} = a_{w,\max} \cdot \left(\frac{k_{v,x}}{k_{vs}}\right)^2_{Xp=\max}.$$

If the calculated quantity is pre-setting n_i for which flow factor k_v is not defined, formula (4.88) should be used, as an extension of relation (4.77):

$$a_{w,x} = a_{w,\max} \cdot \left(\frac{1}{1 + a_{w,I,\min} \cdot \left[\exp[2c \cdot (1 - n_i/n_{\max})] - 1\right]}\right).$$

As indicated by experimental verification, the maximum available range of the closing plug travel totals about $h_{100} = 1.8$ mm, and the valve, according to the manufacturer's specifications [12, 13], was analysed as it co-operated with a thermostatic head with the amplification factor of $k_m = 0.25$ mm/K, which means that the plug lift over the valve seat of $h_x = 0.5$ mm corresponds to proportional range $X_p = 2$ K and the relative lift is about $h_x/h_{max} = 0.294$. The experimental analysis results reveal some significant differences between the parameters (flow factors k_v) compared to the averaged data provided by the valve manufacturer. Therefore, the calculation results obtained using these data may be burdened with a spread if compared to the results obtained for a specific valve specimen. Nonetheless, further considerations presented herein are based on the experimental verification results. It is assumed that the original closing characteristic, beyond the travel range of the plug where the flow channel is shaped due to the requirement that tight closing has to be ensured, is an equal-percentage curve with factor $c = 3$ [12, 13]. The valve flow factor values for the closing plug maximum range of travel h_{max} depending on pre-setting n_i, are listed in Table 6.7. The data are the results of experimental testing performed by the Author. The table also presents the results of calculations of the throttling and the closing element inner authority ($a_{w,I}$ and a_w, respectively).

Table 6.7 Throttling/closing characteristic data of the *Danfoss RTD-N15* valve and calculated values of the throttling and the closing element inner authority ([1] and own testing results)

Number of pre-setting n_i		The valve flow factor k_{vx} [$m^3/(h\ bar^{0.5})$]	Inner authority			
			Throttling element			Closing element
			$[a_{w,I,,min}]_{Xp=2,}$	$a_{w,I,,min}$		a_{wx}
$X_p = 2$ K/ $X_p = \max$	1	0.04/0.057	0.491	0.413	$X_p = \max$	0.000088
	2	0.08/0.076	0.291	0.556		0.000156
	3	0.12/0.164	0.306	0.278		0.00073
	4	0.20/0.247	0.249	0.283		0.00165
	5	0.27/0.37	0.3075	0.273		0.0037
	6	0.36/0.50	0.374	0.31		0.00676
	7	0.45/0.639	0.556	0.358		0.011
	8 ≡ N	0.60/0.784	–	–		0.0166
$X_p = \max$	8 ≡ N	0.784	–	–		–

Because the closing element inner authority a_w was calculated for a single pair of the plug relative lift values and because there is no manufacturer's information about the exact shape of the original closing characteristic, the result may be burdened with a significant error.

The variability in the throttling element inner authority $a_{w,I}$ as a function of pre-setting n_i, for $X_p = 2$ K, can be written using the following polynomial equation:

$$\left[a(n_i)_{w,I,\min}\right]_{Xp} = 0.001772n_i^6 - 0.0436n_i^5 + 0.42559n_i^4 - 2.0897n_i^3 + 5.396n_i^2 - 6.904n_i + 3.704.$$

The valve inner authority a_z is determined using the same calculations as in the previous example. The difference is that it is the throttling element inner authority $a_{w,I}$ that is calculated and the value has to be related to the closing element position corresponding to the required proportional range, i.e. $X_p = 2$ K (in this case $h_x/h_{100} = h_x/h_{max} = 0.294$), and not to full opening h_{max}. This is due to the fact that from the point of view of the throttling element and the thermoregulator design conditions, for a given range X_p, the closing element in this situation constitutes a non-regulated hydraulic resistance.

The valve hydraulic resistance for the throttling element full opening and the set proportional range X_p for which the authority of the throttling element responsible for initial throttling was calculated totals:

$$\left[r_{z,100}\right]_{Xp=2} = \frac{1}{\left[k_{vs}^2\right]_{Xp=2}} = \frac{1}{0.6^2} = 2.778\,\frac{\text{h}^2\,\text{bar}}{\text{m}^6}.$$

For the required value of the medium volume flow, the pipework hydraulic resistance totals:

(a)	$r_{str} = \frac{\Delta p_{str,x}}{\dot{V}_x^2} = \frac{0.2}{0.01265^2} = 1250 \frac{\text{h}^2\,\text{bar}}{\text{m}^6}$
(b)	$r_{str} = \frac{\Delta p_{str,x}}{\dot{V}_x^2} = \frac{0.2}{0.03795^2} = 138.9 \frac{\text{h}^2\,\text{bar}}{\text{m}^6}$
(c)	$r_{str} = \frac{\Delta p_{str,x}}{\dot{V}_x^2} = \frac{0.2}{0.08538^2} = 27.44 \frac{\text{h}^2\,\text{bar}}{\text{m}^6}$
(d)	$r_{str} = \frac{\Delta p_{str,x}}{\dot{V}_x^2} = \frac{0.2}{0.0506^2} = 78.11 \frac{\text{h}^2\,\text{bar}}{\text{m}^6}$

The medium volume flow at the throttling element full opening and proportional range X_{p} totals, respectively:

(a)	$\left[\dot{V}_{100}\right]_{Xp=2} = \sqrt{\frac{\Delta p_{cz}}{\left[r_{z,100}\right]_{Xp=2} + r_{str}}} = \sqrt{\frac{0.3}{2.778+1250}} = 0.01547 \frac{\text{m}^3}{\text{h}}$
(b)	$\left[\dot{V}_{100}\right]_{Xp=2} = \sqrt{\frac{\Delta p_{cz}}{\left[r_{z,100}\right]_{Xp=2} + r_{str}}} = \sqrt{\frac{0.3}{2.778+138.9}} = 0.046 \frac{\text{m}^3}{\text{h}}$
(c)	$\left[\dot{V}_{100}\right]_{Xp=2} = \sqrt{\frac{\Delta p_{cz}}{\left[r_{z,100}\right]_{Xp=2} + r_{str}}} = \sqrt{\frac{0.3}{2.778+27.44}} = 0.0996 \frac{\text{m}^3}{\text{h}}$
(d)	$\left[\dot{V}_{100}\right]_{Xp=2} = \sqrt{\frac{\Delta p_{cz}}{\left[r_{z,100}\right]_{Xp=2} + r_{str}}} = \sqrt{\frac{0.3}{2.778+78.11}} = 0.0609 \frac{\text{m}^3}{\text{h}}$

The pressure losses on the elements for the throttling element full opening and the set proportional range X_{p} are as follows:

(a)	$\left[\Delta p_{z,100}\right]_{Xp=2} = \left[r_{z,100}\right]_{Xp=2} \cdot \left[\dot{V}_{100}^2\right]_{Xp=2} = 2.778 \cdot 0.01547^2 = 0.0006648\ \text{bar}$
(b)	$\left[\Delta p_{z,100}\right]_{Xp=2} = \left[r_{z,100}\right]_{Xp=2} \cdot \left[\dot{V}_{100}^2\right]_{Xp=2} = 2.778 \cdot 0.046^2 = 0.00588\ \text{bar}$
(c)	$\left[\Delta p_{z,100}\right]_{Xp=2} = \left[r_{z,100}\right]_{Xp=2} \cdot \left[\dot{V}_{100}^2\right]_{Xp=2} = 2.778 \cdot 0.0996^2 = 0.02756\ \text{bar}$
(d)	$\left[\Delta p_{z,100}\right]_{Xp=2} = \left[r_{z,100}\right]_{Xp=2} \cdot \left[\dot{V}_{100}^2\right]_{Xp=2} = 2.778 \cdot 0.0609^2 = 0.0103\ \text{bar}$

The valve outer authority at the throttling element full opening and the set proportional range X_p totals:

(a)	$\left[a_{z,\min}\right]_{Xp=2} = \frac{\left[r_{z,100}\right]_{Xp=2}}{\left[r_{z,100}\right]_{Xp=2}+r_{str}} = \frac{2.778}{2.778+1250} = 0.00222$
(b)	$\left[a_{z,\min}\right]_{Xp=2} = \frac{\left[r_{z,100}\right]_{Xp=2}}{\left[r_{z,100}\right]_{Xp=2}+r_{str}} = \frac{2.778}{2.778+138.9} = 0.0196$
(c)	$\left[a_{z,\min}\right]_{Xp=2} = \frac{\left[r_{z,100}\right]_{Xp=2}}{\left[r_{z,100}\right]_{Xp=2}+r_{str}} = \frac{2.778}{2.778+27.44} = 0.0919$
(d)	$\left[a_{z,\min}\right]_{Xp=2} = \frac{\left[r_{z,100}\right]_{Xp=2}}{\left[r_{z,100}\right]_{Xp=2}+r_{str}} = \frac{2.778}{2.778+78.11} = 0.03434$

According to formula (4.139), the required pre-setting values are as follows:

$$n_i = n_{\max} \cdot \left(1 - \frac{\ln \frac{\left[\frac{\Delta p_{cz}}{\left(\frac{\dot{V}_x}{[k_{vs}]_{Xp=2}}\right)^2 + \Delta p_{str,x}} - 1 + \left[a_{z,\min} \cdot a(n_i)_{w,I,\min}\right]_{Xp=2} \right]}{\left[a_{z,\min} \cdot a(n_i)_{w,I,\min}\right]_{Xp=2}}}{2c} \right).$$

(a)	$n_i = 8 \cdot \left(1 - \frac{\ln \frac{\left[\frac{0.3}{\left(\frac{0.01265}{0.6}\right)^2 + 0.2} - 1 + 0.00222 \cdot \left(\begin{array}{l} 0.001772 n_i^6 - 0.0436 n_i^5 + 0.42559 n_i^4 \\ -2.0897 n_i^3 + 5.396 n_i^2 - 6.904 n_i + 3.704 \end{array} \right) \right]}{0.00222 \cdot \left(\begin{array}{l} 0.001772 n_i^6 - 0.0436 n_i^5 + 0.42559 n_i^4 \\ -2.0897 n_i^3 + 5.396 n_i^2 - 6.904 n_i + 3.704 \end{array} \right)}}{2 \cdot 3.5} \right)$ $\underline{n_i = 1.0}$

(continued)

(continued)

(b)

$$n_i = 8 \cdot \left(1 - \frac{\ln \dfrac{\left[\dfrac{0.3}{\left(\frac{0.03795}{0.6}\right)^2 + 0.2} - 1 + 0.0196 \cdot \begin{pmatrix} 0.001772n_i^6 - 0.0436n_i^5 + 0.42559n_i^4 \\ -2.0897n_i^3 + 5.396n_i^2 - 6.904n_i + 3.704 \end{pmatrix} \right]}{0.0196 \cdot \begin{pmatrix} 0.001772n_i^6 - 0.0436n_i^5 + 0.42559n_i^4 \\ -2.0897n_i^3 + 5.396n_i^2 - 6.904n_i + 3.704 \end{pmatrix}}}{2 \cdot 3.5} \right)$$

$\underline{n_i = 3.0}$

(c)

$$n_i = 8 \cdot \left(1 - \frac{\ln \dfrac{\left[\dfrac{0.3}{\left(\frac{0.08538}{0.6}\right)^2 + 0.2} - 1 + 0.0919 \cdot \begin{pmatrix} 0.001772n_i^6 - 0.0436n_i^5 + 0.42559n_i^4 \\ -2.0897n_i^3 + 5.396n_i^2 - 6.904n_i + 3.704 \end{pmatrix} \right]}{0.0919 \cdot \begin{pmatrix} 0.001772n_i^6 - 0.0436n_i^5 + 0.42559n_i^4 \\ -2.0897n_i^3 + 5.396n_i^2 - 6.904n_i + 3.704 \end{pmatrix}}}{2 \cdot 3.5} \right)$$

$\underline{n_i = 5.0}$

(d)

$$n_i = 8 \cdot \left(1 - \frac{\ln \dfrac{\left[\dfrac{0.3}{\left(\frac{0.0506}{0.6}\right)^2 + 0.2} - 1 + 0.03434 \cdot \begin{pmatrix} 0.001772n_i^6 - 0.0436n_i^5 + 0.42559n_i^4 \\ -2.0897n_i^3 + 5.396n_i^2 - 6.904n_i + 3.704 \end{pmatrix} \right]}{0.03434 \cdot \begin{pmatrix} 0.001772n_i^6 - 0.0436n_i^5 + 0.42559n_i^4 \\ -2.0897n_i^3 + 5.396n_i^2 - 6.904n_i + 3.704 \end{pmatrix}}}{2 \cdot 3.5} \right)$$

$\underline{n_i = 3.54}$

The values of pre-setting n_i obtained by means of the two methods differ slightly from each other (an ~0.8% difference) because of different forms of the polynomial functions approximating the curves between the defined values. In the case of the valve under analysis, it is possible to set the pre-setting value with a step of 0.5. Consequently, values such as $n_\text{i} = 3.51$ or $n_\text{i} = 3.54$ cannot be selected and the closest to the required one is the value of $n_\text{i} = 3.5$. It might seem that if the value of 3.5 is selected, the effect will be higher-than-required throttling of the medium, and due to that the radiator heat output will be too low. The relations enabling calculations of the medium flow rate at any value of pre-setting n_i are derived on previous pages. They become indispensable if the selected pre-setting value differs from the required one and if the resultant mass/volume flow of the medium and the difference therein compared to the required value have to be defined (e.g. to establish the radiator heat output). This is possible in the case of manual valves, but for thermostatic ones the situation is different. Firstly, in the case of thermostatic valves, the throttling element is not responsible for current regulation, and under normal operating conditions the user has no access thereto and is unable to use the element to change the flow rate values. Secondly, as already described in the chapter devoted to thermostatic valves (or units thereof), the devices are capable of both decreasing and increasing the medium flow rate compared to the required design value, which is due to the initial intermediate position of the current regulation element for design load conditions. Therefore, in the case under consideration, i.e. selecting the pre-setting value lower than required, the medium flow rate will also be smaller than required and its value can be calculated using the relations presented herein. This, however, will be so only at the initial stage, before the response of the thermostatic head co-operating with the radiator valve, when the closing element is still in the design opening position h_N, corresponding to the assumed proportional range $X_p = 2$ K for the required room temperature value. Later on, the smaller-than-required value of the medium flow rate will result in a shortage of the radiator heat output, and this means too low a level of the room temperature. Due to the fact that the thermostatic unit, as the room temperature regulator, will keep trying to maintain the set temperature value, a change will occur in the closing element position to achieve a state in which the water mass flow ensures the radiator heat output proper to maintain the set room temperature level (considering the deviation resulting from the non-zero range of hysteresis and the principle of the device proportional operation). In this case, therefore, a rise will occur in the opening degree, and the proportional range will thus be increased to the value of $X_p > 2$ K. If too high a value of pre-setting n_i is selected, the opposite will be the case. Although the resultant proportional range X_p will be higher than the design value and the temperature oscillations in the room will be smaller, it should be remembered that the regulation quality will deteriorate due to the smaller range of travel of the closing element. These issues are described in detail in [10, 11]. However, despite the fact that it is possible to increase the flow rate above the design value, this action should not be equated with the possibility of the closing element free compensation for however big effects of errors in the selection of the value of pre-setting n_i. Although the flow rate can always be reduced and a "downward" correction can be made (accepting the fact that the regulation quality is compromised), an increase in the flow rate is limited by a few factors. One of them is the closing plug initial (design) position h_N and the potential increment in the opening degree resulting therefrom. The next is the shape of the original closing characteristic. Yet another important factor is the value of the closing element total authority a_c. This can be illustrated using the earlier-presented charts showing closing characteristics as a function of total authority a_c (cf. Fig. 4.1). For the closing plug set initial position, e.g. $h_x/h_{100} = 0.294$, the smaller the value of total authority a_c, the smaller the increments can be at opening, regardless of the type of the characteristic the situation concerns. It can be seen in this example that if the values of total authority a_c are this low, even full opening of the closing element, exceeding the design value h_N, does not involve big changes in the medium flow rate. If the original characteristic is linear, the changes are practically none. It may happen, therefore, that the closing element will be unable to correct the flow rate being the effect of too low a pre-setting value, and the medium mass/volume flow shortage will occur despite the full opening, corresponding to the proportional range a few times higher than 2 K (which may cause increased temperature oscillations in the room). The flow rate sought value can be found for example from the relation defining the medium flow as a function of active pressure and the sum of the pipework and the valve hydraulic resistances for the closing plug full opening at a given value of pre-setting n_i, according to Eqs. (4.140) and (4.142). This also provides justification for the issues taken up on previous pages of this book, where it is argued that low values of the closing element inner and total authority (a_w and a_c, respectively), especially in combination with the original linear closing characteristic, are very unfavourable and may result in unstable operation of thermoregulators. It can be noticed that the shift in the closing plug position will have a significant effect on the medium flow rate only if the positions are close to full closing; big changes will occur already at very small changes in the opening degree. This is due directly to the fact that current regulation will have a practically on-off character. In the case of the valve under analysis, the phenomenon is illustrated by Fig. 5.17, which was created based on experimental data.

The valve closing element inner authority for the selected pre-setting totals:

(a)	$a_{w,x} = 0.000088$
(b)	$a_{w,x} = 0.00073$
(c)	$a_{w,x} = 0.0037$
(d)	$a_{w,x} = a_{w,\max} \cdot \left(\frac{1}{1 + a(n_i)_{w,I,\min} \cdot \left[\exp[2c \cdot (1 - n_i/n_{\max})] - 1 \right]} \right)$ $= 0.166 \cdot \left(\frac{1}{1 + 0.28 \cdot \left[\exp[2 \cdot 3.5 \cdot (1 - 3.6/8)] - 1 \right]} \right) = 0.0011$

In case (d), the selected pre-setting n_i ($n_i = 3.50$) differs slightly from the required one ($n_i = 3.51$–3.54) because the valve makes it impossible to set the exact calculated value, but the difference is negligibly small. It totals about 1% only. The authority value can be calculated like for the three previous values of pre-setting n_i. In order to do so, the value of flow factor k_v has to be known for the value of pre-setting n_i, at the closing element full range of travel $h_{\max}$. It is $k_{vx} = 0.202$ m^3/h. If the value is unknown, relation (4.88) can be used like in this case (taking account of the computational variability in the throttling element authority depending on pre-setting n_i). Both methods give the same result.

The valve authority a' for the selected pre-setting totals:

(a)	$a' = \frac{\Delta p_{z,x}}{\Delta p_{z,x}+\Delta p_{str,x}} = \frac{0.1}{0.3} = 0.333$
(b)	$a' = \frac{\Delta p_{z,x}}{\Delta p_{z,x}+\Delta p_{str,x}} = \frac{0.1}{0.3} = 0.333$
(c)	$a' = \frac{\Delta p_{z,x}}{\Delta p_{z,x}+\Delta p_{str,x}} = \frac{0.1}{0.3} = 0.333$
(d)	$a' = \frac{\Delta p_{z,x}}{\Delta p_{z,x}+\Delta p_{str,x}} = \frac{0.1}{0.3} = 0.333$

The valve outer authority for the selected pre-setting totals:

(a)	$a_{z,x} = \frac{\Delta p_{z,x}}{\Delta p_{z,x}+\Delta p_{str,x}} = \frac{r_{z,x}}{r_{z,x}+r_{str}} = \frac{1/k_{v,x}^2}{1/k_{v,x}^2+\Delta p_{str,x}/\dot{V}_x^2} = \frac{1/0.057^2}{1/0.057^2+0.2/0.01265^2} = 0.198$
(b)	$a_{z,x} = \frac{\Delta p_{z,x}}{\Delta p_{z,x}+\Delta p_{str,x}} = \frac{r_{z,x}}{r_{z,x}+r_{str}} = \frac{1/k_{v,x}^2}{1/k_{v,x}^2+\Delta p_{str,x}/\dot{V}_x^2} = \frac{1/0.164^2}{1/0.164^2+0.2/0.03795^2} = 0.21$
(c)	$a_{z,x} = \frac{\Delta p_{z,x}}{\Delta p_{z,x}+\Delta p_{str,x}} = \frac{r_{z,x}}{r_{z,x}+r_{str}} = \frac{1/k_{v,x}^2}{1/k_{v,x}^2+\Delta p_{str,x}/\dot{V}_x^2} = \frac{1/0.37^2}{1/0.37^2+0.2/0.08538^2} = 0.21$
(d)	$a_{z,x} = \frac{\Delta p_{z,x}}{\Delta p_{z,x}+\Delta p_{str,x}} = \frac{r_{z,x}}{r_{z,x}+r_{str}} = \frac{1/k_{v,x}^2}{1/k_{v,x}^2+\Delta p_{str,x}/\dot{V}_x^2} = \frac{1/0.202^2}{1/0.202^2+0.2/0.0506^2} = 0.24$

The valve closing element total authority for the selected pre-setting totals:

(a)	$a_{c,x} = a_{w,x} \cdot a_{z,x} = 0.000088 \cdot 0.198 = 0.0000174$
(b)	$a_{c,x} = a_{w,x} \cdot a_{z,x} = 0.00073 \cdot 0.21 = 0.000153$
(c)	$a_{c,x} = a_{w,x} \cdot a_{z,x} = 0.0037 \cdot 0.21 = 0.000777$
(d)	$a_{c,x} = a_{w,x} \cdot a_{z,x} = 0.0011 \cdot 0.24 = 0.000264$

The calculation results are listed in Table 6.8.

Variant 2

Find the required values of the *Danfoss RTD-N DN15* valve pre-setting and total authority using the following data:

- required volume flow of the medium: $\dot{V}_x = 0.03795\ \mathrm{m^3/h}$,
- hydraulic resistance of the pipework: $r_{str} = 134.0\ (\mathrm{h^2\ bar})/\mathrm{m^6}$,
- active (differential) pressure in the circuit (Δp_{cz}):

(a) 0.2 bar,
(b) 0.2127 bar,
(c) 0.293 bar.

Table 6.8 Calculation results for Example 6.16, variant 1

Method	Sought quantity			
	Pre-setting n_i	Inner authority a_W	Authority a'/ Outer authority a_Z	Total authority a_C
The common method	Required (a) $n_i = 1.0$ (b) $n_i = 3.0$ (c) $n_i = 5.0$ (d) $n_i = (3.0; 4,0)$	No computational algorithm No requirements	Required: $a' \geq 0.3$ (a) $a' = 0.333$ (b) $a' = 0.333$ (c) $a' = 0.333$ (d) $a' = 0.333$	No computational algorithm No requirements
Method proposed by *Pyrkov* and *Szaflik* in [12, 13]	No computational algorithm	No requirements (a) a_W = unknown (b) a_W = unknown (c) a_W = unknown (d) a_W = unknown	No requirements (a) a_Z = unknown (b) a_Z = unknown (c) a_Z = unknown (d) a_Z = unknown	Required: $a_C \geq 0.3$ (a) a_C = unknown (b) a_C = unknown (c) a_C = unknown (d) a_C = unknown
Proposed method	(a) $n_i = 1.0$ (b) $n_i = 3.0$ (c) $n_i = 5.0$ (d) $n_i = (3.51;3.54)$	No requirements (a) $a_W = 0.000088$ (b) $a_W = 0.00073$ (c) $a_W = 0.0037$ (d) $a_W = 0.0011$	No requirements (a) $a_Z = 0.198$; $a' = 0.333$ (b) $a_Z = 0.21$; $a' = 0.333$ 9c) $a_Z = 0.21$; $a' = 0.333$ (d) $a_Z = 0.24$; $a' = 0.333$	Required: $a_C \geq 0.3$ (a) $a_C = 0.0000174$ (b) $a_C = 0.000153$ (c) $a_C = 0.000777$ (d) $a_C = 0.000264$

Solution

This time the set value is not the pipework pressure loss Δp_{str} but the pipework hydraulic resistance r_{str}, i.e. a characteristic parameter of the pipework. Moreover, different pressure values are set for which the same value of the medium volume flow has to be maintained. In practice for example, this corresponds to a change in the pressure supplied by the pump to the installation or the pump replacement with a different one, or the use of local pressure-stabilizing systems. The calculations are performed using the same formulae as in the previous variant. The final relations and results are presented below.

Calculations performed by means of the common method

The same as in Example 6.15, variant 2.

Calculations performed by means of the method proposed by *Pyrkov* and *Szaflik*

The same as in Example 6.15, variant 2 and in Example 6.16, variant 1.

Calculations performed by means of the proposed method

Method #1:

$$\dot{V}_x - \sqrt{\frac{\Delta p_{cz}}{\frac{1}{\left([k_v(n_i)]_{Xp=2}\right)^2} + r_{str}}} = 0.$$

(a)	$0.03795 - \sqrt{\frac{0.2}{\frac{1}{\left(\begin{array}{l}-0.0004167n_i^6+0.010167n_i^5-0.09792n_i^4+\\0.47083n_i^3-1.16667n_i^2+1.424n_i-0.6\end{array}\right)^2}+134}} = 0$ $\underline{n_i = 7.0}$
(b)	$0.03795 - \sqrt{\frac{0.2127}{\frac{1}{\left(\begin{array}{l}-0.0004167n_i^6+0.010167n_i^5-0.09792n_i^4+\\0.47083n_i^3-1.16667n_i^2+1.424n_i-0.6\end{array}\right)^2}+134}} = 0$ $\underline{n_i = 5.0}$
(c)	$0.03795 - \sqrt{\frac{0.293}{\frac{1}{\left(\begin{array}{l}-0.0004167n_i^6+0.010167n_i^5-0.09792n_i^4+\\0.47083n_i^3-1.16667n_i^2+1.424n_i-0.6\end{array}\right)^2}+134}} = 0$ $\underline{n_i = 3.0}$

The valve authority a' for the selected pre-setting totals:

(a)	$a' = \frac{\Delta p_{z,x}}{\Delta p_{z,x}+\Delta p_{str,x}} = \frac{r_{z,x}}{r_{z,x}+r_{str}} = \frac{1/\left[k_{v,x}^2\right]_{Xp=2}}{1/\left[k_{v,x}^2\right]_{Xp=2}+r_{str}} = \frac{1/0.45^2}{1/0.45^2+134} = 0.0355$
(b)	$a' = \frac{\Delta p_{z,x}}{\Delta p_{z,x}+\Delta p_{str,x}} = \frac{r_{z,x}}{r_{z,x}+r_{str}} = \frac{1/\left[k_{v,x}^2\right]_{Xp=2}}{1/\left[k_{v,x}^2\right]_{Xp=2}+r_{str}} = \frac{1/0.27^2}{1/0.27^2+134} = 0.093$
(c)	$a' = \frac{\Delta p_{z,x}}{\Delta p_{z,x}+\Delta p_{str,x}} = \frac{r_{z,x}}{r_{z,x}+r_{str}} = \frac{1/\left[k_{v,x}^2\right]_{Xp=2}}{1/\left[k_{v,x}^2\right]_{Xp=2}+r_{str}} = \frac{1/0.12^2}{1/0.12^2+134} = 0.341$

The valve outer authority for the selected pre-setting totals:

(a)	$a_{z,x} = \frac{\Delta p_{z,x}}{\Delta p_{z,x}+\Delta p_{str,x}} = \frac{r_{z,x}}{r_{z,x}+r_{str}} = \frac{1/k_{v,x}^2}{1/k_{v,x}^2+r_{str}} = \frac{1/0.64^2}{1/0.64^2+134} = 0.0179$
(b)	$a_{z,x} = \frac{\Delta p_{z,x}}{\Delta p_{z,x}+\Delta p_{str,x}} = \frac{r_{z,x}}{r_{z,x}+r_{str}} = \frac{1/k_{v,x}^2}{1/k_{v,x}^2+r_{str}} = \frac{1/0.37^2}{1/0.37^2+134} = 0.0517$
(c)	$a_{z,x} = \frac{\Delta p_{z,x}}{\Delta p_{z,x}+\Delta p_{str,x}} = \frac{r_{z,x}}{r_{z,x}+r_{str}} = \frac{1/k_{v,x}^2}{1/k_{v,x}^2+r_{str}} = \frac{1/0.164^2}{1/0.164^2+134} = 0.217$

The valve closing element inner authority for the selected pre-setting totals:

(a)	$a_{w,x} = 0.011$
(b)	$a_{w,x} = 0.0037$
(c)	$a_{w,x} = 0.00073$

The valve total authority for the selected pre-setting totals:

(a)	$a_{c,x} = a_{w,x} \cdot a_{z,x} = 0.011 \cdot 0.0179 = 0.000197$
(b)	$a_{c,x} = a_{w,x} \cdot a_{z,x} = 0.0037 \cdot 0.0517 = 0.000191$
(c)	$a_{c,x} = a_{w,x} \cdot a_{z,x} = 0.00073 \cdot 0.217 = 0.000158$

The calculation results are listed in Table 6.9.

Example 6.17
Variant 1

Find the required values of pre-setting and total authority of the *Herz TS-FV DN15* valve of a linear figure co-operating with a thermostatic head using the following data:

- active (differential) pressure in the circuit: $\Delta p_{cz} = 0.3$ bar,
- hydraulic resistance of the pipework: $r_{str} = 29.585$ bar/(m^3/h)^2,
- required volume flow of the medium:

Table 6.9 Calculation results for Example 6.16, variant 2

Method	Sought quantity			
	Pre-setting n_i	Inner authority a_w	Authority a'/outer authority a_z	Total authority a_c
The common method	No computational algorithm	No computational algorithm No requirements	Required: a' ≥ 0.3 No computational algorithm	No computational algorithm No requirements
Method proposed by *Pyrkov* and *Szaflik* in [12, 13]	No computational algorithm	No requirements (a) a_w = unknown (b) a_w = unknown (c) a_w = unknown	No requirements (a) a_z = unknown (b) a_z = unknown (c) a_z = unknown	Required: $a_c \geq 0.3$ (a) a_c = unknown (b) a_c = unknown (c) a_c = unknown
Proposed method	Required (a) $n_i = 7.0$ (b) $n_i = 5.0$ (c) $n_i = 3.0$	No requirements (a) $a_w = 0.011$ (b) $a_w = 0.0037$ (c) $a_w = 0.00073$	No requirements (a) $a_z = 0.0179$; $a' = 0.0355$ (b) $a_z = 0.0517$; $a' = 0.093$ (c) $a_z = 0.217$; $a' = 0.341$	Required: $a_c \geq 0.3$ (a) $a_c = 0.000197$ (b) $a_c = 0.000191$ (c) $a_c = 0.000158$

Table 6.10 Throttling characteristic data of the *Herz TS-FV 7523* valve [1]

Number of pre-setting n_i		The valve flow factor $k_{v,x}$ [m³/(h bar$^{0.5}$)]
$X_p = 2$ K	1	0.019
	2	0.043
	3	0.089
	4	0.17
	5	0.26
	6	0.3
$X_p = \max$	6	0.39

(a) $\dot{V}_x = 0.01035\,\mathrm{m}^3/\mathrm{h}$,
(b) $\dot{V}_x = 0.04388\,\mathrm{m}^3/\mathrm{h}$,
(c) $\dot{V}_x = 0.08222\,\mathrm{m}^3/\mathrm{h}$,
(d) $\dot{V}_x = 0, .0842\,\mathrm{m}^3/\mathrm{h}$.

Solution

The case concerns a thermostatic, double-regulation valve with two adjustable sections of the medium flow, with the throttling-closing element structure as presented in Fig. 2.29e. The data of flow factor k_v as a function of pre-setting n_i for the analysed valve co-operating with a thermostatic head are listed in Table 6.10.

This time the set value is not the pipework pressure loss Δp_{str} but the pipework hydraulic resistance r_{str}, i.e. a characteristic parameter of the pipework. It follows that the situation concerns selection of a single pipework and not four different ones generating the same pressure losses for different values of the medium mass/volume flow, like in the previous examples. The variant thus concerns, for instance, finding the correct value of pre-setting n_i for a single heating circuit depending on the required flow rate of the medium. The previous examples could be solved using either hydraulic resistances (obtaining an indirect result in the form of the values of pre-setting n_i—Method #1) or the pressure drops corresponding thereto (Method #2). In this case, Method #1 is more convenient.

Calculations performed by means of the common method

In order to carry out the calculations in this variant, it is necessary to find the pressure that will have to be throttled on the valve for the sought i-th pre-setting n_i and the nominal opening degree of the closing plug corresponding to proportional range $Xp = 2$ K. Because there is no direct information on the pressure value, it is impossible to determine the required value of pre-setting n_i by means of the commonly applied calculation method. The situation, therefore, is the same as in Example 6.15, variant 2.

Calculations performed by means of the method proposed by *Pyrkov* and *Szaflik*

The same as in Example 6.15, variant 2 and in Example 6.16, variant 1.

Calculations performed by means of the proposed method

Method #1:

Based on the data from the valve throttling/closing characteristic measurements presented in Table 6.10, function $k_v(n_i)_{\mathrm{Xp=2}}$ for pre-settings $n_i = 1.0$–6.0 can be written as:

$$[k_v(n_i)]_{Xp=2} = 0.00005\,n_i^5 - 0.002375\,n_i^4 + 0.022667\,n_i^3 - 0.07012\,n_i^2 + 0.1098\,n_i - 0.041.$$

According to formula (4.141a), the required pre-setting values are as follows:

$$\dot{V}_x - \sqrt{\frac{\Delta p_{cz}}{\frac{1}{\left([k_v(n_i)]_{Xp=2}\right)^2} + r_{str}}} = 0.$$

(a)	$0.01035 -$ $\underline{n_i = 1.0}$	$\sqrt{\frac{0.3}{\frac{1}{\left(0.00005n_i^5 - 0.002375n_i^4 + 0.022667n_i^3 - 0.07012n_i^2 + 0.1098n_i - 0.041\right)^2} + 29.585}} = 0$
(b)	$0.04388 -$ $\underline{n_i = 3.0}$	$\sqrt{\frac{0.3}{\frac{1}{\left(0.00005n_i^5 - 0.002375n_i^4 + 0.022667n_i^3 - 0.07012n_i^2 + 0.1098n_i - 0.041\right)^2} + 29.585}} = 0$
(c)	$0.08222 -$ $\underline{n_i = 5.0}$	$\sqrt{\frac{0.3}{\frac{1}{\left(0.00005n_i^5 - 0.002375n_i^4 + 0.022667n_i^3 - 0.07012n_i^2 + 0.1098n_i - 0.041\right)^2} + 29.585}} = 0$
(d)	$0.0842 -$ $\underline{n_i = 5.29}$	$\sqrt{\frac{0.3}{\frac{1}{\left(0.00005n_i^5 - 0.002375n_i^4 + 0.022667n_i^3 - 0.07012n_i^2 + 0.1098n_i - 0.041\right)^2} + 29.585}} = 0$

Like in the previous examples, the value of pre-setting n_i is calculated by means of the other method, after relevant quantities are determined. It is assumed that, in the range of the defined values of pre-setting n_i, the original throttling characteristic is an equal-percentage curve with factor $c = 4.9$ and the original closing characteristic

Table 6.11 Data of the *Herz TS-FV7523* valve throttling characteristic and calculated values of the throttling and the closing element inner authority ([2] and own testing results)

Number of pre-setting n_i		The valve flow factor $k_{v,x}$ [m^3/(h bar$^{0.5}$)]	Inner authority			
			Throttling element			Closing element
			$[a_{w,I,min}]_{Xp=2,}$	$a_{w,I,,min}$		$a_{w,x}$
$X_p = 2$ K/ $X_p = \max$	1	0.019/0.02	0.0705	0.108	$X_p = \max$	0.00039
	2	0.043/0.045	0.0694	0.108		0.0020
	3	0.089/0.094	0.0777	0.122		0.0087
	4	0.17/0.192	0.0838	0.124		0.036
	5	0.26/0.32	0.0804	0.118		0.10
	6	0.3/0.39	–	–		0.15
$X_p = \max$	6	0.39	–	–		0.15

is linear. As indicated by experimental verification, the closing plug maximum travel range (lift) is $h_{100} = h_{max} = \sim 1.3$ mm. The values of flow factor k_v declared by the manufacturer for the set proportional range $X_p = 2$ K are obtained at the plug lift over the valve seat of $h_x = \sim 0.55$ mm. This gives the relative lift at the level of $h_x/h_{100} = h_x/h_{max} = 0.423$. Using the above information, the throttling and the closing element inner authority is calculated like in Example 6.16. The calculation results are listed in Table 6.11.

The variability in the throttling element inner authority $a_{w,I}$ as a function of pre-setting n_i, for $X_p = 2$ K, can be written using the following equation:

$$\left[a(n_i)_{w,I,\min}\right]_{Xp} = 0.00018384 n_i^4 - 0.003781 n_i^3 + 0.0228 n_i^2 - 0.0458 n_i + 0.0971.$$

The valve hydraulic resistance at the throttling element full opening and the set proportional range X_p totals:

$$\left[r_{z,100}\right]_{Xp=2} = \frac{1}{\left[k_{vs}^2\right]_{Xp=2}} = \frac{1}{0.3^2} = 11.11 \, \frac{\text{h}^2\,\text{bar}}{\text{m}^6}$$

The valve outer authority at the throttling element full opening and the set proportional range X_p totals:

$$\left[a_{z,\min}\right]_{Xp=2} = \frac{\left[r_{z,100}\right]_{Xp=2}}{\left[r_{z,100}\right]_{Xp=2} + r_{str}} = \frac{11.11}{11.11 + 29.585} = 0.273 -$$

According to formula (4.139), the required pre-setting values are as follows:

$$n_i = n_{\max} \cdot \left(1 - \frac{\ln \frac{\left[\frac{\Delta p_{cz}}{\dot{V}_x^2 \cdot \left([r_{z,100}]_{Xp=2} + r_{str}\right)} - 1 + [a_{z,\min} \cdot a(n_i)_{w,I,\min}]_{Xp=2}\right]}{[a_{z,\min} \cdot a(n_i)_{w,I,\min}]_{Xp=2}}}{2c}\right).$$

(a)	$n_i = n_{\max} \cdot \left(1 - \frac{\ln \frac{\left[\frac{0.3}{0.01035^2 \cdot (11.11+29.585)} - 1 + 0.273 \cdot \left(0.00018384 n_i^4 - 0.003781 n_i^3 + 0.0228 n_i^2 - 0.0458 n_i + 0.0971\right)\right]}{0.273 \cdot \left(0.00018384 n_i^4 - 0.003781 n_i^3 + 0.0228 n_i^2 - 0.0458 n_i + 0.0971\right)}}{2 \cdot 4.9}\right)$ $\underline{\underline{n_i = 1.0}}$
(b)	$n_i = n_{\max} \cdot \left(1 - \frac{\ln \frac{\left[\frac{0.3}{0.04388^2 \cdot (11.11+29.585)} - 1 + 0.273 \cdot \left(0.00018384 n_i^4 - 0.003781 n_i^3 + 0.0228 n_i^2 - 0.0458 n_i + 0.0971\right)\right]}{0.273 \cdot \left(0.00018384 n_i^4 - 0.003781 n_i^3 + 0.0228 n_i^2 - 0.0458 n_i + 0.0971\right)}}{2 \cdot 4.9}\right)$ $\underline{\underline{n_i = 3.0}}$

(continued)

(continued)

(c)	$n_i =$ $$n_{\max}\cdot\left(1-\frac{\ln\dfrac{\left[\frac{0.3}{0.08222^2\cdot(11.11+29.585)}-1+0.273\cdot\begin{pmatrix}0.00018384n_i^4-0.003781n_i^3+\\0.0228n_i^2-0.0458n_i+0.0971\end{pmatrix}\right]}{0.273\cdot\begin{pmatrix}0.00018384n_i^4-0.003781n_i^3+\\0.0228n_i^2-0.0458n_i+0.0971\end{pmatrix}}}{2\cdot4.9}\right)$$ $\underline{\underline{n_i = 5.0}}$
(d)	$n_i =$ $$n_{\max}\cdot\left(1-\frac{\ln\dfrac{\left[\frac{0.3}{0.0842^2\cdot(11.11+29.585)}-1+0.273\cdot\begin{pmatrix}0.00018384n_i^4-0.003781n_i^3+\\0.0228n_i^2-0.0458n_i+0.0971\end{pmatrix}\right]}{0.273\cdot\begin{pmatrix}0.00018384n_i^4-0.003781n_i^3+\\0.0228n_i^2-0.0458n_i+0.0971\end{pmatrix}}}{2\cdot4.9}\right)$$ $\underline{\underline{n_i = 5.34}}$

Due to its structure, the analysed valve has to be set to the total value of pre-setting n_i. The situation is similar to the valve in Example 6.16, and the selection of the pre-setting value lower or higher than the required one will cause the same effects. Selecting pre-setting $n_i = 6$, it is certain that the obtained value of the medium volume flow will not be too low. In the steady state, the valve, co-operating with the head by adjusting the closing plug position and proportional range X_p ($X_p < 2$ K), will set the required hydraulic resistance corresponding to the required pre-setting $n_i = 5.29$–5.34 for $X_p = 2$ K.

The valve closing element inner authority for the selected pre-setting totals:

(a)	$a_{w,x} = 0.00039$
(b)	$a_{w,x} = 0.0087$
(c)	$a_{w,x} = 0.1$
(d)	$a_{w,x} = 0.15$

The valve authority a' for the selected pre-setting totals:

(a)	$a' = \frac{\Delta p_{z,x}}{\Delta p_{z,x}+\Delta p_{str,x}} = \frac{r_{z,x}}{r_{z,x}+r_{str}} = \frac{1/[k^2_{v,x}]_{Xp=2}}{1/[k^2_{v,x}]_{Xp=2}+r_{str}} = \frac{1/0.019^2}{1/0.019^2+29.585} = 0.99$
(b)	$a' = \frac{\Delta p_{z,x}}{\Delta p_{z,x}+\Delta p_{str,x}} = \frac{r_{z,x}}{r_{z,x}+r_{str}} = \frac{1/[k^2_{v,x}]_{Xp=2}}{1/[k^2_{v,x}]_{Xp=2}+r_{str}} = \frac{1/0.089^2}{1/0.089^2+29.585} = 0.81$
(c)	$a' = \frac{\Delta p_{z,x}}{\Delta p_{z,x}+\Delta p_{str,x}} = \frac{r_{z,x}}{r_{z,x}+r_{str}} = \frac{1/[k^2_{v,x}]_{Xp=2}}{1/[k^2_{v,x}]_{Xp=2}+r_{str}} = \frac{1/0.26^2}{1/0.26^2+29.585} = 0.333$

In case (d), in order to establish the valve authority a', it is necessary to find the valve hydraulic resistance r_z, flow factor k_v, or the pressure throttled on the element that will stabilize for the required flow rate. The following is then obtained:

(d)	$a' = \frac{\Delta p_{z,x}}{\Delta p_{ob}} = \frac{\Delta p_{z,x}}{\Delta p_{z,x}+\Delta p_{str,x}} = \frac{\Delta p_{cz}-r_{str}\cdot \dot{V}^2_x}{\Delta p_{cz}} = \frac{0.3-29.585\cdot 0.0842^2}{0.3} = 0.3$

The valve outer authority for the selected pre-setting totals:

(a)	$a_{z,x} = \frac{\Delta p_{z,x}}{\Delta p_{z,x}+\Delta p_{str,x}} = \frac{r_{z,x}}{r_{z,x}+r_{str}} = \frac{1/k^2_{v,x}}{1/k^2_{v,x}+r_{str}} = \frac{1/0.02^2}{1/0.02^2+29.585} = 0.988$
(b)	$a_{z,x} = \frac{\Delta p_{z,x}}{\Delta p_{z,x}+\Delta p_{str,x}} = \frac{r_{z,x}}{r_{z,x}+r_{str}} = \frac{1/k^2_{v,x}}{1/k^2_{v,x}+r_{str}} = \frac{1/0.094^2}{1/0.094^2+29.585} = 0.793$
(c)	$a_{z,x} = \frac{\Delta p_{z,x}}{\Delta p_{z,x}+\Delta p_{str,x}} = \frac{r_{z,x}}{r_{z,x}+r_{str}} = \frac{1/k^2_{v,x}}{1/k^2_{v,x}+r_{str}} = \frac{1/0.32^2}{1/0.32^2+29.585} = 0.248$
(d)	$a_{z,x} = \frac{\Delta p_{z,x}}{\Delta p_{z,x}+\Delta p_{str,x}} = \frac{r_{z,x}}{r_{z,x}+r_{str}} = \frac{1/k^2_{v,x}}{1/k^2_{v,x}+r_{str}} = \frac{1/0.39^2}{1/0.39^2+29.585} = 0.182$

The valve closing element total authority for the selected pre-setting totals:

(a)	$a_{c,x} = a_{w,x} \cdot a_{z,x} = 0.00039 \cdot 0.988 = 0.000385$
(b)	$a_{c,x} = a_{w,x} \cdot a_{z,x} = 0.0087 \cdot 0.793 = 0.0069$
(c)	$a_{c,x} = a_{w,x} \cdot a_{z,x} = 0.1 \cdot 0.248 = 0.0248$
(d)	$a_{c,x} = a_{w,x} \cdot a_{z,x} = 0.15 \cdot 0.182 = 0.0273$

The results are listed in Table 6.12.

Table 6.12 Calculation results for Example 6.17, variant 1

Method	Sought quantity			
	Pre-setting n_i	Inner authority a_w	Authority a'/outer authority a_z	Total authority a_c
The common method	No computational algorithm	No computational algorithm No requirements	Required: a' ≥ 0.3 No computational algorithm	No computational algorithm No requirements
Method proposed by *Pyrkov* and *Szaflik* in [12, 13]	No computational algorithm	No requirements (a) a_w = unknown (b) a_w = unknown (c) a_w = unknown (d) a_w = unknown	No requirements (a) a_z = unknown (b) a_z = unknown (c) a_z = unknown (d) a_z = unknown	Required: $a_c \geq 0.3$ (a) a_c = unknown (b) a_c = unknown (c) a_c = unknown 9d) a_c = unknown
Proposed method	Required (a) $n_i = 1.0$ (b) $n_i = 3.0$ (c) $n_i = 5.0$ (d) $n_i =$ (5.29;5.34)	No requirements (a) $a_w = 0.00039$ (b) $a_w = 0.0087$ (c) $a_w = 0.1$ (d) $a_w = 0.15$	No requirements (a) $a_z = 0.988$; $a' = 0.99$ (b) $a_z = 0.793$; $a' = 0.81$ (c) $a_z = 0.248$; $a' = 0.333$ (d) $a_z = 0.182$; $a' = 0.3$	Required: $a_c \geq 0.3$ (a) $a_c = 0.000385$ (b) $a_c = 0.0069$ (c) $a_c = 0.0248$ (d) $a_c = 0.0273$

Variant 2

Find the required values of pre-setting and total authority of the *Herz TS-FV7523 DN15* valve of a linear figure co-operating with a thermostatic head using the following data:

- hydraulic resistance of the pipework: $r_{str} = 34.515$ bar/(m^3/h)^2,
- required volume flow of the medium: $\dot{V}_x = 0.04388\,\text{m}^3/\text{h}$,
- active (differential) pressure in the circuit (Δp_{cz}):

(a) 0.095 bar,
(b) 0.133 bar,
(c) 0.31 bar.

Solution
The calculations are performed using the same formulae as in the previous variant. The final relations and results are presented below.

Calculations performed by means of the common method

The same as in Example 6.15, variant 2.

Calculations performed by means of the method proposed by *Pyrkov* and *Szaflik*

The same as in Example 6.15, variant 2 and in Example 6.16, variant 1.

Calculations performed by means of the proposed method

Method #1:

$$\dot{V}_x - \sqrt{\frac{\Delta p_{cz}}{\frac{1}{\left([k_v(n_i)]_{Xp=2}\right)^2} + r_{str}}} = 0.$$

(a)	$0.04388 - \sqrt{\frac{0.095}{\frac{1}{\left(0.00005n_i^5 - 0.002375n_i^4 + 0.022667n_i^3 - 0.07012n_i^2 + 0.1098n_i - 0.041\right)^2} + 34.515}} = 0$ $\underline{n_i = 5.0}$
(b)	$0.04388 - \sqrt{\frac{0.133}{\frac{1}{\left(0.00005n_i^5 - 0.002375n_i^4 + 0.022667n_i^3 - 0.07012n_i^2 + 0.1098n_i - 0.041\right)^2} + 34.515}} = 0$ $\underline{n_i = 4.0}$
(c)	$0.04388 - \sqrt{\frac{0.31}{\frac{1}{\left(0.00005n_i^5 - 0.002375n_i^4 + 0.022667n_i^3 - 0.07012n_i^2 + 0.1098n_i - 0.041\right)^2} + 34.515}} = 0$ $\underline{n_i = 3.0}$

The valve closing element inner authority for the selected pre-setting totals:

(a)	$a_{w,x} = 0.1$
(b)	$a_{w,x} = 0.036$
(c)	$a_{w,x} = 0.0087$

The valve authority a' for the selected pre-setting totals:

(a)	$a' = \frac{\Delta p_{z,x}}{\Delta p_{z,x}+\Delta p_{str,x}} = \frac{r_{z,x}}{r_{z,x}+r_{str}} = \frac{1/[k_{v,x}^2]_{Xp=2}}{1/[k_{v,x}^2]_{Xp=2}+r_{str}} = \frac{1/0.26^2}{1/0.26^2+34.515} = 0.3$
(b)	$a' = \frac{\Delta p_{z,x}}{\Delta p_{z,x}+\Delta p_{str,x}} = \frac{r_{z,x}}{r_{z,x}+r_{str}} = \frac{1/[k_{v,x}^2]_{Xp=2}}{1/[k_{v,x}^2]_{Xp=2}+r_{str}} = \frac{1/0.17^2}{1/0.17^2+34.515} = 0.5$
(c)	$a' = \frac{\Delta p_{z,x}}{\Delta p_{z,x}+\Delta p_{str,x}} = \frac{r_{z,x}}{r_{z,x}+r_{str}} = \frac{1/[k_{v,x}^2]_{Xp=2}}{1/[k_{v,x}^2]_{Xp=2}+r_{str}} = \frac{1/0.089^2}{1/0.089^2+34.515} = 0.785$

The valve outer authority for the selected pre-setting totals:

(a)	$a_{z,x} = \frac{\Delta p_{z,x}}{\Delta p_{z,x}+\Delta p_{str,x}} = \frac{r_{z,x}}{r_{z,x}+r_{str}} = \frac{1/k_{v,x}^2}{1/k_{v,x}^2+r_{str}} = \frac{1/0.32^2}{1/0.32^2+34.515} = 0.22$
(b)	$a_{z,x} = \frac{\Delta p_{z,x}}{\Delta p_{z,x}+\Delta p_{str,x}} = \frac{r_{z,x}}{r_{z,x}+r_{str}} = \frac{1/k_{v,x}^2}{1/k_{v,x}^2+r_{str}} = \frac{1/0.192^2}{1/0.192^2+34.515} = 0.44$
(c)	$a_{z,x} = \frac{\Delta p_{z,x}}{\Delta p_{z,x}+\Delta p_{str,x}} = \frac{r_{z,x}}{r_{z,x}+r_{str}} = \frac{1/k_{v,x}^2}{1/k_{v,x}^2+r_{str}} = \frac{1/0.094^2}{1/0.094^2+34.515} = 0.766$

The valve closing element total authority for the selected pre-setting totals:

(a)	$a_{c,x} = a_{w,x} \cdot a_{z,x} = 0.1 \cdot 0.22 = 0.022$
(b)	$a_{c,x} = a_{w,x} \cdot a_{z,x} = 0.036 \cdot 0.44 = 0.0158$
(c)	$a_{c,x} = a_{w,x} \cdot a_{z,x} = 0.0087 \cdot 0.766 = 0.00666$

The calculation results are listed in Table 6.13.

Example 6.18
Variant 1

Find the required values of pre-setting and total authority of the *Herz TS-90-V DN15* valve of a linear figure co-operating with a thermostatic head using the following data:

Table 6.13 Calculation results for Example 6.17, variant 2

Method	Sought quantity			
	Pre-setting n_i	Inner authority a_w	Authority a'/outer authority a_z	Total authority a_c
The common method	No computational algorithm	No computational algorithm No requirements	Required: $a' \geq 0.3$ No computational algorithm	No computational algorithm No requirements
Method proposed by *Pyrkov* and *Szaflik* in [12, 13]	No computational algorithm	No requirements (a) a_w = unknown (b) a_w = unknown (c) a_w = unknown (d) a_w = unknown	No requirements (a) a_z = unknown (b) a_z = unknown (c) a_z = unknown (d) a_z = unknown	Required: $a_c \geq 0.3$ (a) a_c = unknown (b) a_c = unknown (c) a_c = unknown (d) a_c = unknown
Proposed method	Required (a) $n_i = 5.0$ (b) $n_i = 4.0$ (c) $n_i = 3.0$	No requirements (a) $a_w = 0.1$ (b) $a_w = 0.036$ (c) $a_w = 0.0087$	No requirements (a) $a_z = 0.22$; $a' = 0.3$ (b) $a_z = 0.44$; $a' = 0.5$ (c) $a_z = 0.766$; $a' = 0.785$	Required: $a_c \geq 0.3$ (a) $a_c = 0.022$ (b) $a_c = 0.0158$ (c) $a_c = 0.00666$

- active (differential) pressure in the circuit: $\Delta p_{cz} = 0.3$ bar,
- hydraulic resistance of the pipework: $r_{str} = 22.786$ bar/(m^3/h)^2,
- required volume flow of the medium:

(a) $\dot{V}_x = 0.0453\,\text{m}^3/\text{h}$,
(b) $\dot{V}_x = 0.07923\,\text{m}^3/\text{h}$,
(c) $\dot{V}_x = 0.096\,\text{m}^3/\text{h}$,
(d) $\dot{V}_x = 0.093\,\text{m}^3/\text{h}$.

Solution

The case concerns a thermostatic, double-regulation valve with one adjustable section of the medium flow, as presented in Fig. 2.28. The data of flow factor k_v as a function of pre-setting n_i for the analysed valve co-operating with a thermostatic head are listed in Table 6.14.

In terms of the computational methodology presented in this book, the valve is rather unique. As it can be noticed in Fig. 2.28, the structure of the throttling and the closing elements, their position in relation to the valve seat, like in the case of the *Danfoss MSV*-I valve, implies that the original throttling characteristic is approximately linear. But the original throttling characteristic of the valve tested under the requirements of the EN 215:2004 standard [4] and co-operating with a thermostatic head will not be linear, which results directly from the structure of the plugs, their mutual position and the principle of the operation of valves in relation to the requirements set out for thermostatic units. According to standard EN 215:2004 [4], the flow factor is to be determined for every value of pre-setting n_i, but at a set constant value of proportional range X_p. The closing plug, setting the value of X_p, is permanently in a single position, whereas the throttling plug moves setting different values of throttling. Because the closing element is placed inside the throttling one, the total hydraulic resistance value varies only due to a change in the surface area of the cross-section through which the liquid leaves the mechanism. By contrast,

Table 6.14 Throttling/closing characteristic data of the *Herz TS-90-V* valve [2]

Number of pre-setting n_i		The valve flow factor k_{vx} [m^3/(h bar$^{0.5}$)]
$X_p = 2$ K	1	0.03
	2	0.05
	3	0.09
	4	0.15
	5	0.2
	6	0.25
	7	0.32
	8	0.5
	9	0.55
$X_p = \max$	max.	1.1

the surface area of the cross-section of the inflow under the mechanism (not to be confused with the valve seat orifice inlet), which is determined by the plug, is constant. This is due to the fact that as the throttling plug rises, only one of the two hydraulic resistances of the medium outflow section in the regulated section decreases. Consequently, the resultant hydraulic resistance changes less than it would if both elements moved at the same time. As a result, the original closing characteristic will in this situation be characterized by downward convexity. If, like in the case of manual valves with a double coaxial plug, the two elements moved simultaneously, the original throttling characteristic would be linear. There is yet another reason for which the original closing characteristic is not linear. The closing plug inherent part is a seal and the plug does not reach the throttling plug edge. As it comes out, it causes additional local swirling of the liquid and big increments in hydraulic resistances in its initial travel range. The closing characteristics obtained by means of experiments are presented in Fig. 5.9. The flow factor maximum value is determined for the full lift of both plugs. The lift is much bigger than for the maximum pre-setting defined for the valve. Apart from that, the manufacturer specified the pre-setting numbers relating them in some ranges to the plug lift by means of a non-linear dependence, i.e. a doubled pre-setting value does not mean a doubled opening degree for example. An approximately linear relation holds in the pre-setting range of $n_i = 3–7$.

Considering all the above issues, it is generally impossible to determine the valve inner authority a_w by means of Method #2 using the data provided by the manufacturer. Instead, appropriate quantities have to be measured, i.e. flow factor k_v depending on the regulated section opening degree (both plugs jointly). This is a typical situation for this kind of thermostatic valves. However, the required value of pre-setting n_i can be calculated analytically by means of Method #1 proposed in this book. The method requires the knowledge of neither the value of inner authority a_w nor the mathematical description of the original closing characteristic.

Calculations performed by means of the common method

The same as in Example 6.15, variant 2.

Calculations performed by means of the method proposed by *Pyrkov* and *Szaflik*

The same as in Example 6.15, variant 2 and in Example 6.16, variant 1.

Calculations performed by means of the proposed method

Method #1:

Based on the data from the valve throttling/closing characteristic measurements presented in Table 6.14, for the pre-setting range of $n_i = 1–7$, function $k_v(n_i)_{\mathrm{I}}$ can be expressed as:

$$[k_v(n_i)]_{Xp=2} = -0.000138889n_i^6 + 0.0035n_i^5 - 0.0343055n_i^4 + 0.164166n_i^3 - 0.39055n_i^2 + 0.45733n_i - 0.17.$$

According to formula (4.141a), the required pre-setting values are as follows:

$$\dot{V}_x - \sqrt{\frac{\Delta p_{cz}}{\frac{1}{\left([k_v(n_i)]_{Xp=2}\right)^2} + r_{str}}} = 0.$$

(a)	$0.0453 - \sqrt{\frac{0.3}{\frac{1}{\left(\begin{array}{l} -0.000138889n_i^6 + 0.0035n_i^5 - 0.0343055n_i^4 + \\ 0.164166n_i^3 - 0.39055n_i^2 + 0.45733n_i - 0.17 \end{array}\right)^2} + 22.786}} = 0$ $\underline{\underline{n_i = 3.0}}$
(b)	$0.07923 - \sqrt{\frac{0.3}{\frac{1}{\left(\begin{array}{l} -0.000138889n_i^6 + 0.0035n_i^5 - 0.0343055n_i^4 + \\ 0.164166n_i^3 - 0.39055n_i^2 + 0.45733n_i - 0.17 \end{array}\right)^2} + 22.786}} = 0$ $\underline{\underline{n_i = 5.0}}$
(c)	$0.096 - \sqrt{\frac{0.3}{\frac{1}{\left(\begin{array}{l} -0.000138889n_i^6 + 0.0035n_i^5 - 0.0343055n_i^4 + \\ 0.164166n_i^3 - 0.39055n_i^2 + 0.45733n_i - 0.17 \end{array}\right)^2} + 22.786}} = 0$ $\underline{\underline{n_i = 7.0}}$
(d)	$0.093 - \sqrt{\frac{0.3}{\frac{1}{\left(\begin{array}{l} -0.000138889n_i^6 + 0.0035n_i^5 - 0.0343055n_i^4 + \\ 0.164166n_i^3 - 0.39055n_i^2 + 0.45733n_i - 0.17 \end{array}\right)^2} + 22.786}} = 0$ $\underline{\underline{n_i = 6.58}}$

The valve authority a' for the selected pre-setting totals:

(a)	$a' = \frac{\Delta p_{z,x}}{\Delta p_{z,x} + \Delta p_{str,x}} = \frac{r_{z,x}}{r_{z,x} + r_{str}} = \frac{1/[k_{v,x}^2]_{Xp=2}}{1/[k_{v,x}^2]_{Xp=2} + r_{str}} = \frac{1/0.09^2}{1/0.09^2 + 22.786} = 0.844$
(b)	$a' = \frac{\Delta p_{z,x}}{\Delta p_{z,x} + \Delta p_{str,x}} = \frac{r_{z,x}}{r_{z,x} + r_{str}} = \frac{1/[k_{v,x}^2]_{Xp=2}}{1/[k_{v,x}^2]_{Xp=2} + r_{str}} = \frac{1/0.2^2}{1/0.2^2 + 22.786} = 0.523$
(c)	$a' = \frac{\Delta p_{z,x}}{\Delta p_{z,x} + \Delta p_{str,x}} = \frac{r_{z,x}}{r_{z,x} + r_{str}} = \frac{1/[k_{v,x}^2]_{Xp=2}}{1/[k_{v,x}^2]_{Xp=2} + r_{str}} = \frac{1/0.32^2}{1/0.32^2 + 22.786} = 0.3$

In case (d), in order to establish the valve authority a', it is necessary to find the valve hydraulic resistance, the flow factor or the pressure throttled on the element that will stabilize for the required flow rate. The following is then obtained:

(d)	$a' = \frac{\Delta p_{z,x}}{\Delta p_{z,x}+\Delta p_{str,x}} = \frac{\Delta p_{z,x}}{\Delta p_{ob}} = \frac{\Delta p_{cz}-r_{str}\cdot\dot{V}_x^2}{\Delta p_{cz}} = \frac{0.3-22.786\cdot0.093^2}{0.3} = 0.343$

Table 6.15 Throttling/closing characteristic data of the *Herz TS-90-V* valve and calculated inner authority

Relative opening degree	The valve flow factor k_{vx} [m^3/(h bar$^{0.5}$)]	Inner authority a_w	Pre-setting	The valve flow factor k_{vx} [m^3/(h bar$^{0.5}$)]	Inner authority a_w
0.1	0.174	0.472	1	0.041	0.963
0.2	0.364	0.413	2	0.047	0.955
0.3	0.547	0.378	3	0.085	0.818
0.4	0.686	0.394	4	0.174	0.472
0.5	0.803	0.413	5	0.237	0.415
0.6	0.923	0.39	6	0.31	0.389
0.7	1.037	0.329	7	0.379	0.376
0.8	1.113	0.294	8	0.496	0.344
0.9	1.17	0.233	9	0.885	0.339
1.0	1.202	–	Max	1.202	–

In order to find the value of the valve total authority a_c, it is necessary to know the values of flow factor k_v for the throttling element given pre-settings n_i, at the full lift of the closing element, and the mathematical description of the original throttling characteristic, which is linear. The other data and the calculation results are listed in Table 6.15. The valve inner authority is found using relation (4.46). The mean value, for the valve operation from the full range of the regulating element travel, totals $a_{w,min} \approx 0.368$.

The valve closing element inner authority for the selected pre-setting totals:

(a)	$a_{w,x} = 0.818$
(b)	$a_{w,x} = 0.415$
(c)	$a_{w,x} = 0.376$
(d)	$a_{w,x} = 0.381$

The valve enables infinitely variable pre-setting adjustment, so it is possible to select (approximately) the value of $n_i = 6.58$ and calculate the inner and the outer authority values for it.

The valve outer authority for the selected pre-setting totals:

(a)	$a_{z,x} = \frac{r_{z,x}}{r_{z,x}+r_{str}} = \frac{1/k_{v,x}^2}{1/k_{v,x}^2+r_{str}} = \frac{1/0.085^2}{1/0.085^2+22.786} = 0.86$
(b)	$a_{z,x} = \frac{r_{z,x}}{r_{z,x}+r_{str}} = \frac{1/k_{v,x}^2}{1/k_{v,x}^2+r_{str}} = \frac{1/0.237^2}{1/0.237^2+22.786} = 0.44$
(c)	$a_{z,x} = \frac{r_{z,x}}{r_{z,x}+r_{str}} = \frac{1/k_{v,x}^2}{1/k_{v,x}^2+r_{str}} = \frac{1/0.379^2}{1/0.379^2+22.786} = 0.234$
(d)	$a_{z,x} = \frac{r_{z,x}}{r_{z,x}+r_{str}} = \frac{1/k_{v,x}^2}{1/k_{v,x}^2+r_{str}} = \frac{1/0.342^2}{1/0.342^2+22.786} = 0.273$

The valve closing element total authority for the selected pre-setting totals:

(a)	$a_{c,x} = a_{w,x} \cdot a_{z,x} = 0.818 \cdot 0.86 = 0.7$
(b)	$a_{c,x} = a_{w,x} \cdot a_{z,x} = 0.415 \cdot 0.44 = 0.183$
(c)	$a_{c,x} = a_{w,x} \cdot a_{z,x} = 0.376 \cdot 0.234 = 0.088$
(d)	$a_{c,x} = a_{w,x} \cdot a_{z,x} = 0.381 \cdot 0.273 = 0.104$

The calculation results are listed in Table 6.16.

Variant 2

Find the required values of pre-setting and total authority of the *Herz TS-90-V DN15* valve of a linear figure co-operating with a thermostatic head using the following data:

- hydraulic resistance of the pipework: $r_{str} = 58.33$ bar/(m^3/h)^2,
- required volume flow of the medium: $\dot{V}_x = 0.0453\,\text{m}^3/\text{h}$,
- active (differential) pressure in the circuit (Δp_{cz}):

(a) 0.171 bar,
(b) 0.211 bar,
(c) 0.373 bar.

Solution
The calculations are performed using the same formulae as in the previous variant. The final relations and results are presented below.

Calculations performed by means of the common method

The same as in Example 6.15, variant 2.

Calculations performed by means of the method proposed by *Pyrkov* and *Szaflik*

The same as in Example 6.15, variant 2 and in Example 6.16, variant 1.

Table 6.16 Calculation results for Example 6.18, variant 1

Method	Sought quantity			
	Pre-setting n_i	Inner authority a_w	Authority a'/outer authority a_z	Total authority a_c
The common method	No computational algorithm	No computational algorithm No requirements	Required: $a' \geq 0.3$ No computational algorithm	No computational algorithm No requirements
Method proposed by *Pyrkov* and *Szaflik* in [12, 13]	No computational algorithm	No requirements (a) a_W = unknown (b) a_W = unknown (c) a_W = unknown (d) a_W = unknown	No requirements (a) a_Z = unknown (b) a_Z = unknown (c) a_Z = unknown (d) a_Z = unknown	Required: $a_c \geq 0.3$ (a) a_C = unknown (b) a_C = unknown (c) a_C = unknown (d) a_C = unknown
Proposed method	Required (a) $n_i = 3.0$ (b) $n_i = 5.0$ (c) $n_i = 7.0$ (d) $n_i = 6.58$	No requirements (a) $a_W = 0.818$ (b) $a_W = 0.415$ (c) $a_W = 0.376$ (d) $a_W = 0.381$	No requirements (a) $a_Z = 0.86$; $a' = 0.844$ (b) $a_Z = 0.44$; $a' = 0.523$ (c) $a_Z = 0.234$; $a' = 0.3$ (d) $a_Z = 0.273$; $a' = 0.343$	Required: $a_c \geq 0.3$ (a) $a_c = 0.7$ (b) $a_c = 0.183$ (c) $a_c = 0.088$ (d) $a_c = 0.104$

Calculations performed by means of the proposed method

Method #1:

$$\dot{V}_x - \sqrt{\frac{\Delta p_{cz}}{\frac{1}{\left([k_v(n_i)]_{Xp=2}\right)^2} + r_{str}}} = 0.$$

(a)	$0.0453 - \sqrt{\frac{0.171}{\frac{1}{\left(\begin{array}{l}-0.000138889n_i^6 + 0.0035n_i^5 - 0.0343055n_i^4 + \\ 0.164166n_i^3 - 0.39055n_i^2 + 0.45733n_i - 0.17\end{array}\right)^2} + 58.33}} = 0$ $\underline{n_i = 5.0}$
(b)	$0.0453 - \sqrt{\frac{0.211}{\frac{1}{\left(\begin{array}{l}-0.000138889n_i^6 + 0.0035n_i^5 - 0.0343055n_i^4 + \\ 0.164166n_i^3 - 0.39055n_i^2 + 0.45733n_i - 0.17\end{array}\right)^2} + 58.33}} = 0$ $\underline{n_i = 4.0}$

(continued)

(continued)

(c)	$0.0453-\sqrt{\frac{0.373}{\frac{1}{\left(\begin{array}{l}-0.000138889n_i^6+0.0035n_i^5-0.0343055n_i^4+\\0.164166n_i^3-0.39055n_i^2+0.45733n_i-0.17\end{array}\right)^2}+58.33}}=0$ $\underline{n_i=3.0}$

The valve authority a' for the selected pre-setting totals:

(a)	$a'=\frac{\Delta p_{z,x}}{\Delta p_{z,x}+\Delta p_{str,x}}=\frac{r_{z,x}}{r_{z,x}+r_{str}}=\frac{1/\left[k_{v,x}^2\right]_{Xp=2}}{1/\left[k_{v,x}^2\right]_{Xp=2}+r_{str}}=\frac{1/0.2^2}{1/0.2^2+58.33}=0.3$
(b)	$a'=\frac{\Delta p_{z,x}}{\Delta p_{z,x}+\Delta p_{str,x}}=\frac{r_{z,x}}{r_{z,x}+r_{str}}=\frac{1/\left[k_{v,x}^2\right]_{Xp=2}}{1/\left[k_{v,x}^2\right]_{Xp=2}+r_{str}}=\frac{1/0.15^2}{1/0.15^2+58.33}=0.432$
(c)	$a'=\frac{\Delta p_{z,x}}{\Delta p_{z,x}+\Delta p_{str,x}}=\frac{r_{z,x}}{r_{z,x}+r_{str}}=\frac{1/\left[k_{v,x}^2\right]_{Xp=2}}{1/\left[k_{v,x}^2\right]_{Xp=2}+r_{str}}=\frac{1/0.09^2}{1/0.09^2+58.33}=0.68$

The valve closing element inner authority for the selected pre-setting totals:

(a)	$a_{w,x}=0.415$
(b)	$a_{w,x}=0.472$
(c)	$a_{w,x}=0.818$

The valve outer authority for the selected pre-setting totals:

(a)	$a_{z,x}=\frac{r_{z,x}}{r_{z,x}+r_{str}}=\frac{1/k_{v,x}^2}{1/k_{v,x}^2+r_{str}}=\frac{1/0.237^2}{1/0.237^2+58.33}=0.234$
(b)	$a_{z,x}=\frac{r_{z,x}}{r_{z,x}+r_{str}}=\frac{1/k_{v,x}^2}{1/k_{v,x}^2+r_{str}}=\frac{1/0.174^2}{1/0.174^2+58.33}=0.362$
(c)	$a_{z,x}=\frac{r_{z,x}}{r_{z,x}+r_{str}}=\frac{1/k_{v,x}^2}{1/k_{v,x}^2+r_{str}}=\frac{1/0.085^2}{1/0.085^2+58.33}=0.7$

The valve closing element total authority for the selected pre-setting totals:

(a)	$a_{c,x}=a_{w,x}\cdot a_{z,x}=0.415\cdot 0.234=0.097$
(b)	$a_{c,x}=a_{w,x}\cdot a_{z,x}=0.472\cdot 0.362=0.171$
(c)	$a_{c,x}=a_{w,x}\cdot a_{z,x}=0.818\cdot 0.7=0.573$

The calculation results are listed in Table 6.17.

Table 6.17 Calculation results for 6.18, variant 2

Method	Sought quantity			
	Pre-setting n_i	Inner authority a_w	Authority a'/outer authority a_z	Total authority a_c
The common method	No computational algorithm	No computational algorithm No requirements	Required: $a' \geq 0.3$ No computational algorithm	No computational algorithm No requirements
Method proposed by *Pyrkov* and *Szaflik* in [12, 13]	No computational algorithm	No requirements (a) a_W = unknown (b) a_W = unknown (c) a_W = unknown (d) a_W = unknown	No requirements (a) a_Z = unknown (b) a_Z = unknown (c) a_Z = unknown (d) a_Z = unknown	Required: $a_c \geq 0.3$ (a) a_C = unknown (b) a_C = unknown (c) a_C = unknown (d) a_C = unknown
Proposed method	Required (a) $n_i = 5.0$ (b) $n_i = 4.0$ (c) $n_i = 3.0$	No requirements (a) $a_w = 0.415$ (b) $a_w = 0.472$ (c) $a_w = 0.818$	No requirements (a) $a_z = 0.234$; $a' = 0.3$ (b) $a_z = 0.362$; $a' = 0.432$ (c) $a_z = 0.7$; $a' = 0.68$	Required: $a_c \geq 0.3$ (a) $a_c = 0.097$ (b) $a_c = 0.171$ (c) $a_c = 0.573$

Table 6.18 Throttling/closing characteristic data of the *Danfoss MSV-C DN15* valve [1]

Number of pre-setting n_i	The valve flow factor k_{vx} [m^3/(h bar$^{0.5}$)]	Number of pre-setting n_i	The valve flow factor k_{vx} [m^3/(h bar$^{0.5}$)]
2	0.6	5	2.1
3	1.1	6	2.8
4	1.6	7	3.2

Example 6.19

Find the required pre-setting values of the *Danfoss MSV-C DN15* manual valve, version with no measuring orifice, using the following data:

- active (differential) pressure in a given part of the circuit: $\Delta p_{cz} = 0.4$ bar,
- pipework pressure loss for the required volume flow of the medium: $\Delta p_{str,x} = 0.2$ bar,
- required volume flow of the medium:

(a) $\dot{V}_x = 0.268\,\text{m}^3/\text{h}$,
(b) $\dot{V}_x = 0.492\,\text{m}^3/\text{h}$,
(c) $\dot{V}_x = 0.6\text{m}^3/\text{h}$.

The valve throttling/closing characteristic data are listed in Table 6.18.

Solution

Calculations performed by means of the common method

In order to carry out the calculations in this variant, it is necessary to find the pressure that will have to be throttled on the valve for the sought x-th opening degree of the valve corresponding to the i-th pre-setting. By contrast to thermostatic valves, for a manual valve this always corresponds to the maximum relative travel range h_x/h_{100} of the closing element responsible for regulation. According to formula (6.14), the pressure will total:

$$\Delta p_{z,x} = \Delta p_{cz} - \Delta p_{str,x} = 0.2\text{bar}.$$

Knowing this pressure value, the required flow factor $k_{v,x}$ can be calculated and the pre-setting n_i corresponding thereto can be assigned to it according to Table 6.18. The following is then obtained:

(a)	$k_{v,x} = \frac{\dot{V}_x}{\sqrt{\Delta p_{z,x}}} = \frac{0.268}{\sqrt{0.2}} = 0.6$	$n_i = 2.0$
(b)	$k_{v,x} = \frac{\dot{V}_x}{\sqrt{\Delta p_{z,x}}} = \frac{0.492}{\sqrt{0.2}} = 1.1$	$n_i = 3.0$
(c)	$k_{v,x} = \frac{\dot{V}_x}{\sqrt{\Delta p_{z,x}}} = \frac{0.6}{\sqrt{0.2}} = 1.34$	$n_i = (3.0;4.0)$

Calculations performed by means of the proposed method

Method #1:

Based on the data presented in Table 6.18, function $k_v(n_i)$ can be written as:

$$k_v(n_i) = -0.0075n_i^5 + 0.15833n_i^4 - 1.27917n_i^3 + 4.9417n_i^2 - 8.6133n_i + 6.$$

Naturally, solutions should be looked for in the set of real numbers and in the interval of pre-settings which are available for a given valve and which are used to create function $k_v(n_i)$.

According to formula (4.141b), the required pre-setting values are as follows:

$$\dot{V}_x - \sqrt{\frac{\Delta p_{cz}}{\frac{1}{(k_v(n_i))^2} + \frac{\Delta p_{str,x}}{\dot{V}_x^2}}} = 0.$$

(a)	$0.268-\sqrt{\dfrac{0.4}{\dfrac{1}{\begin{pmatrix}-0.0075n_i^5+0.15833n_i^4-1.27917n_i^3\\+4.9417n_i^2-8.6133n_i+6\end{pmatrix}^2}+\dfrac{0.2}{0.268^2}}}=0$ $\underline{\underline{n_i=2.0}}$
(b)	$0.492-\sqrt{\dfrac{0.4}{\dfrac{1}{\begin{pmatrix}-0.0075n_i^5+0.15833n_i^4-1.27917n_i^3\\+4.9417n_i^2-8.6133n_i+6\end{pmatrix}^2}+\dfrac{0.2}{0.492^2}}}=0$ $\underline{\underline{n_i=3.0}}$
(c)	$0.6-\sqrt{\dfrac{0.4}{\dfrac{1}{\begin{pmatrix}-0.0075n_i^5+0.15833n_i^4-1.27917n_i^3\\+4.9417n_i^2-8.6133n_i+6\end{pmatrix}^2}+\dfrac{0.2}{0.6^2}}}=0$ $\underline{\underline{n_i=3.45}}$

Calculations performed by means of the method proposed by *Pyrkov* and *Szaflik*

This method is described in detail in [12, 13]. Only the indispensable computational relations will be presented here, together with the results they produce.

The pressure losses on the elements for the valve full opening are as follows:

(a)	$\Delta p_{z,100}=\left(\frac{\dot{V}_x}{k_{vs}}\right)^2=\left(\frac{0.268}{3.9}\right)^2=0.00472\,\text{bar}$
(b)	$\Delta p_{z,100}=\left(\frac{\dot{V}_x}{k_{vs}}\right)^2=\left(\frac{0.492}{3.9}\right)^2=0.0159\,\text{bar}$
(c)	$\Delta p_{z,100}=\left(\frac{\dot{V}_x}{k_{vs}}\right)^2=\left(\frac{0.6}{3.9}\right)^2=0.0237\,\text{bar}$

The valve outer authority at full opening totals:

(a)	$a_{z,100}\equiv a_{z,\min}=\frac{\Delta p_{z,100}}{\Delta p_{z,100}+\Delta p_{str,100}}=\frac{0.00472}{0.00472+0.2}=0.0231$
(b)	$a_{z,100}\equiv a_{z,\min}=\frac{\Delta p_{z,100}}{\Delta p_{z,100}+\Delta p_{str,100}}=\frac{0.0159}{0.0159+0.2}=0.0736$
(c)	$a_{z,100}\equiv a_{z,\min}=\frac{\Delta p_{z,100}}{\Delta p_{z,100}+\Delta p_{str,100}}=\frac{0.0237}{0.0237+0.2}=0.106$

The valve inner authority for the selected pre-setting totals:

(a)	$a_{w,x} = a_{w,\min} = 0.1385$
(b)	$a_{w,x} = a_{w,\min} = 0.1385$
(c)	$a_{w,x} = a_{w,\min} = 0.1385$
(d)	$a_{w,x} = a_{w,\min} = 0.1385$

The required pre-setting values are as follows:

$$n_i = n_{\max} \cdot \left[1 - \frac{\ln\left(1 - \frac{1}{a_{z,min} \cdot \bar{a}_{w,min}} + \frac{\Delta p_{cz}}{\bar{a}_{w,min} \cdot \left(\frac{\dot{V}_x}{k_{VS}}\right)^2}\right)}{2c} \right]$$

(a)	$n_i = 8 \cdot \left[1 - \frac{\ln\left(1 - \frac{1}{0.0231 \cdot 0.1385} + \frac{0.4}{0.1385 \cdot 0.00472}\right)}{2 \cdot 4}\right] = 2.3$
(b)	$n_i = 8 \cdot \left[1 - \frac{\ln\left(1 - \frac{1}{0.0736 \cdot 0.1385} + \frac{0.4}{0.1385 \cdot 0.0159}\right)}{2 \cdot 4}\right] = 3.56$
(c)	$n_i = 8 \cdot \left[1 - \frac{\ln\left(1 - \frac{1}{0.106 \cdot 0.1385} + \frac{0.4}{0.1385 \cdot 0.0237}\right)}{2 \cdot 4}\right] = 4.0$

The calculation results are listed in Table 6.19.

The above computational examples concerning selection of pre-setting n_i make use of polynomial functions describing the required quantities changeability. Finding the relations using a popular spreadsheet, a smaller number of significant digits was set each time compared to their maximum number offered in the program. It is assumed that the accuracy of the calculations from selected types of functions and so-posed condition is sufficient to present the computational algorithms (cf. the results of the pre-setting calculations in Example 6.16). Moreover, it is much higher than the minimum one required for example in practical calculations of the value of pre-setting n_i. If a need should arise, it is possible to achieve even higher accuracy of results either by changing the type of the function describing a given quantity changeability (e.g. using spline functions) or increasing the number of significant digits in relevant notations. In some cases, it may also be reasonable to use for the calculations a polynomial function of a degree lower than the one resulting from the number of the points the function is based on. Naturally, the curve will then not pass through all nodal points (i.e. defined values of pre-setting n_i) and the values returned by it will differ slightly from the required ones (the nodal points). It may

Table 6.19 Calculation results for Example 6.19

Method	Pre-setting, n_i
The common method	Required
	(a) $n_i = 2.0$
	(b) $n_i = 3.0$
	(c) $n_i = (3.0;4.0)$
Method proposed by *Pyrkov* and *Szaflik* in [12, 13]	(a) $n_i = 2.3$
	(b) $n_i = 3.56$
	(c) $n_i = 4.0$
Proposed method	(a) $n_i = 2.0$
	(b) $n_i = 3.0$
	(c) $n_i = 3.45$

then be better to approximate in between the points. In some situations, such an action will even be more justified than trying to achieve convergence of the values in nodal points because as the degree of the polynomial gets higher, the number of the function inflection points increases, and the function values between the nodal points may vary considerably. In other words, the interpolation accuracy decreases, which is referred to as the *Runge effect*.

Each of the first four computational examples concerning selection of the required value of pre-setting n_i is presented in two variants. The first variant comprises four and the second—three cases. The characteristic feature of the first variant is that for each circuit it represents, the same value of active pressure Δp_{cz} is set, but the medium volume flow values are different. The pipework hydraulic resistance r_{str} in each circuit is either different (Examples 6.15 and 6.16) or the same (Examples 6.17 and 6.18). The set identical value of active pressure Δp_{cz} corresponds practically to the situation where a pump generating a certain pressure value is incorporated into the installation. The pipework constant pressure loss Δp_{str}, at different required values of the medium flow rate, corresponds to the situation where pipes differing in terms of hydraulic resistance r_{str} are selected in each circuit where the radiators heat output to individual rooms is different. A constant value of the pipework hydraulic resistance r_{str}, on the other hand, corresponds to the occurrence of circuits made of identical pipes—with the same lengths and local obstacles.

In the first computational variant, the parameters in the first three cases are defined to be satisfied if a defined and known value is selected of pre-setting n_i, resulting from experimental measurements of each of the valves considered in the examples. This makes it possible to compare and verify the calculation results by means of the herein-presented methods of selection of the value of pre-setting n_i. The fourth case in every example is described by such values of the required parameters that calculating the required value of pre-setting n_i by means of the commonly applied method should be impossible.

Using the commonly applied computational method, the necessary value of pre-setting n_i cannot be calculated directly. It is only possible to calculate the required value of flow factor k_v and then select the pre-setting using a relevant table. It follows that if the required value of k_v lies between two defined values of the parameter, the appropriate value of pre-setting n_i is selected at the discretion of the user choosing an intermediate value of k_v and generating a bigger or a smaller error. This is what happens in practice almost all the time. Usually, the higher value of pre-setting n_i is selected, which in the case of manual valves (Examples 6.15 and 6.19) requires additional adjustment of the valve on an already operating installation to ensure that the measured values of the flow rate and the system operation parameters are kept. This implies additional time and investment outlays. Apart from that, in the case of manual valves, the commonly applied method enables calculations of the valve outer authority a_z only, and not its total authority a_c—the value responsible for resultant closing characteristics of valves installed in the pipework. The situation is even worse in the case of thermostatic valves, where the valve is to operate at an intermediate position of the closing element. Not even outer authority a_z is determined but only the quantity resulting from the share in pressure losses arising on the valve in the total pressure losses in the circuit (herein marked as a'). The quantity is thus calculated for any position of the valve closing element over the seat that results from design assumptions and it may therefore vary from 0 to 1, despite the fact that neither the value of pre-setting n_i, and the throttling element hydraulic resistance resulting therefrom, nor the pipework hydraulic resistances r_{str} change. As it has already been shown and proved, the valve authority is the picture of the original distribution of those quantities. Therefore in this situation, the commonly applied method does not make it possible to calculate either outer authority a_z or, much less, total authority a_c, which means that using it, it is impossible to determine final static (thermal) characteristics that the radiator will operate with. For this reason, the obtained results cannot be related to the 0.3–0.7 condition, which in the radiator regulation practice is defined for the parameter that ultimately deforms the valve original closing characteristic, i.e. total authority a_c. Moreover, the method does not enable calculation of changes in the medium volume flow if changes occur in the value of pre-setting n_i, or at current regulation through changes in the closing element position over the valve seat. It is impossible then to define the radiator heat output at any position of the valve closing and/or throttling element, e.g. if other than the required value of pre-setting n_i is selected.

In [12, 13], *Pyrkov* and *Szaflik* propose a computational method of selecting the value of pre-setting n_i for manual valves. But there is no such procedure for thermostatic valves and for valves with two adjustable sections of the medium flow. They give relations that make it possible to calculate inner authority a_w of a thermostatic valve with two adjustable sections of the medium flow which are the same as the ones proposed herein. However, *Pyrkov* and *Szaflik* make use of the flow factor values corresponding to the closing element partial opening h_N and not to the element maximum range of travel h_{max}. Such an approach illustrates the distribution of pressures for a given position of the closing element, which may open beyond the position changing the said distribution. Consequently, it does not reflect the deformation of the original closing characteristic, i.e. inner authority a_w, as illustrated by the computational example presented earlier in this book and confirmed experimentally.

In the case of manual valves, the method put forward by *Pyrkov* and *Szaflik* offers a possibility of direct calculation of the required pre-setting n_i and of determination of the value of total authority a_c. However, the results obtained by means of the formulae used in the method fail to agree with experimental data, as indicated by relevant comparisons. Analysing the results of calculations of the required pre-setting n_i in Example 6.15, it can be noticed that in each of the first three cases (*a, b, c*) it is slightly different from the required value resulting from the valve experimental measurements. This is the effect of the issues discussed herein earlier. In their computational relations, *Pyrkov* and *Szaflik* make use of the mean value of inner authority a_w, which results in values of pre-setting n_i differing from the required ones because values of the valve inner authority a_w for calculated values of pre-setting n_i differ from the mean value of a_w. The agreement would be achieved if the calculated value of pre-setting n_i was characterized by the calculated value of inner authority a_w equal to the mean one adopted for the sought pre-setting calculations. Moreover, the method proposed by *Pyrkov* and *Szaflik* requires the knowledge of the values of a number of parameters, such as inner authority a_w, outer authority a_z or the mathematical description of the original closing/throttling characteristic.

While calculating the parameter responsible for the final deformation of the closing curve, *Pyrkov* and *Szaflik* are right to use total authority a_c, i.e. the product of inner authority a_w and outer authority a_z. But in each case under consideration, they suggest that the value of the valve outer authority a_z should be calculated for the valve full opening and the regulating element maximum range of travel h_{max}, i.e. at the maximum value of pre-setting n_i, despite the fact that the selected pre-setting values are different from those that the valve actually operates with. The value of outer authority a_z should be calculated for the selected values of pre-setting n_i. The approach presented by *Pyrkov* and *Szaflik* would be right if pre-setting n_i was not selected by reducing the plug maximum travel range (lift) h_{max} and by the rise in the valve hydraulic resistance resulting therefrom, which is the case here, and if current regulation started each time from the maximum available travel range h_{max}, i.e. if the valve could open more than the set opening degree. Then, however, the case would not be setting the pre-setting value but determining flow factor k_v as a function of pre-setting n_i in the same manner as for thermostatic valves, where the closing element operates without a reduction in the maximum available range of travel h_{max}, but the computational position corresponds to an intermediate lift h_N. This is discussed in detail in this book and this is the situation that Example 6.16 refers to.

For inner authority a_w in valves with one adjustable section of the medium flow, the researchers assume that the parameter value is constant, irrespective of the selected pre-setting n_i. This means that they do not take account of the fact that the regulated section hydraulic resistance in such valves increases as the regulating element maximum available travel range h_{max} is reduced. As a result, the value of inner authority a_w increases and the regulation characteristic curvature changes. The value calculated by means of the method proposed by *Pyrkov* and *Szaflik* corresponds only to the case where the maximum available value of pre-setting n_{max} is selected and—like in the case of outer authority a_z—it will be correct only if the valve operates with this pre-setting value.

If pre-setting n_i is calculated using the methods proposed in Sect. 4.2, the obtained results agree with the required values resulting from experimental measurements of the valves. The condition is satisfied by both computational methods presented in this book. This concerns the calculations performed for the manual and the thermostatic valve with both one and two adjustable sections of the medium flow. In Method #1, defined by formulae (4.127) and (4.141), no parameters need to be known except the ones normally provided by manufacturers. There is therefore no need to determine the values of inner authority a_w, outer authority a_z or the mathematical description of the original closing/throttling characteristic. Method #2 makes use of inner authority a_w as a function of pre-setting n_i and not the parameter mean value because if the calculated value of a_w varies depending on n_i, such an approach enables correct determination of the required value of pre-setting n_i. Created functions $a_w(n_i)$ should reflect the variability of inner authority a_w as accurately as possible, i.e. they should be based on a possibly high number of computational and, consequently, measuring points of the required hydraulic characteristic of the valve because the calculation results accurate convergence with the required values is obtained for pre-setting values for which the required quantities are defined (measured flow factor k_v, calculated inner authority a_w). If the calculated pre-setting n_i lies between the defined values, the value of inner authority a_w is found based on the numbers characterizing the other values of pre-setting n_i and the assumed type of the function approximating the sought curve. Therefore, the higher the number of measuring points, the more accurately the shape of the function $a_w(n_i)$ can be approximated, in between them, compared to the actual one, and the more accurate the determination of the pre-setting value sought in these intervals.

The two proposed methods of the pre-setting value determination differ substantially. As it can be noticed, relations (4.127) and (4.141) from Method #1, where functions $k_v(n_i)$ are used, do not require the knowledge of the valve original throttling or closing characteristic. Nor is it necessary to know the values of inner or outer authority (a_w and a_z, respectively) whether for the closing or the throttling element. All these quantities are indispensable, however, to determine the sought value of pre-setting n_i using the relations expressing the parameter directly, like in formulae (4.117), (4.118), (4.131), (4.132), (4.138) and (4.139), according to Method #2. Therefore, it is less complicated and less time-consuming to perform the calculations by means of the method that does not require the knowledge of the quantities, i.e. Method #1. However, compared to Method #2, Method #1 has a certain drawback. In order to find the required value of pre-setting n_i, it is necessary to know the flow factors (or e.g. the hydraulic resistances) for each value of pre-setting n_i available for the valve. Otherwise, it is impossible to create function $k_v(n_i)$. In the case of Method #2 in the version expressed by formulae (4.131), (4.132), (4.138) and (4.139), the function is not determined, but a proposal is made for determination of function $a(n_i)_w$, which means that some quantities also need to be known for all values of pre-setting n_i. If, however, the exact mathematical description of the original closing/throttling characteristic is known, the calculated values of inner authority a_w for a given value of pre-setting n_i will be constant and function $a(n_i)_w$ will not have to be defined, which will simplify the notation, as presented in formulae (4.117) and (4.118). Then, in order to define the mathematical description of the original closing/throttling characteristic, in the most favourable case, i.e. if the closing/throttling element is designed and shaped a for a single type of characteristic (without dividing into two different types, e.g. linear and equal-percentage), only two measuring points are required: the point for the full opening and any other (in the case of an original characteristic that ensures that a non-zero flow is cut off completely). It is therefore unnecessary to know the entire shape of the resultant closing and/or throttling characteristic plotted for all available values of pre-setting n_i, which is an advantage of the method in this simplification (constancy of $a(n_i)_w$). In practice, however, such measurements are as a rule performed for every regulation valve, which means that both the methods, making use of functions $k_v(n_i)$ and $a(n_i)_w$, are characterized by similar functionality. In the case of the method assuming constancy of $a(n_i)_w$, the difference between the calculation results and the actual values will get bigger the bigger the difference between the original closing/throttling characteristic adopted for the calculations and the real one. Analysing the results of the calculations performed for the *Danfoss MSV-I* valve, it can be seen that the differences are significant, even though the mathematical description of the characteristic is known, which is the effect of the issues discussed during the description of the *M-3176* valve measuring data in Sect. 5.3.1. But it should be noted that the data provided by the manufacturer are averaged and concern a specific lot of a given valve model. In practice, it may turn out that the error in the calculated pre-setting n_i resulting from the assumption of the inner authority constancy will be smaller than the spread of the parameters for the given valve different specimens. Nonetheless, the mathematical description of the original characteristic is usually unknown, or it requires a combination of a few functions, which is the case for example for the *Danfoss RTD-N15* or the *Danfoss MSV-C15* valve. In such a situation, using a method based on a single description of a function and a constant value of inner authority a_w may lead to big errors in the results of calculations of the required pre-setting n_i. The errors could reach up to about fifteen per cent or even more, as illustrated by the computational case presented in Example 6.19, and they get bigger with a rise in the valve outer authority n_z.

In the proposed methodology of the parameter determination, outer authority a_z is calculated taking account of the original distribution of hydraulic resistances in the circuit. In the case of a manual valve, the pressure loss arising on the regulating element is then calculated for the actually selected pre-setting n_i that ensures that the excess of pressure in the circuit will be throttled as required. In the case of thermostatic valves, this is done for the actually selected pre-setting n_i and the closing element maximum travel range h_{max}. Inner authority a_w of a manual valve with one adjustable section of the medium flow is calculated taking account of the variability of the closing element hydraulic resistance as a function of a reduction in the element maximum range of travel h_{max} and, thereby, taking account of the changeability of the distribution of hydraulic resistances within the valve. The lower the value of pre-setting n_i, the smaller the plug lift over the valve seat and range of travel, which means a smaller surface area of the medium flow, a bigger hydraulic resistance of the element and a higher value of inner authority a_w. Inner authority a_w of a thermostatic valve with one or two adjustable sections of the medium flow is calculated taking account of the changeability of the throttling element hydraulic resistance. The lower the value of pre-setting n_i, the smaller the throttling section surface area of the medium flow, the bigger the element hydraulic resistance and the smaller the inner authority of the closing element with a defined initial resistance. Total authority a_c calculated by means of the proposed method thus illustrates the actual deformation of the closing characteristic of both types of the valve installed in the pipework and operating with a defined value of pre-setting n_i.

The comparison between the results of the valve authority calculations obtained by means of the commonly applied method and the method proposed herein indicates that the parameter can take substantially different values. If the calculations are based on the so-far applied methodology, the 0.3–0.7 condition can be satisfied easily. Analysing the computational examples, it can be seen that the condition is met in every case, for each valve. However, if the value of total authority a_c is calculated, i.e. the parameter that should in fact be confronted with this particular interval, it turns out that the condition is usually not satisfied. What is more, the obtained value is by a few orders smaller than the minimum required value. In Example 6.15, concerning calculations of the *Danfoss MSV-I DN20* valve, the assumption is satisfied only in one of the four cases under analysis, i.e. if a very low pre-setting value is selected ($n_i = 0.2$), which results in the regulating element travel range up to $h_x/h_{max} = 6.25\%$, for which the value of total authority totals $a_c = 0.31$. The condition under consideration is met, but it should be noted that in the case of a valve with one adjustable section of the medium flow, such a big reduction in the closing element maximum range of travel h_{max} naturally involves deterioration in the quality of regulation (as previously described herein). For the *Herz TS-90-V* valve, on the other hand, the pipework parameters are set so that even in the least favourable case from the point of view of the valve authority, the condition set so far should be satisfied. Like previously, however, it turns out that after inner authority a_w is taken into consideration, the resultant quantity, i.e. total authority a_c, has a satisfactory value only for low values of pre-setting n_i, corresponding again to a big reduction in the current regulation element maximum range of travel h_{max}. In the case of the *Danfoss RTD-N15* and the *HERZ TS-FV* valve, the condition set so far is also always met. Still, the value of total authority a_c is in every case very low, by a few orders lower than the minimum required value. From the point of view of the regulation process, this situation is very unfavourable, although it is a common occurrence for valves with two adjustable sections of the medium flow, which are nowadays the most popular types of valves. Additionally, in contrast to valves with one adjustable section of the medium flow (Examples 6.15 and 6.19), as the value of pre-setting n_i gets smaller, the value of inner authority a_w decreases. Despite the fact then that this solution of double regulation makes it possible for the closing element to operate each time with a constant travel range h_{max} that is not reduced to very small values, irrespective of the selected value of pre-setting n_i, it does not ensure favourable shapes of closing characteristics.

Variant 2 in the examples models a slightly different situation of the circuit operation. For the pipework constant hydraulic resistance r_{str} and the required value of the medium volume flow, different values of active pressure Δp_{cz} are set, which is a frequent occurrence in practice. The pipework selected for the circuit is characterized by a certain value of hydraulic resistance r_{str} and the required volume flow of the medium results from the need for the radiator required heat output. In such a situation, different values of active pressure Δp_{cz} may be the effect of the installation of regulators and stabilizers of the pressure difference in risers or branches, or of the change in the pump throttling characteristic, the pump type, size, etc. In each case, however, a specific value of the medium volume flow has to be kept, and a value of the radiator valve pre-setting n_i has to be found that will satisfy the postulate.

The calculations performed for each of the valves by means of the algorithms presented in this book are the same as in Variant 1. No comparison is made between the results of calculations and the two confronted methods because the data defined in the variant are not taken into account in the methods, and using them, it is impossible to carry out the calculations, which is described in the comments supplementing each case. If the results of the valve authority calculations obtained in this variant are examined, a certain regularity can be noticed. A rise in the circuit active pressure Δp_{cz} naturally translates into a decrease in the required value of pre-setting n_i of the valve that has to throttle ever increasing pressure to ensure a constant value of the medium volume flow. Due to that, the value of the valve outer authority a_z rises. It does not mean, however, that the value of total authority a_c also rises because the latter can decrease as well, as illustrated in Examples 6.16 and 6.17. Therefore, a rise in the installation active pressure Δp_{cz}, for valves with two adjustable sections of the medium flow, will cause deterioration in the valves regulation characteristics. The final (thermohydraulic) regulation characteristics will also deteriorate.

A range is frequently mentioned in this book in which, according to the commonly accepted criterion, the value of the valve total authority a_c should be included, i.e. $a_c = 0.3\text{–}0.7$. However, as indicated by the results of the calculations presented in Sect. 5.3 and in this chapter, it is usually impossible to satisfy the condition in practice. This results from the fact that the regulation valve maximum values of inner authority a_w usually do not exceed the value of $a_w = 0.4$. They are thus close to the lower allowable value required for total authority a_c, even without taking account of the additional drop arising due to hydraulic resistances of the pipes and devices connected to the system. In the case of the radiator typical double-regulation valves (two adjustable sections of the medium flow), the condition mentioned above is impossible to meet ($a_w \ll 0.4$). The conclusion is that in the case of the heating installation regulation valves, the radiator regulation valves in particular, the valve total authority a_c should not be the only parameter to assess the valve regulation capacity and its impact on the final regulation characteristics. It is only one of the two valve-related factors that decide about regulation curves. The other is the valve original regulation characteristic. Each time the values of inner and outer authority (and thereby total authority) are decreased, the characteristic is deformed, which creates the risk of on-off operation (cf. Chap. 4 and the valves regulation characteristics presented therein). But valves can be designed taking this phenomenon into account, and their closing elements can be shaped appropriately (the solution is discussed in [6, 7]) to obtain a curve with a sufficient downward curvature. However, the problem here is that it is impossible to define a universal characteristic of a given valve. This is due to the fact that the valve may operate with receivers with different characteristics and in different installations with different values of hydraulic resistance r_{str} and very different impact on the original regulation characteristic. The $a_c = 0.3–0.7$ condition is justified if the valve original regulation characteristic is shaped taking account of a specific characteristic of the regulated object and the required final regulation characteristic. In this case, obtaining the value of $a_c = 0.3–0.7$ does not produce considerable deviations from the required final curves (cf. Figs. 4.1 and 4.10). However, such an approach is not used. There are now no requirements concerning regulation characteristics of valves intended for heating installations, and freedom in this field is unlimited. Moreover, regulated objects may differ noticeably in terms of the shape of regulation characteristics (e.g. convector and panel radiators, convector and underfloor radiators, etc.). In the light of this, formulating a number condition for the valve total authority a_c (and for authority in general) has no practical grounds because it does not provide the information on the installation actual regulation characteristic or on the regulation quality. The joint impact of inner authority a_w, outer authority a_z and of the shape of the valve original regulation characteristic should be considered. This requires the knowledge of actual values of the parameters which can be determined by means of the computational relations proposed in this book.

Example 6.20
Based on the installation known structure, find the required pre-setting value of the radiator regulation valve in circuit 5 and the error in the circuit balance using the following data:

- temperature parameters of the installation operation: $t_z/t_p = 80/60$ °C,
- active (differential) pressure in the installation: $\Delta p_{cz} = 8$ kPa,
- copper pipes with the wall absolute roughness at the level of $k = 0.01$ mm,
- water computational density: $\rho = 972$ kg/m^3,
- analysed radiator regulation valves:

(a) Herz *TS-90-V*
(b) *Danfoss RTD-N 15*

- the diagram of the installation is presented in Fig. 6.19; the data concerning the pipes inner diameters, lengths and local pressure loss coefficients are gathered in Table 6.20.

Solution
In order to solve the problem, it is necessary to find the required value of the water pressure loss in the valve and the volume/mass flow. If nomograms illustrating pressure losses as a function of the medium flow rate are used for the valves under analysis and not tables listing the values of flow factor k_v as a function of pre-setting n_i, determination of the k_v value is not absolutely necessary. The volume flow value

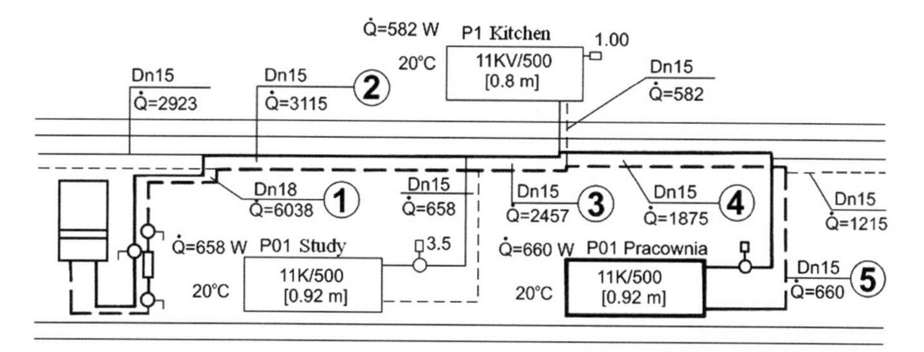

Fig. 6.19 Diagram of the installation under analysis

Table 6.20 Installation parameters

Branch No	Thermal power	Inner diameter	Total length	Sum of coefficients of local pressure losses
	$\dot{Q}$, [W]	d_w, [m]	l, [m]	$\sum \zeta$
1	6038	0.016	5.0	14
2	3115	0.013	4.8	10
3	2457	0.013	7.8	4
4	1875	0.013	8.4	1
5	660	0.013	5.0	15

can be found knowing the required thermal power and the temperature parameters of the circuit operation. Due to the fact that the medium flows to the valve through a few upstream-located common branches (distributing sections) (from 1 to 4), the pressure on the valve will be the active pressure value at the beginning of the system, reduced by pressure losses arising in upstream-located common branches, according to formula (6.14):

$$\Delta p_z = \Delta p_{cz} - \Delta p_{str}.$$

The pipework pressure losses Δp_{str} have to be established. They constitute the sum of losses arising on local obstacles and in straight sections, according to formula (3.5):

$$\Delta p_{str} = \sum_{i=1}^{n} (R \cdot l + Z)_i,$$

where l_i is the sum of the lengths of a given branch return and supply pipes. Because the value of absolute roughness k is known, the pressure losses in the installation subsequent branches can be found from the relations given in Sect. 3.1. But it is also possible to use the nomograms created based on those relations. This will proceed in the manner described in Sect. 3.2 using a simplifying assumption that the read-out value is the water mean temperature in the circuit and that a nomogram created for this particular temperature value is available. In this case, the temperature totals (80 + 60)/2 = 70 °C. Figure 6.20 presents a nomogram concerning the pipes considered in the computational example. By means of the nomogram, approximate readouts can be made directly for the medium set flow velocity w or for its mass flow $\dot{m}$. Alternatively, the hydraulic resistance value r_{str} (the line slope) of a pipe with diameter d can be determined and then, based on it, the linear pressure loss R can be found. This procedure is presented in Example 6.7. In this example, the target value is read out from the nomogram.

The local pressure losses of the medium in the i-th section of the pipework can be determined from formula (3.3):

$$\Delta p_{m,i} \equiv Z_{\mathrm{i}} = \frac{1}{2} \cdot \rho_{\mathrm{i}} \cdot w_{\mathrm{i}}^2 \cdot \Sigma\zeta_{\mathrm{i}}.$$

The calculations of pressure losses Δp, the medium velocity w and the medium mass flow $\dot{m}$/volume flow $\dot{V}$ for each of the branches will run using the same computational relations. The medium mass flow can be calculated from relation (6.11):

$$\dot{m}_i = \frac{\dot{Q}_i}{c_w \cdot \left(t_{\mathrm{z}} - t_{\mathrm{p}}\right)}.$$

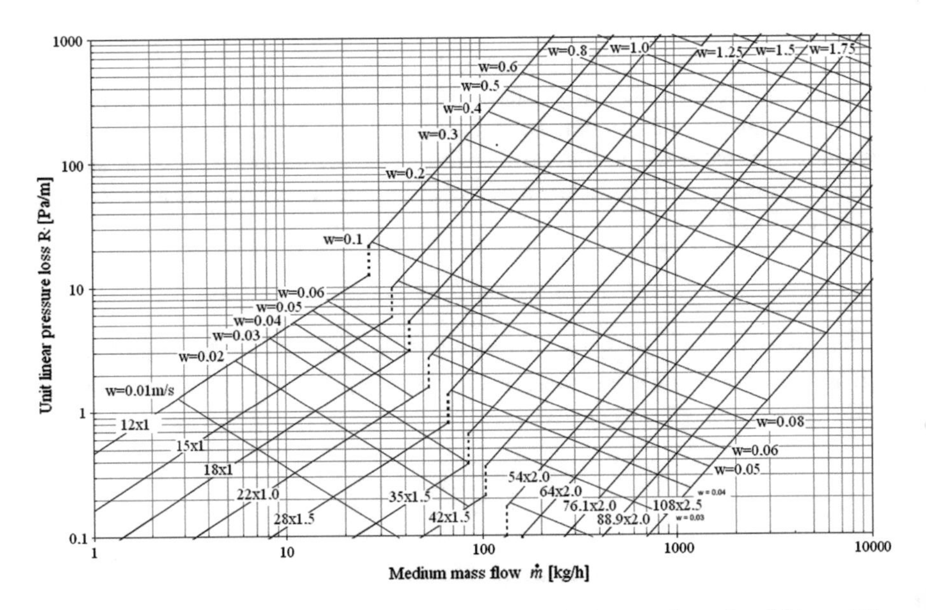

Fig. 6.20 Unit linear pressure loss R in copper pipes with absolute roughness $k = 0.01$ mm [5]

The medium volume flow can be calculated from the following relation:

$$\dot{V}_\mathrm{i} = \frac{\dot{m}_i}{\rho}. \tag{6.15}$$

Knowing the pipe diameter, the medium velocity in a given section of the pipework can be found based on the volume flow value determined earlier or using the initial heat flux equation. The following values are then obtained:

$$w_\mathrm{i} = \frac{\dot{V}_\mathrm{i}}{A_\mathrm{i}} = \frac{4 \cdot \dot{V}_\mathrm{i}}{\pi \cdot d_\mathrm{i}^2}, \tag{6.16}$$

$$\begin{aligned} \dot{Q}_\mathrm{i} &= \dot{m}_\mathrm{i} \cdot c_\mathrm{W} \cdot \left(t_\mathrm{z} - t_\mathrm{p}\right) = \left(w_\mathrm{i} \cdot A_\mathrm{i} \cdot \rho_\mathrm{i}\right) \cdot c_\mathrm{W} \cdot \left(t_\mathrm{z} - t_\mathrm{p}\right) \\ &= w_i \cdot \pi \cdot \frac{d_\mathrm{i}^2}{4} \cdot \rho_\mathrm{i} \cdot c_\mathrm{W} \cdot \left(t_\mathrm{z} - t_\mathrm{p}\right) \Rightarrow w_\mathrm{i} = \frac{4 \cdot \dot{Q}_\mathrm{i}}{\pi \cdot d_\mathrm{i}^2 \cdot \rho_\mathrm{i} \cdot c_\mathrm{W} \cdot \left(t_\mathrm{z} - t_\mathrm{p}\right)}. \end{aligned} \tag{6.17}$$

Substituting the data for subsequent branches, the result is:

Branch 1

$$\dot{m}_1 = \frac{\dot{Q}_1}{c_w \cdot \left(t_\mathrm{z} - t_\mathrm{p}\right)} = \dot{m}_1 = \frac{6038}{4186 \cdot (80 - 60)} = 0.0721\,\mathrm{kg/s} = 259.6\,\mathrm{kg/h},$$

$$\dot{V}_1 = \frac{\dot{m}_1}{\rho_1} = \frac{0.0721}{972} = 0.0000742\,\mathrm{m^3/s} = 0.277\,\mathrm{m^3/h},$$

$$w_1 = \frac{4 \cdot \dot{V}_1}{\pi \cdot d_\mathrm{i}^2} = \frac{4 \cdot 0.0000742}{\pi \cdot 0.016^2} = 0.37\,\mathrm{m/s},$$

$$(R \cdot l)_1 = 115 \cdot 5 = 575 \text{ Pa},$$

$$Z_1 = \frac{1}{2} \cdot \rho_1 \cdot w_1^2 \cdot \Sigma\zeta_1 = \frac{1}{2} \cdot 972 \cdot 0.37^2 \cdot 14 = 931.5\,\mathrm{Pa}.$$

Branch 2

$$\dot{m}_2 = \frac{\dot{Q}_2}{c_w \cdot \left(t_\mathrm{z} - t_\mathrm{p}\right)} = \dot{m}_2 = \frac{3115}{4186 \cdot (80 - 60)} = 0.0372\,\mathrm{kg/s} = 134\,\mathrm{kg/h},$$

$$\dot{V}_2 = \frac{\dot{m}_2}{\rho_2} = \frac{0.0372}{972} = 0.0000383\,\mathrm{m^3/s} = 0.138\,\mathrm{m^3/h},$$

$$w_2 = \frac{4 \cdot \dot{V}_2}{\pi \cdot d_2^2} = \frac{4 \cdot 0.0000383}{\pi \cdot 0.013^2} = 0.2885\,\mathrm{m/s},$$

$(R \cdot l)_2 = 100 \cdot 4.8 = 480\ \text{Pa},$

$$Z_2 = \frac{1}{2} \cdot \rho_2 \cdot w_2^2 \cdot \Sigma\zeta_2 = \frac{1}{2} \cdot 972 \cdot 0.2885^2 \cdot 10 = 404.5\,\text{Pa}.$$

Branch 3

$$\dot{m}_3 = \frac{\dot{Q}_3}{c_w \cdot (t_z - t_p)} = \frac{2457}{4186 \cdot (80 - 60)} = 0.02935\,\text{kg/s} = 105.6\,\text{kg/h},$$

$$\dot{V}_3 = \frac{\dot{m}_3}{\rho_3} = \frac{0.02935}{972} = 0.0000302\,\text{m}^3/\text{s} = 0.109\,\text{m}^3/\text{h},$$

$$w_3 = \frac{4 \cdot \dot{V}_3}{\pi \cdot d_3^2} = \frac{4 \cdot 0.0000302}{\pi \cdot 0.013^2} = 0.227\,\text{m/s},$$

$(R \cdot l)_3 = 70 \cdot 7.8 = 546\ \text{Pa},$

$$Z_3 = \frac{1}{2} \cdot \rho_3 \cdot w_3^2 \cdot \Sigma\zeta_3 = \frac{1}{2} \cdot 972 \cdot 0.227^2 \cdot 4 = 100\,\text{Pa}.$$

Branch 4

$$\dot{m}_4 = \frac{\dot{Q}_4}{c_w \cdot (t_z - t_p)} = \frac{1875}{4186 \cdot (80 - 60)} = 0.0224\,\text{kg/s} = 80.6\,\text{kg/h},$$

$$\dot{V}_4 = \frac{\dot{m}_4}{\rho_4} = \frac{0.0224}{972} = 0.000023\,\text{m}^3/\text{s} = 0.08295\,\text{m}^3/\text{h},$$

$$w_4 = \frac{4 \cdot \dot{V}_4}{\pi \cdot d_4^2} = \frac{4 \cdot 0.000023}{\pi \cdot 0.013^2} = 0.173\,\text{m/s},$$

$(R \cdot l)_4 = 40 \cdot 8.4 = 336\ \text{Pa},$

$$Z_4 = \frac{1}{2} \cdot \rho_4 \cdot w_4^2 \cdot \Sigma\zeta_4 = \frac{1}{2} \cdot 972 \cdot 0.173^2 \cdot 1 = 14.5\,\text{Pa}.$$

Branch 5

$$\dot{m}_5 = \frac{\dot{Q}_5}{c_w \cdot (t_z - t_p)} = \frac{660}{4186 \cdot (80 - 60)} = 0.00788\,\text{kg/s} = 28.4\,\text{kg/h},$$

$$\dot{V}_5 = \frac{\dot{m}_5}{\rho_5} = \frac{0.00788}{972} = 0.0000081\,\text{m}^3/\text{s} = 0.0292\,\text{m}^3/\text{h},$$

$$w_5 = \frac{4 \cdot \dot{V}_5}{\pi \cdot d_5^2} = \frac{4 \cdot 0.0000081}{\pi \cdot 0.013^2} = 0.061\,\mathrm{m/s},$$

$(R \cdot l)_5 = 5 \cdot 5 = 25$ Pa,

$$Z_5 = \frac{1}{2} \cdot \rho_5 \cdot w_5^2 \cdot \Sigma\zeta_5 = \frac{1}{2} \cdot 972 \cdot 0.061^2 \cdot 15 = 27.15\,\mathrm{Pa}.$$

The above parameters calculation results are listed in Table 6.21.
The required pressure drop on the valve totals:

$$\Delta p_z = \Delta p_{cz} - \Delta p_{str} = 8000 - 3440 = 4560\,\mathrm{Pa}.$$

This means that the required value of the valve flow factor, according to formula (3.57) is:

$$k_v = \frac{\dot{V}_5}{\sqrt{\Delta p_z}} = \frac{0.0292}{\sqrt{0.0456}} = 0.137\,\mathrm{m^3/h}.$$

The required value of pre-setting n_i of the valves under analysis can be read out using the data provided by the manufacturer.

(a) For the *Herz TS-90-V* valve.

For this valve the manufacturer provides both the nomogram illustrating the pressure loss dependence on the flow rate, which makes it possible to read out the required value of pre-setting n_i, and a table with the pre-setting values together with corresponding values of flow factor k_v, as presented in Fig. 6.21. The figure presents the lines of the required pressure loss Δp_z and the medium mass flow $\dot{m}$. The point where the lines intersect is the required working point of the valve. It can be seen that it lies between the pressure loss lines for two neighbouring pre-setting values, i.e. $n_i = 3$ and $n_i = 4$. The values of flow factor k_v corresponding to them are $k_{v,3} = 0.09$ and $k_{v,4} = 0.15$, respectively. This is a valve with infinitely variable adjustment of the pre-setting value, which means that an exact, intermediate value thereof can be selected. Such a procedure is presented in earlier computational examples. The value required in this case, which results from the use of an interpolating polynomial, is $n_i = 3.78$.

In practice, the nomogram makes it also possible to read out the value of the valve flow factor k_v directly for a given value of pre-setting n_i, and it is not necessary for the manufacturer to provide a table with any relevant data. The definition of the flow factor defined for the pressure loss of 1 bar = 100 kPa will suffice. If, therefore, a point of intersection of the line corresponding to a given pre-setting n_i and the line of the pressure loss of 100 kPa is read from the chart, the medium volume flow for the point will be directly the flow factor value if the flow rate in the chart is expressed in $\mathrm{m^3/h}$. In the case of other units, appropriate conversion is necessary. This regularity can be analysed based on Figs. 6.21 and 6.22.

Table 6.21 Results of the calculations of the pipework pressure losses

Branch No.	Thermal power	Inner diameter	Length	Sum of local pressure losses	Volume flow	Coefficient of linear pressure losses	**Pressure drop due to linear resistances**	Water velocity	**Pressure drop due to local resistances**	**Total pressure drop on the branch**
	$\dot{Q}$ [W]	d_W [m]	L [m]	$\sum \zeta$ -	$\dot{V}$ [m^3/h]	R [Pa/m]	$\boldsymbol{R \cdot l}$ [Pa]	w [m/s]	$\boldsymbol{Z}$ [Pa]	$\boldsymbol{R \cdot l + Z}$ [Pa]
1	6038	0.016	5.0	14	0.277	115	575	0.37	931.5	1506.5
2	3115	0.013	4.8	10	0.138	100	480	0.2885	404.5	884.5
3	2457	0.013	7.8	4	0.109	70	546	0.227	100.2	646.2
4	1875	0.013	8.4	1	0.0829	40	336	0.173	14.5	350.5
5	660	0.013	5.0	15	0.0292	5	25	0.061	27.15	52.15
									$\Sigma(R \cdot l + Z)_i$ = 3440 Pa	

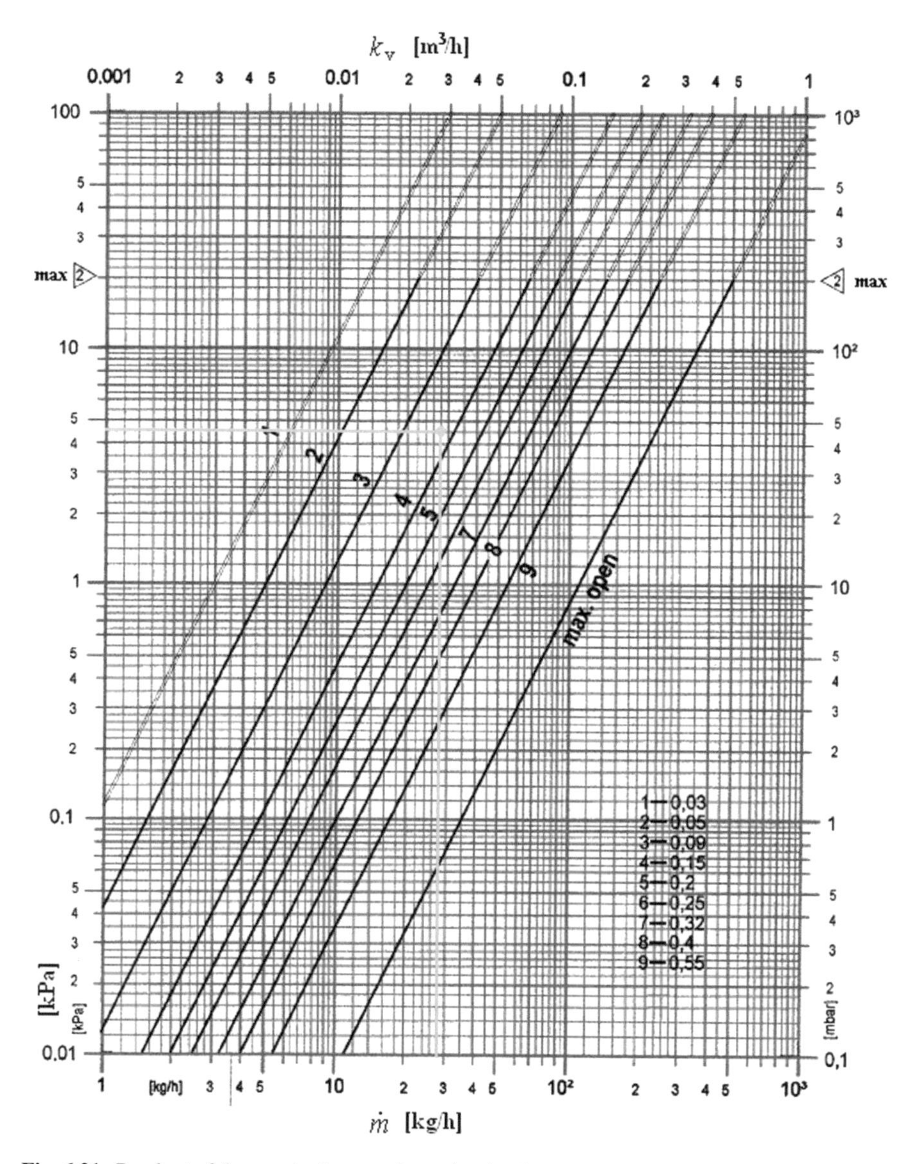

Fig. 6.21 Readout of the required pre-setting value for the *Herz TS-90-V* valve [2]

(b) For the *Danfoss RTD-N 15* valve.

Like for the previous valve, the manufacturer provides both the nomogram illustrating the pressure loss dependence on the flow rate, which makes it possible to read out the required value of pre-setting n_i, and a table with the pre-setting values together with corresponding values of flow factor k_v, as presented in Fig. 6.22. The figure presents the lines of the required pressure loss Δp_z and the medium volume flow $\dot{V}$.

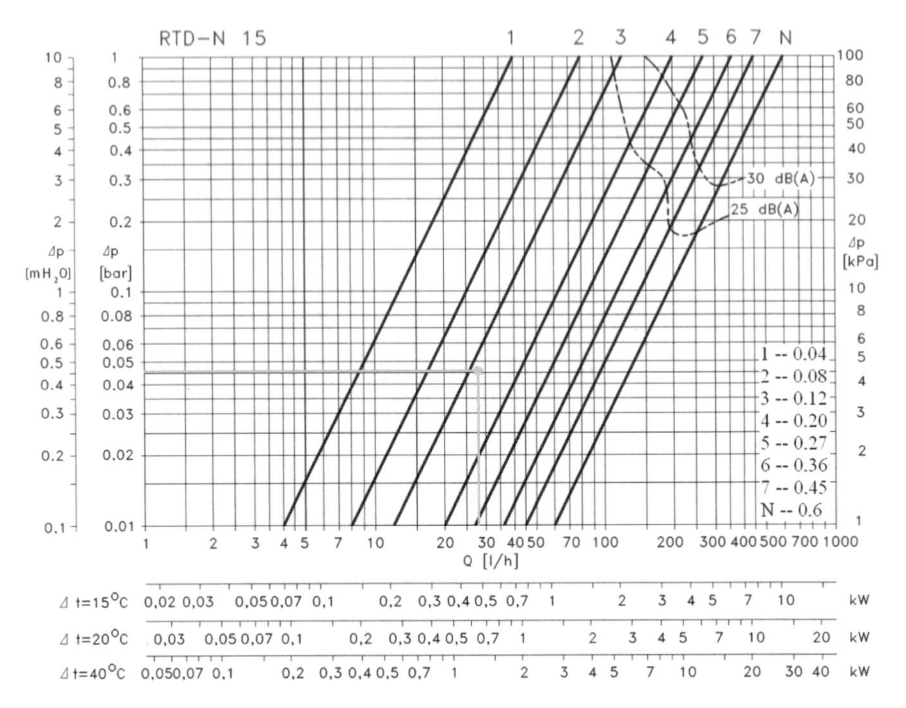

Fig. 6.22 Readout of the required pre-setting value for the *Danfoss* RTD-N 15 valve [1]

The point where the lines intersect is the required working point of the valve. It can be seen that it lies between the pressure loss lines for two neighbouring pre-setting values, i.e. $n_i = 3$ and $n_i = 4$. The values of flow factor k_v corresponding to them are $k_{v,3} = 0.12$ and $k_{v,4} = 0.2$, respectively. This valve does not enable infinitely variable adjustment of pre-setting n_i. A choice can be made between pre-settings expressed by integers and halves (even though the manufacturer does not present pressure drop lines for the latter), which creates a problem. If pre-setting $n_i = 3$ is selected, the resulting value of flow factor k_v is too low. The current regulation element may not be able to compensate for this phenomenon even if it opens entirely above the partial position h_N for $X_p = 2$ K, for which the flow factor value is defined. But if it does, proportional range X_p will be increased significantly and bigger oscillations in the room temperature may occur. If pre-setting $n_i = 4$ is selected, the effect is a substantially excessive value of flow factor k_v. The current regulation element will then have to close significantly—from position h_N for $X_p = 2$ K. In other words, it will operate within a very limited range of travel, and the regulation stability will deteriorate. A solution is to choose pre-setting $n_i = 3.5$, which—as indicated by the calculations performed in Example 6.16—is characterized by the flow factor value $k_{v,3.5} = 0.16$. Nevertheless, the value is still much higher than required. In practice, using a different valve should be considered.

The parameter referred to as the *error in the circuit balance* is usually calculated using the following relation:

$$\delta = \frac{|\Delta p_{cz} - \Delta p_{str} - \Delta p_z|}{\Delta p_{cz}} \cdot 100\%. \tag{6.18}$$

It informs about the percentage difference between active pressure Δp_{cz} and pressure losses Δp_{str} in the circuit. If a pre-setting is selected that results in the flow factor equal to the required value, the numerator in the formula is zeroed and the error in the circuit balance totals 0%. Otherwise, the error is non-zero and has to be calculated using the new value of pressure loss Δp_z arising on the valve, for a given value of the selected pre-setting n_i. Transforming formula (3.573), the following is obtained:

$$\Delta p_z = \left(\frac{\dot{V}}{k_v}\right)^2. \tag{6.19}$$

As described earlier, in the case of the *Herz TS-90-V* valve it is possible to select a pre-setting which is exactly equal to the required value, and the error in the circuit balance will be none. The situation is different for the *Danfoss RTD-N15* valve. Using formulae (6.19) and (6.20), for the two pre-settings under consideration, $n_i = 3$ and $n_i = 3.5$, the following is obtained, respectively:

– for pre-setting $n_i = 3$:

$$\Delta p_z = \left(\frac{\dot{V}}{k_v}\right)^2 = \left(\frac{0.0292}{0.12}\right)^2 = 0.0592\,\text{bar} = 5920\,\text{Pa},$$

$$\delta = \frac{|8000 - 3440 - 5920|}{8000} \cdot 100\% = 17\%.$$

– for pre-setting $n_i = 3.5$:

$$\Delta p_z = \left(\frac{\dot{V}}{k_v}\right)^2 = \left(\frac{0.0292}{0.16}\right)^2 = 0.0333\,\text{bar} = 3330\,\text{Pa},$$

$$\delta = \frac{|8000 - 3440 - 3330|}{8000} \cdot 100\% = 15.4\%.$$

It is most often specified that the percentage error calculated in this manner should not exceed 10%.

However, this computational procedure requires a few words of comment. It should first be considered whether the numerator should be calculated as an absolute value of the pressure difference, or perhaps without the module. The latter seems to be a better-grounded approach because the result then indicates clearly if the selected value of pre-setting n_i may cause a shortage or excess of the flow rate. For the radiator thermoregulators, the first case, as described in Sect. 2.1.2, is more unfavourable than the second because it may involve obtaining permanently too low a value of the room temperature. For this reason, the sign of the result of the percentage error in the circuit balance is so important. Taking this condition into account, formula (6.18) is expressed as:

$$\delta = \frac{\Delta p_{cz} - \Delta p_{str} - \Delta p_z}{\Delta p_{cz}} \cdot 100\%. \tag{6.20}$$

For the data under consideration, it gives the following values:

$$\delta = \frac{8000 - 3440 - 5920}{8000} \cdot 100\% = -17\%,$$

$$\delta = \frac{8000 - 3440 - 3330}{8000} \cdot 100\% = 15.4\%.$$

Apart from that, the expression functional form itself should be considered. According to *Newton's* first law, it is impossible for the driving and the braking force in a system operating in the steady state not to balance each other. If one of them was stronger, according to *Newton's* second law, the system would either accelerate infinitely (driving force stronger than the braking one) or it would either not change its initial state or the sense of its action would reverse (driving force weaker than the braking one). In reality, the driving forces and the braking ones (here: the circuit active pressure Δp_{cz} and the circuit pressure losses Δp_{str}) balance each other and the error in the circuit balance totals zero, except that the equilibrium is obtained for a different flow velocity and mass/volume flow of the medium compared to the initial/assumed values. If the percentage error in the circuit balance δ calculated according to formula (6.20) is negative, the medium mass/volume flow value is smaller than required. If it is positive, the value is higher than required. In reality, therefore, the results obtained by means of formula (6.18) or (6.19) will not be correct quantitatively if a pre-setting value is selected that results in the value of flow factor k_v other than the required one. Firstly, always the same initial value of active pressure Δp_{cz} in the system is substituted in the formula. As described in Sect. 2.3.4, the parameter is a function of the pumped medium flow rate. Due to that, changes in the valve setting will involve changes in active pressure Δp_{cz}. Moreover, the pressure loss Δp_z arising on the valve is calculated, according to formula (6.19), for the initial value of the medium mass/volume flow, which is right assuming that the selected value of pre-setting n_i results in the value of flow factor k_v equal to the required one—a condition that as a rule is not satisfied if the said calculations are to be carried out (because it is impossible to select the appropriate exact pre-setting). If the valve flow factor k_v is different from the required value, the medium mass/volume flow will also be different. In addition, the pipework pressure loss Δp_{str} is substituted as the initial value, for the required value of the medium mass/volume flow, whereas in reality it will be different. Another aspect that needs considering is the specificity of the radiator thermoregulator operation, being the effect of the function the device serves. The regulator will try to compensate for any deviations from the room temperature resulting from deviations in the medium mass/volume flow and the radiator heat output values by closing or opening beyond the design value h_N. This generates additional oscillations in the values of the parameters mentioned above. In reality, then, all the values in formula (6.18) will differ from the calculated ones and the value of the percentage error in the circuit balance will also be different. The parameter should be treated as an indicator of quality and the values of the results of calculations performed using it can only be directly compared within the operating parameters of a given heating circuit.

The calculations presented herein are based on some simplifying assumptions made in practice for calculations performed by hand to complete them fast and minimize the risk of errors. Linear pressure losses R are determined for the mean temperature of water in the system. However, as discussed in Sect. 3.2, the losses depend on the assumed temperature of water because water viscosity is strongly dependent on temperature. Moreover, the quantity does not depend on temperature linearly. For this reason, linear pressure losses R are not a linear function of temperature and assuming its arithmetic mean for the calculations will cause some differences compared to actual results. Anyway, considering that the temperature difference under discussion is small (20 °C), the differences will be slight.

Moreover, it is assumed in the example that in every circuit the temperatures at the supply and return are equal to the required values. But in reality this is seldom the case. Firstly, flowing through no matter how well thermally insulated pipes, water is subject to cooling. Secondly, its cooling degree in the radiator is also different from the one assumed for the calculations. This happens for at least two reasons. Firstly, flowing through the radiator, water with a temperature lower than required will cool down by a smaller value of temperature. Secondly, the radiator size is selected to ensure that the device heat output is as required and that the medium temperature drops by a required value. Due to the fact that radiators are manufactured in specific series of types with regard to dimensions, it is impossible to ensure that the radiator size will be exactly as required in every situation. Consequently, the cooling of water will also differ from the required value even if the radiator supply temperature complies with the initial assumptions.

Hydraulic calculations performed in isolation from the circuit thermal calculations, i.e. assuming that the assumptions made in the process are satisfied, will thus generate errors in comparison with real operating parameters achieved in the installation. This can be noticed especially in large, extensive installations. Only a full analysis, i.e. the system conjugate thermal and hydraulic balancing that takes account of the phenomena discussed herein, will give correct results. The fact that such algorithms are rather complex is of no practical importance because nowadays the computations are carried out by means of specialist computer programs capable of performing a full conjugate analysis. A supplement discussion of the issue of the installation thermal balancing and the conjugate thermal and hydraulic analysis can be found in [8–11]. A full mathematical model enables a detailed and precise analysis of the heating installation operation.

References

1. Catalogue information of Danfoss
2. Catalogue information of Herz
3. European Standard EN 12831:2003: Heating systems in buildings. Method for calculation of the design heat load
4. European Standard EN 215:2004: Thermostatic radiator valves—Requirements and test methods
5. Górecki, A., Fedorczyk, Z., Płachta, J., Płuciennik, M., Rutkiewicz, A., Stefański, W., Zimmer, J.: Instalacje wodociągowe, ogrzewcze i gazowe, na paliwo gazowe, wykonane z rur miedzianych. Wytyczne stosowania i projektowania (Water, heating and gas installations, for gaseous fuel, made of copper pipes. Application and design guidelines). Biblioteka Polskiego Centrum Promocji Miedzi (electronic edition) (2009)
6. Kołodziejczyk, W.: Armatura regulacyjna w ogrzewaniach wodnych (Control armature in hydronic heating systems). Arkady, Warszawa (1985)
7. Mielnicki, J.S.: Centralne ogrzewanie. Regulacja i eksploatacja (Central heating. Regulation and exploitation). Arkady, Warszawa (1985)
8. Muniak, D.: Grzejniki w wodnych instalacjach grzewczych. Dobór, konstrukcja i charakterystyki cieplne (Radiators in hydronic heating installations. Structure, selection and thermal characteristics). WNT/PWN, Warszawa (2015)
9. Muniak, D.: Radiators in hydronic heating installations. Structure, selection and thermal characteristics. SPRINGER (2017)
10. Muniak, D.: Wspomaganie komputerowe równoważenia hydraulicznego instalacji centralnego ogrzewania. Część I (Computer-Aided Hydraulic Balancing in Central Heating Installations. Part I). District Heating, Heating, Ventilation **42**, 352–355 (Sept 2011)
11. Muniak, D.: Wspomaganie komputerowe równoważenia hydraulicznego instalacji centralnego ogrzewania. Część II (Computer-Aided Hydraulic Balancing in Central Heating Installations. Part II), District Heating, Heating, Ventilation **43**, 480–486 (Nov 2012)
12. Pyrkov, V.: Gidrawliczeskoje regulirowanije sistem otoplenija i ochłażdjenija. Teorija i praktika, Danfoss, Kijów (2005)
13. Pyrkov, V.: Regulacja hydrauliczna systemów ogrzewania i chłodzenia. Teoria i praktyka (Hydraulic regulation of heating and cooling systems. Theory and practice). Systherm Serwis, Poznań (2007)

Printed in the United States
By Bookmasters